Theoretical and Applied Approaches to Economic Geography and Spatial Planning

George M. Korres
https://orcid.org/0000-0002-4073-977X
University of Aegean, Greece

Maria Michailidis
https://orcid.org/0000-0003-1315-225X
University of Nicosia, Cyprus

Anastasios K. Sepetis
University of West Attica, Greece

Spyros Anagnostou
University of Aegean, Greece

IGI Global
Scientific Publishing
Publishing Tomorrow's Research Today

Vice President of Editorial	Melissa Wagner
Director of Acquisitions	Mikaela Felty
Director of Book Development	Jocelynn Hessler
Production Manager	Mike Brehm
Cover Design	Jose Rosado

Published in the United States of America by
IGI Global Scientific Publishing
701 East Chocolate Avenue
Hershey, PA, 17033, USA
Tel: 717-533-8845 | Fax: 717-533-7115
Website: https://www.igi-global.com E-mail: cust@igi-global.com

Library of Congress Cataloging-in-Publication Data

LCCN: 2025016768 (CIP Data Pending)
ISBN13: 9798337315546
Isbn13Softcover: 9798337315553
EISBN13: 9798337315560

British Cataloguing in Publication Data
A Cataloguing in Publication record for this book is available from the British Library.

Table of Contents

Detailed Table of Contents

Abstract: The extremely favorable framework provided with the implementation of a series of "reorganization reforms" (Tanzimat) in the Ottoman Empire was associated with the onset of a hundred-year period of prosperity for the Island of Lesvos. The economic growth, mainly driven by the thriving olive oil and soap-making industry, was accompanied by an unprecedented socio-economic transformation and gradually gave rise to a flourishing and diversified industrial development and to a burgeoning export trade. The regional economic cycle of expansion covered the period between 1840 and 1940, even though the Interwar years of 1923-1940 could be considered as a transitional phase between the two cycles. In the post-War period, the new, contraction cycle was marked by a continuing trend of decline, at an economic, social and demographic level, which transformed the once innovative and thriving trade hub in the Eastern Mediterranean to a hard-border zone in the Eastern Aegean, geopolitically and economically marginalized.

Unemployment is one of the most important macroeconomic indicators for both developed and developing countries, serving as a key indicator of economic performance and social well-being. Economic crises, structural problems within economies, and regional disparities may increase unemployment. This study uses spatial econometric techniques to examine the determinants of regional unemployment in Turkey. Based on the NUTS-2 classification, 26 subregions were selected for empirical analysis for 2009-2021. The study considers income levels, general prices, education levels, youth population growth, and out-migration variables at the regional level as explanatory variables. The findings show that unemployment rates in Turkey are spatially dependent, while the random effect spatial error model results indicate that higher income levels and out-migration reduce unemployment. Conversely, general price levels, education levels, and a growing youth population increase regional unemployment rates.

Sustainable local development is important for the success of sustainable development. Local governments have become important actors in ensuring local development. As a basic service provider, local governments are the closest units to citizens in solving local problems and are easily accessible by citizens. This section aims to analyze the roles of local governments in sustainable local development. In this study, what are the roles and strategies for local governments in sustainable local development? In which areas are they successful or unsuccessful? What should local governments do to eliminate these failures? Answers were sought to the research questions. In order to find answers to the research questions, systematic literature review from qualitative research methods was used. This research examines the roles of local governments, one of the actors of sustainable local development, in local development and the strategies expected to be implemented by local governments.

The development of smart strategies in small island territories contributes to environmental protection and the improvement of characteristics related to the permanent phenomenon of geographical discontinuity, aiming at sustainable development and eco-friendly tourism. This is achieved by reducing environmental, social, and economic burdens through an integrated, holistic approach. The main environmental and socio-economic benefits rely on the implementation of innovative projects focused on energy and natural resource management, waste management, sustainable mobility, and energy-efficient building upgrades. Furthermore, expanding research in this area can contribute to the further development and application of solutions not only on small and medium-sized European islands but across all island regions, making the sector more efficient and resilient.

The definition of the term inclusive growth can be found in different political and scientific concepts, as several scientific and institutional articles define inclusive growth as a combination of social views with economic development and equitable management of natural resources through policy approaches. Literature research on the policies of international organizations reveals a general and growing recognition that economic growth alone does not provide equal opportunities to different segments of society and it is not possible without the development of international policies to reduce social inequalities, optimal management of natural resources and intergenerational justice. Deepening that study of the definition of the concept of inclusive sustainable development by international organizations, we find that it is defined in a broader concept of development, including various non-financial aspects of life, such as health, employment prospects and environmental issues with a focus on local communities and the Sustainable Development Goals.

This chapter examines how spatial planning contributes to the realization of sustainable development goals by balancing development demands and environmental conservation. It examines the impacts of spatial planning, highlighting its potential to address economic, social, and environmental challenges while ensuring value for money in development projects. This chapter draws insights to identify the concept of spatial planning and its importance in integrating it into city development planning in order to optimize resources, reduce life cycle costs, and enhance investment value. It emphasizes the need for environmental and economic assessments to mitigate risk associated with improper spatial arrangements. It also examines the historical significance of spatial planning in mitigating environmental impacts, increasing resilience to natural hazards, and safeguarding investments. By providing theoretical foundations and practical approaches, this chapter serves as a comprehensive resource for policymakers interested in implementing effective spatial planning for sustainable development.

This paper empirically analyzes the impact of high-speed rail (HSR) on migrants' settlement intentions, using data from the 2016 and 2017 China Migrants Dynamic Survey (N=294,425). The results of the linear probability model (LPM) suggest that cities with HSR increase migrants' settlement intentions by 5.58%. Mechanism analysis shows that the more relatives accompanying the migrants, the smaller the effect of HSR on the settlement intention. If migrants move across province or with their own land assets in hometown, the settlement intention is easily affected by HSR. In addition, HSR cities can provide higher income and business environment for migrants. Regional heterogeneity analysis shows that the effect of HSR cities in Northeast China on the settlement intention of migrants is not significant, while it is significant in other regions of China. This study linking transport infrastructure development with micro-level migration decisions, offering insights for urban planning and labor mobility policies.

Natural heritage sites and areas are vital for preserving biodiversity and ecological integrity, especially on islands with limited size and isolation. This study investigates the impact of uncontrolled built-up area sprawl on protected areas and surrounding forest ecosystems in the Southern Aegean Region, where development pressures like tourism and infrastructure expansion threaten fragile ecosystems. The analysis highlights the importance of forest and protected areas in maintaining biodiversity. Using geospatial datasets from the Copernicus Land Monitoring Service and the Ministry for the Environment, the study examines spatial relationships between urban expansion and Natura 2000 sites and forested regions. The findings reveal significant conflicts from built-up dispersion, highlighting the inadequacy of current management frameworks in protecting these natural resources. This underscores the urgent need for integrated spatial planning that balances ecological preservation with development, ensuring the sustainability of biodiversity on islands amidst ongoing urban pressures.

The traditional settlement of Spetses is a rich palimpsest of historical and architectural value, on which the social, political and economic details of the society are reflected throughout the last centuries. The seafaring character of the Spetsian society is clearly visible in the 18th century, when the traditional character of the settlement was formed, its transformation into a high society tourist resort in the 19th century, as well as its subsequent inseparable connection with tourism until this day. The research examines how the urban structure of the settlement, the typologies and the morphological characteristics of the traditional architecture are affected by the phenomenon of tourism, as it changes its character and form. Also, the research focuses on the strict policies that targets the conservation of the traditional settlement, arising from the great influx of the mass tourism of the 20th century and the consequent uncontrolled mass building construction.

Over the past few decades, tourism has become a key economic activity, with the crises of the last 15 years making many peripheral economies in Europe and beyond increasingly dependent on it. However, empirical research on touristification at the regional level remains limited. To address this gap, we introduce a composite index that measures the territorial and social pressures from tourism's supply and demand, as well as regions' dependence on tourism in terms of employment and output. Our analysis covers all EU NUTS-2 regions from 2009 to 2022 and identifies three main zones of touristification: the sunlust insular/seaside EU South, the Italian and Austrian Alps, and various urban destinations across Central and Northern EU. Ultimately, two key asymmetries emerge: one between tourism-oriented Southern EU and production-oriented Northern EU, and within the former, another between tourism-dependent Greece, Portugal, and Croatia, and the economically diverse Spain and Italy.

This chapter examines the role of tourism in the development of coastal zones, focusing on the legislative frameworks shaping tourism-related infrastructure. Since tourism depends on quality infrastructure such as accommodation, transport, and recreational facilities, the chapter analyzes laws that either support or restrict such development in coastal areas. A comparative approach is applied, reviewing frameworks from various Mediterranean countries to highlight similarities, differences, and the presence or absence of environmental safeguards. Special emphasis is given to Greece, where coastal zones are both ecologically sensitive and economically strategic for tourism. The analysis assesses the Greek legal framework's ability to balance environmental protection with tourism growth, examining how current laws and planning instruments influence the scale, location, and type of infrastructure. Overall, the chapter contributes to understanding how tourism infrastructure can align with sustainability and spatial justice goals.

There is an increasing need to develop new holistic approaches that address the underlying causes, behavior and impacts of natural disasters that appear in rural and residential areas, threatening people, property and heritage, as well as other values at risk. An example of integrated risk management and socioecological perspective is thus offered as a case study in south Lesvos Island, Greece. Two successive natural disasters "hurt" and drastically limited the local environmental, residential, productive, social, economic and tourist capital of the area. It was considered of utmost importance to generally monitor and analyze the post-disaster situation, in order for authorities and public services to plan a series of actions for the implementation of the necessary short-term and medium-to-the-long term rehabilitation measures, and overall reconstruction and development of the affected areas. All in all, we focused on the socioeconomic and biophysical causes and effects of the pathology of natural hazards (i.e. natural hazards geography) rather than trying to treat the symptoms.

Preface

INTRODUCTION

Economic geography and spatial planning stand at the heart of contemporary debates on sustainable development, territorial cohesion, and the governance of regional transformation. Across the world, regions grapple with uneven development patterns, demographic pressures, shifting labor dynamics, and the profound spatial implications of globalization. As scholars and practitioners seek to understand and manage these complexities, there is a growing need for interdisciplinary approaches that recognize the interdependence of economic structures, environmental systems, and spatial organization. This edited volume responds to that need by offering a collection of theoretical reflections and applied research grounded in diverse regional contexts.

The contributing chapters reveal the dynamic ways in which territories evolve—economically, socially, and environmentally—through both endogenous forces and external shocks. Islands, urban centers, peripheral regions, and environmentally sensitive areas are examined not only as geographic spaces but also as socio-economic systems shaped by policy choices, institutional frameworks, and historical pathways. Through these case studies, the volume highlights the significance of spatial planning instruments, local governance, and strategic development policies in mediating the pressures of growth, decline, and transformation.

At a time when sustainability, resilience, and spatial justice have become guiding principles for public policy, this volume offers insights into how spatial planning and economic geography can inform more inclusive and forward-looking development strategies. By integrating empirical methodologies with conceptual advancements, the book seeks to foster dialogue between academia and practice, contributing to a deeper understanding of how territories can adapt to contemporary challenges while preserving their social and environmental integrity.

CHAPTER OVERVIEW

Chapter 1 examines the long-term economic cycles of Lesvos, tracing a historical trajectory from 1840 to 2024. It demonstrates how institutional reforms catalyzed industrial expansion before prolonged post-war contraction shaped the island's modern socio-economic reality.

Chapter 2 analyzes regional unemployment in Turkey through spatial panel data, providing evidence of spatial dependence and identifying key determinants—including income, migration, and demographic factors—that influence employment patterns across subregions.

Chapter 3 focuses on the roles and responsibilities of local governments in sustainable local development. Using a systematic literature review, it evaluates strategies, performance gaps, and opportunities for strengthening local governance in sustainability initiatives.

Chapter 4 explores smart strategies for sustainable development on small and medium-sized European islands. It emphasizes innovative solutions for resource management, mobility, energy efficiency, and ecosystem protection.

Chapter 5 reviews policies of international organizations addressing inclusive sustainable development. It synthesizes conceptual frameworks and policy trends that integrate economic growth with equity, natural resource stewardship, and intergenerational justice.

Chapter 6 discusses spatial planning as a mechanism for balancing development needs with environmental conservation. It presents both theoretical foundations and practical approaches for integrating spatial planning into city development and risk mitigation.

Chapter 7 investigates how high-speed rail infrastructure in China influences migrants' settlement intentions, demonstrating that improved connectivity shapes labor mobility and regional development outcomes.

Chapter 8 analyzes the impact of built-up area dispersion on natural heritage sites in the South Aegean. It highlights threats posed by uncontrolled development to protected ecosystems and underscores the need for integrated spatial planning.

Chapter 9 assesses the effects of tourism on the traditional settlement of Spetses. The chapter examines architectural change, spatial transformation, and conservation pressures linked to mass tourism.

Chapter 10 introduces a composite indicator to measure touristification across EU regions, revealing deepening core-periphery asymmetries between tourism-dependent southern regions and more diversified northern economies.

Chapter 11 examines legislative frameworks governing tourism infrastructure in Mediterranean coastal zones, with an emphasis on Greece's efforts to reconcile economic development with ecological sensitivity.

Chapter 12 presents a natural hazards geography assessment for South Lesvos, offering a holistic evaluation of disaster impacts and outlining post-disaster reconstruction and resilience strategies.

Chapter 13 highlights the role of ecomuseums as catalysts for territorial development through community-led initiatives that strengthen regional identity and promote sustainable cultural and economic activities.

Chapter 14 investigates how spatial planning affects traffic safety, showing how land use patterns, urban morphology, transport strategies, and road network design shape accident frequency and severity.

CONCLUSION

The chapters collected in this volume underscore the profound interconnections between spatial dynamics and economic development. Whether examining long-term island economies, regional unemployment structures, cultural landscapes, or the pressures of tourism and natural hazards, each contribution illuminates facets of contemporary spatial challenges. Together, they reveal that economic geography and spatial planning are not discrete fields but mutually reinforcing lenses essential for understanding regional trajectories.

Several recurring themes emerge across the case studies: the importance of evidence-based policymaking, the growing relevance of sustainability and resilience, and the need for inclusive, community-centered approaches to territorial development. The research demonstrates that spatial inequalities, environmental vulnerabilities, and development pressures cannot be adequately addressed without a holistic view of the social, economic, and ecological systems that shape territorial outcomes.

We hope that this volume will serve as a valuable resource for scholars, practitioners, policymakers, and students seeking deeper insights into the complexities of spatial development. By bridging theoretical perspectives with empirical applications, the book aspires to contribute to ongoing debates about how regions can navigate contemporary challenges, foster equitable growth, and build sustainable futures. It is our hope that the insights presented here will inspire further research and inform innovative spatial planning practices across diverse geographic contexts.

Chapter 1
Island Development and Regional Economic Cycles:
The Case of Lesvos, 1840–2024

Spyros Anagnostou
https://orcid.org/0009-0001-3930-3223
Department of Geography, University of the Aegean, Greece

ABSTRACT

Abstract: The extremely favorable framework provided with the implementation of a series of "reorganization reforms" (Tanzimat) in the Ottoman Empire was associated with the onset of a hundred-year period of prosperity for the Island of Lesvos. The economic growth, mainly driven by the thriving olive oil and soap-making industry, was accompanied by an unprecedented socio-economic transformation and gradually gave rise to a flourishing and diversified industrial development and to a burgeoning export trade. The regional economic cycle of expansion covered the period between 1840 and 1940, even though the Interwar years of 1923-1940 could be considered as a transitional phase between the two cycles. In the post-War period, the new, contraction cycle was marked by a continuing trend of decline, at an economic, social and demographic level, which transformed the once innovative and thriving trade hub in the Eastern Mediterranean to a hard-border zone in the Eastern Aegean, geopolitically and economically marginalized.

DOI: 10.4018/979-8-3373-1554-6.ch001

1. INTRODUCTION

Regional economic cycles refer to the cyclical patterns of growth and contraction in the economy of a specific region, such as a state, country, or a smaller geographic area. These cycles are shaped by local factors, such as natural resources, industrial potential, demographics, geopolitical context, policies, etc. The modern history of Lesvos Island, between 1840 and 2024 features a typical sequence of two major regional economic cycles.

The year 1840 could be considered as the start line for a major, hundred-year long growth cycle, as the implementation of the "reorganization reforms" (Tanzimat) in the Ottoman Empire is associated with the onset of a long period of economic prosperity for Lesvos. The abolition of the monopoly of trading in olive oil (1839) that enjoyed the local Ottoman governor, by the first major Tanzimat edict (*Edict of* Gülhane) represents a decisive triggering factor in the industrialization process of the island. The edict, promulgated in November 1839, though first implemented in 1840-41, guaranteed, for the first time, the security of persons and property. An even more favorable framework was provided with the reform Edict of 1856 (*Hatt-ı hümayun*) which continued and expanded the Tanzimat policies, particularly by guaranteeing the equality of citizens, regardless of religion and class, in terms of taxation, justice, administration and education. Even though, in real life, property was never as secure as the Edict mandated, the Tanzimat reforms boosted the property of the Greeks in the Empire.

While the substantive strengthening of the economic power of Greeks resulted in all parts of the Empire, this was definitively more effective in certain "model provinces", such as the Province of the Islands (*Cezair-i Bahr-i Sefid Vilayeti*) (Strohmeier, 2010). The remarkable economic upturn of Lesvos in particular, which the Ottomans of that period had called *"The Golden Island"* and *"The Garden of the Ottoman Empire"* (Conze, 1865), was probably the most unique of all. As Evridiki Sifneou (1996a:68) notes, quoting a relevant report of 1858 by the Austro-Hungarian vice-consul in Mytilene, Theodore Bargigli, "in no other part of the Empire were favorable legislative concessions implemented with such ease".

2. A CENTURY OF GROWTH: 1840-1940

During the decades that followed the Tanzimat reforms (mid-19th century) the unfettered entrepreneurship took the role of locomotive in Lesvos, and the thriving industrialization process gradually verticalized the processing of olive oil and of its various derivatives, such as soap, kernel oil, green soap, glycerin, refined oils etc. The first big soap industries were built in the 1880s, and in the following years, soap became, after olive oil, the major industrial product and export item of Lesvos -and the main driver of growth, for the next half century. The comparison with Crete, made for the year 1890 by the French Consul in Smyrna Vital Cuinet (1892), is revealing of the importance of the production of olive oil in the island - and, even more so, of the production of soap: As the provided data shows, Lesvos produced at the time approximately the same quantity of olive oil with Crete, while the soap production in the island was more than the double of that in Crete (Fig. 2). Between 1894 and 1896, a major innovative step in the island's soap industry was inaugurated, with the establishment of the first five olive-pit factories, which produced kernel oil used for the manufacture of green soap. It is significant that these were the first industries of this kind in the entire Ottoman Empire (Sifneou 1996b). This development marked a new stage for the local trade, with green soap appearing in Lesvos' exports after 1898. The next major innovative process in the island's industry came some thirty years later, in 1928, with the establishment of the first refineries, for the refining of high-acidity olive oils.

Figure 1. Map of Lesvos by Richard Kiepert, 1915

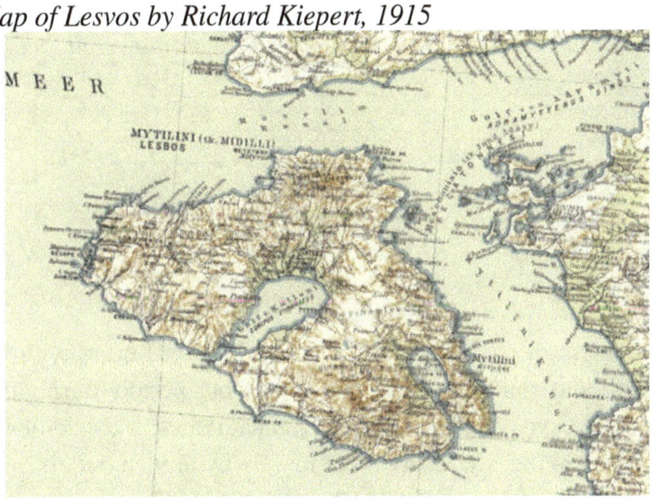

Source: Kiepert R., Second edition Karte von Kleinasien -Aivalyk, 1915 (detail)

By 1890, the city of Mytilene alone counted 14 steam engine factories (Agriantoni, 1986), which made it the fourth Greek city (though still under Ottoman rule) for its steam potential, behind Piraeus, Ermoupolis and Patras (Fig. 3). For the introduction of steam, olive pit was used as biofuel, which was abundant on the island and ensured reduced energy costs for businesses. In parallel with the verticalization of the production of olive oil, a significant diversification of the industrial activity also marked the end of the 19th century. By 1894, the industrial potential of the island consisted of: 200 olive mills, 32 steam-powered olive oil factories, 2 steam-powered olive-pit factories, 47 soap factories, 16 tanneries, 34 flour mills, 9 steam-powered flour mills and 17 spinning mills. Two decades later, in 1913, the accumulating profits had led to new business investments. Steam-powered factories had reached the number of 144, with an even greater diversification of the industrial activity in the island, which now included, among others, several building material factories, industrial tanneries, spinning mills and two steam engine machine-shops (Sifneou, 1996a).

Figure 2. Comparison of average production of olive oil and soap, for the islands of Lesvos, Crete and Samos, in 1890

	Lesvos	Crete	Samos
Olive oil	10.000 tons	9.000-13.000 tons	2.000 tons
Soap	3.800 tons	1.800 tons	250 tons

Source: Cuinet, 1892; Sifneou 1996b

Figure 3. Greek cities with the biggest number of steam-powered factories in 1890

Piraeus	52
Ermoupolis	23
Patras	20-24
Mytilene	14
Athens	12-13

Source: Agriantoni, 1986

The development of trade followed a parallel upward trend with that of local agricultural and industrial production. In the previous period (during the 18th century), the trade of Lesvos had very special characteristics: It concerned exclusively olive oil for soap-making, was carried out mainly by French traders, and had as its exclusive market that of Marseille (Boulanger, 1991). The new geopolitical landscape however, after the Napoleonic Wars, had already reoriented the markets of Lesvos to a new trade axis: Alexandria-Smyrna-Mytilene-Constantinople-Southern Russia (Vafeas & Mandamadiotou, 2023). In the first decades of the great changes that followed 1840, the class that took control of the olive oil trade was the "local aristocracy" (Newton, 1865) actually, the bourgeois landowners, who initially began to trade the olive oil they produced on their own estates. However, the industrial development of the island that followed, soon brought about the gradual diversification of exports, which were no longer limited to olive oil. The export trade of Lesvos became extremely prosperous, particularly in the last three decades of the 19th century. The conditions for this had become favorable through a series of factors:

The main factor was again that of the Ottoman Tanzimat reforms of 1839, 1856 and 1876, which had allowed the strengthening of the economic power of Greeks through the massive acquisition of land and through the free exercise of trade. Besides, the new liberalized framework introduced by the breakthrough Reforms found a more fertile ground in Lesvos, as the industriousness and the merchant and entrepreneurial spirit of the Greeks of the island had always been praised by most foreign visitors of that period (Newton, 1865; von Scherzer, 1873; Rougon, 1892; Launay, 1897).

Curiously enough, the severe natural disasters that occurred during the second half of the 19th century had also a mid-term positive impact on the island's economy and trade. The 'Great Frost' of January 1850, which devastated the olive groves, and the two ravaging earthquakes of 1867 and 1889 (Kampouris, 1978) were the three worst calamities. Among the three, the 'Great Frost' was the one with the greatest impact. The island's farmers worked hard in the following years, replanting the olive groves with new trees and new varieties -and managing, over the following decades, not only to restore olive oil production to pre-disaster levels, but to increase it significantly, thus contributing to the launch of the island's exports. At the same time, the unforeseen annihilation of olive oil production in 1850, led many among the most entrepreneurial inhabitants to leave the island for Smyrna, Constantinople, Alexandria, Marseilles, Varna, Braila, Odessa and other places, where they formed powerful merchant communities. These wealthy communities and, primarily, the largest among them, that of Alexandria, through the large profits of the 55 Lesvian trading and financial companies that were active there, played the prime role in the accumulation of capital, that was invested both in improving agricultural production and in the industrialization of Lesvos (Sifneou 1996b). All of these communities formed a wide trading network, often reinforced by family ties, which significantly supported the island's trade take off in the second half of the 19th century. Most of the merchants and businessmen eventually returned in the island, where they built splendid manor-houses and invested their fortunes in olive farming and in the local industry. Many of them even purchased cultivated land (olive groves, but also cereal crops) on the coastal region of Asia Minor, between Edremit and Dikili.

Figure 4. Major political and geopolitical milestones and natural events in the history of Lesvos, between 1839 and 1939

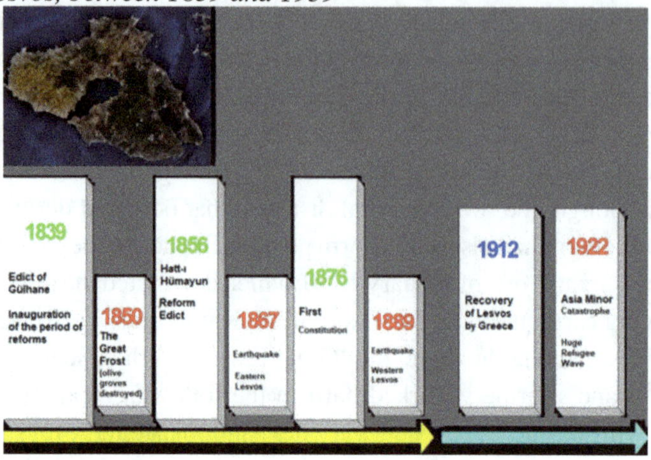

Source: Own elaboration

According to the French Consul in Smyrna, Firmin Rougon (1892:494) the trade activity of Lesvos exceeded, in 1889, 30,000,000 french francs, an amount that was equivalent to the sum of the trade of all the rest of the Ottoman islands in the Aegean (with the exception of the semi-autonomous island of Samos). This comparison becomes even more significant if one considers that the Province of the Islands represented, at that time, the province of the Empire with the highest per capita income (Karpat, 1985). Presenting a clear surplus, the exports represented that same year (1889) 16,000,000 francs, and it is noteworthy that nearly half of these were directed to the Ottoman empire, with Austria holding the second place.

Continued growth gradually brought an ever-expanding business diversification that became significant toward the end of the century. One of the most iconic businessmen of the island's flourishing economy, Panos Kourtgis, founded in 1883 the *Aegean Steam Navigation*. The passenger shipping company possessed a fleet of 11 steamships, between 300 and 3,200 GT, in direct competition with the all-powerful Austrian Lloyd and French Fraissinet companies, which were based in Trieste and Marseilles, respectively. For the next three decades, the maritime lines linked Lesvos to the major ports of the Aegean, the Eastern Mediterranean and the Black Sea, such as Piraeus, Heraklion, Trieste, Alexandria, Constantinople, Samsun, Trabzon, Varna, Constanta, Braila and many others (Kaitatzidis, 2009). Eight years later, in 1891, Kourtgis also founded the *Banque de Mételin* (Bank of Mytilene), with branches in 10 cities -among which Constantinople, Smyrna, Athens, Piraeus and Thessaloniki. The *Banque de Mételin* made also use of the vast network of 195 agencies of the *Aegean Steam Navigation*, covering the entire Eastern Mediterranean and Black Sea regions. The shipping company greatly contributed to the emergence of the port of Mytilene as a major crossroads in the Eastern Mediterranean and, in association with the bank's activities, it favored the further expansion of the island's trade networks. However, the prosperity of Lesvos, on the verge of the two centuries, was not reflected only in the business activities of the local entrepreneurs, but also in the unprecedented socio-economic transformation of the island, in a remarkable urbanization process, in new road infrastructures and, even more so, in the establishment of exemplary schools, exclusively funded with the resources of local communities and wealthy contributors. The president of the French Academy of Sciences, Louis de Launay (1897:85) who visited the island in 1889, even asserts that, "the development of schools is almost comparable to that in France".

An element that was particularly important for the unprecedented flourishing of the island's trade activity, was the role of the port of Mytilene as a commercial hub, not only for the island but for a much wider region (von Scherzer, 1873). About a third of the imports in Lesvos were eventually re-exported to Asia Minor, by the big importers of Mytilene, but also by small merchants who traveled far in the Anatolian mainland to make their products available in the markets of numerous towns and villages (Sifneou, 2007). This was also due to the fact that, unlike the port of Mytilene, the Asia Minor ports opposite the island were shallow-water ports that were not accessible to big -or even medium-sized- vessels and, at the same time, there was practically no rail and road network connecting the towns of that region with the big mainland cities. This made the port of Mytilene, at that period, the main port in the entire *Vilayet* (Province) of the Aegean, for its commercial activity (Cuinet, 1892). In 1909, the annual traffic of the port reached the number 1,749 steamships and 3,352 sailing ships (Vogiatzis, 2012). A particularly interesting element of the nature of the economic ties between Lesvos and Asia Minor is that a large part of the island's exports consisted of industrial products, such as textile, or machinery for olive mills. However, the contribution of Lesvos in the industrialization process in Asia Minor was not limited to the supply of machinery, but also involved transfer of industrial know-how and of capitals, which led, before 1912, to a dependence of a part of the north-western Asia Minor economy on Lesvos' businessmen and their companies, (Sifneou, 2007).

In 1912, all the 42 soap factories of Lesvos belonged to Greek Christians, in total contrast with Crete, where, since the 18th century, the majority of the island's soap factories belonged to Ottoman families. According to the 1913 report of A. Sophianopoulos (Vogiatzis, 2012) the annual value of soap exports represented that year 11,595,100 drachmas, which corresponded to 32% of the island's income from exportable industrial products. The sector continued to thrive in the following years, and, by the end of the decade, Lesvos accounted for more than half of the country's soap exports (Fig. 5).

Figure 5. Greek exports of common soaps by area (1919)

Area	Exports (%)
Mytilene	54%
Thessaloniki	17%
Heraklion	15%
Zakynthos	11%
Other	3%

Source: Sifneou, 2002; Hellenic Statistical Authority, Trade statistics for the year 1919

3. THE INTERWAR ADVERSITIES IN A STILL FLOURISHING ECONOMY AND THE POST-WAR CYCLE OF DECLINE

Even though the Interwar period could be considered as the last phase of the island's growth cycle, it also represents a period during which the extremely subversive geopolitical and political developments created the first turbulence and setbacks in the local economy. (Due to the great Greco-Turkish war of 1919-1922, and Greece's late entry into World War II, the Interwar period in Greek history covers the span between 1923-1940).

By the mid-1920s, the soap industry of Lesvos had reached a peak. As notes Sifneou (2002:95), in the years 1920 to 1926, the average annual production of white soap in the island reached 2,700 tons, compared with the average production in all types of soap for all the rest of the country of 1,700 tons.

It was during this period that huge extra export taxes (such as added tax, tax for the refugees, for the war invalids, for the orphanage, for the Agricultural Bank, for the water-supply etc.) had been imposed on foreign trade commodities. In 1926, at a time when the market had showed a slight recovery, the export taxes were raised once again, a measure that had only local effect and resulted, in 1927, in the closing of all but 6 of the island's 58 soap factories and in the plummeting of its soap exports (Fig. 6). The disastrous policy of the Greek state, which aimed at tax-collection, didn't make any use of the geographical advantages and the established commercial networks of Lesvos, while the business world of the island was seeking equality of taxation with "Old Greece" (Sifneou, 2002).

Figure 6. White soap exports of Lesvos (1923-1930)

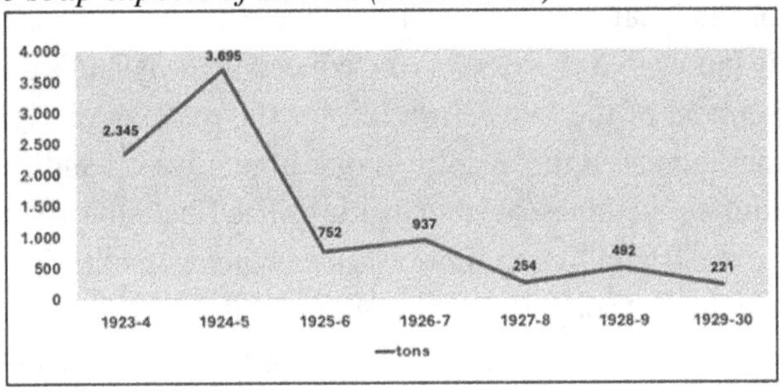

Source: Sifneou, 2002 – Bulletin of the Commercial and Industrial Chamber of Lesvos, Jan. 1930 – Sept. 1931

The grinding taxation on foreign trade, with specific local effect in Lesvos, was certainly not the only adversity for the local economy in the mid-1920s. Many of the long established markets for soaps, as well as for olive oil and other local products, were situated in the Ottoman Empire. The sweeping geopolitical changes in the early 20th century, such as the Italo-Turkish War of 1911, the recovery of the islands of the Aegean by Greece, in 1912, the First World War (1914-1918), the first wave of Asia Minor refugees, in 1914, and -most important- the Asia Minor Catastrophe of 1922, led to the eclipse of many of the island's traditional markets and to the shrinkage of its commercial networks (Sifneou, 2002).

The brief boom of a new sector, namely the tobacco sector, during the Interwar period, compensated to some extent for the losses from the adversities in the sector of soap making. Tobacco cultivation in Lesvos officially started around 1910, though its cultivation was taking place at least since 1886 (Bakalis, 2016). With the arrival of the first wave of refugees from Asia Minor, in 1914, tobacco cultivation spread even further on the island, and the first commercial and industrial companies began to operate. The arrival of the second and largest wave of Asia Minor refugees in 1922, gave a further boost to the sector, providing low cost and skilled labor. Exports were maximized in the mid-1920s, as the production of tobacco leaves reached 1,200 tons in 1925 (Mantzaris, 1928). By 1924, four tobacco factories were operating in the city of Mytilene, employing around 650 workers. These factories also processed a significant part of the tobacco production of Chios, as Lesvian traders were buying tobacco quantities from the nearest island, which lacked warehouses and processing factories (Bakalis, 2016). Due to the negative impact of the 'Great Depression' in international trade, as well as the drought of the summer of 1932, the tobacco sector stopped following an upward trend in the 1930's and significantly declined with the turmoil and devastation of World War II. Even though the tobacco sector never reached in Lesvos the development and the importance it had in some regions of

Northern and Western Greece, it nevertheless formed, for the two Interwar decades, the third pillar of the local economy, behind olive growing and soap making.

In 1928 a new innovative process was introduced in the island's olive oil industry with the establishment of the first refineries, for the refining of high-acidity olive oils. This new fact increased the quantities of edible olive oil and correspondingly reduced the quantities available for soap making. The final blow to the soap exports of Lesvos -and to the island's economy- was nevertheless dealt after World War II, with the global decline of soap demand. Commercial detergent products with fatty alcohol sulphates had already entered the German market in 1932, and the American market in 1933. The conclusive decline in soap exports, however, and the definitive end of this key sector of the economy of Lesvos came in the post-war years, when surfactants had become widespread in developed countries, largely replacing soap in cleaning and laundry. Thus, the 1940s, that is the ten-year devastating period of World War II and the Greek civil war that followed, marked the abrupt end of the hundred-year cycle of growth.

The geopolitical, political and economic context that emerged after the end of the Greek civil war (1949) was extremely adverse for the island: The local industrial infrastructure was destroyed and the main export product, soap, had become obsolete as its demand in the international markets was annihilated. With Marshall Plan funds having been invested almost exclusively in the metropolitan areas of Athens and Thessaloniki, no national or foreign aid investments were directed to the geopolitically marginalized island, for several decades. Ironically, the post-war period of rapid growth for the European economies had been the most adverse period for Lesvos. The context slightly changed only after 1981, when the over-centralized Greek state started evolving toward a more decentralized development model.

4. THE EVOLUTION OF THE POPULATION, FOLLOWING THE CYCLES OF ECONOMIC GROWTH AND DECLINE

The population boom that accompanied the century-long period of economic thriving of Lesvos, took end with the outbreak of World War II, thus marking the beginning of a new cycle. The population curve between 1840 and 2021 (Fig. 7) shows that there have actually been two main cycles -the first one being that of population growth (1840 to 1940), followed by a cycle of population collapse (1940 to 2024). Each cycle can be subdivided in two distinct phases:

- Within the cycle of growth, the first phase, ranging between 1840 and 1911, shows an impressive near-doubling of the population, over the 71-year period, from 72,776 to 138,538. The developments that shaped the population boom of this period are however quite complex: The substantial population surplus created by the positive natural balance of the period could have been essentially cancelled by the significant wave of migration of the island's inhabitants to the nearby coasts of Asia Minor, but also to a number of major ports of the Mediterranean and the Black Sea. However, the influx of a very large number of Greek citizens during the second half of the 19th century far exceeded the numbers of the island's residents who emigrated. This very significant influx of Greek nationals was not directed exclusively to Lesvos, but also to other islands of the *Vilayet* (Province) of the Islands, as well as to many areas of western Anatolia, as a result of the favorable conditions created by the Tanzimat reforms of 1839 and 1856 (Karpat, 1985) Moreover, the economic boom and the vast market of the Ottoman Empire offered much greater opportunities for the development of business activities than those in the small Greek kingdom of the time (Strohmeier, 2010).

- In the second phase of the cycle of growth, ranging between 1911 and 1940, the population of the island remains practically stable, with a very slight decrease, from 138,538 to 134,058. A curious exception is however observed in the short period between 1911 and 1920, where a temporary reversal of the population growth trend is recorded, with the island losing 27,478 residents (-19,8%) in just 9 years. This paradoxical "interval" has puzzled several authors, as it raises the assumption that it could be due to a possible overestimation of the population for 1911 (Kolodny & Darques, 2004). While this case should not be ruled out, there are still two important components which could have greatly contributed to this temporary though significant population shrinkage over the 9-year period: The first component is the mass exodus of a large part of the island's Muslim population (approximately 11,000 out of a total of 18,000), immediately after the integration of Lesvos in the Greek State, in 2012-13. A second component, that is often overlooked, could be that of the severe blood toll of the youth of Lesvos during the Second Balkan War and the First World War. It is characteristic that, at the great Battle of Skra-di-Legen alone (May 1918) 338 out of the 441 dead (Anagnostou, 2024) as well as the vast majority of the 164 missing and the 2,204 seriously injured on the Greek side were conscripts from Lesvos, who served in the newly formed Archipelago Division. After 1923, the mass settlement of refugees from Asia Minor brought the population of the island back to 1911 levels. Thus, the 1928 census recorded a population of 137,160, of whom around 30,500 (or 22.3%) were Asia Minor refugees (Madison Kolodny, 1974; Kolodny & Darques, 2004; Paraskevaidis, 2013). It is quite interesting to note that, the difference in the numbers between 1920 and 1928 corresponds to the influx of approximately 30,500 refugees, minus the outflow of the last group of (approximately) 7,000 Muslims who left the island after 1923, with the enforced exchange of population agreed in the Treaty of Lausanne. This in turn suggests that the strongly positive natural balance of the period, which ranged from +7‰ to +14‰, was completely counterweighted by a significant migratory outflow (which however involved less the local population than the recently settled Asia Minor refugees).

- Within the cycle of economic decline, the first four decades, between 1940 and 1981, corresponding to the most adverse economic period for the island, are characterized by a severe population collapse of -33,9%. This phase saw a mass population outflow (directed mainly to Athens -but also to Australia, South Africa, Canada and various other oversea destinations) while the reversal of the natural balance and the significant aging of the population also occured during this period.
- In the second phase of the cycle of decline (1981 to 2024) the island's population first showed stabilization trends, thanks to a slightly positive migration balance, which were then followed by a (relatively smooth) decrease, after 2001, mainly due, this time, to the strongly negative natural balance of -6‰ to -3‰ for this period. For the last ten year period, the actual negative natural balance is however even more unfavorable, (ranging between -8‰ and -5‰) if we do not take into account the population of illegal immigrants who temporarily reside on the island, an extremely fluctuating population which however significantly affects the crude birth rate.

Figure 7. Evolution of the population of Lesvos, 1840-2021

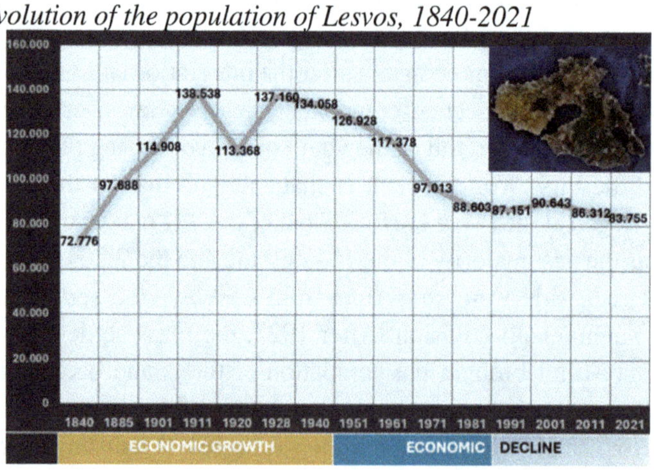

Source: Karpat, 1985; Anagnostou, 2004; Hellenic Statistical Authority–Population Censuses, 1920-2021 – own elaboration

5. CONCLUSION

World War II, and the Greek civil war that followed, have marked a clear turning point between the two distinct cycles of growth and decline, for both the economy and the population of Lesvos. The once innovative island and thriving trade hub in the Eastern Mediterranean was transformed in the post-War period to a hard-border zone in the Eastern Aegean, geopolitically and economically marginalized. The obsolescence of the traditional sectors that had been the source of the island's prosperity could only have been addressed with bold public investment programs, which have never been implemented by the Greek state. However, in the last 40 years, developments such as the establishment of the University of the Aegean and (more importantly) the establishment, in 1986, of the country's 13 Regional Authorities (which brought about the administrative upgrade of the city of Mytilene, from a prefectural capital to a regional capital) have given some impetus to the local tertiary sector.

In the view of Strauss & Howe's (1991) 'generational theory', the local generations of the coming saeculum period could once again play a leading role in a new cycle of innovation, growth and development for the Island of Lesvos. The main challenge they have to overcome today is however the dramatic surge in illegal immigration flows, over the last ten years, and the related policies of the EU for the permanent and definitive relegation of Lesvos into a major hub for the so called 'Aegean Route', something that would create new setbacks for future development prospects, as it would definitely favor the development of all kinds of parasitic, distorted and unsustainable activities, while, at the same time, would act as a deterrent to the island's tourism development prospects.

REFERENCES

Agriantoni Ch., (1986), *The beginnings of industrialization in Greece in the 19th century*, Athens [Αγριαντώνη Χ., (1986), *Οι απαρχές της εκβιομηχάνισης στην Ελλάδα του 19° αιώνα,* Αθήνα] (in Greek)

Anagnostou, S. (2004), *The residential development of Lesvos (1462-1912)*, PhD Thesis, Department of Geography, University of the Aegean, Mytilene [Αναγνώστου Στρατής, (2004), *Η οικιστική εξέλιξη της Λέσβου (1462-1912),* Διδακτορική Διατριβή, Τμήμα Γεωγραφίας, Πανεπιστήμιο Αιγαίου, Μυτιλήνη] (in Greek)

Anagnostou S., (2024), *Concise History of Lesvos, 1800-1950*, Mytilene, unpublished [Αναγνώστου Στρατής, (2024), *Συνοπτική Ιστορία της Λέσβου, 1800-1950*, Μυτιλήνη, αδημοσίευτο] (in Greek)

Bakalis, C. (2016). Tobacco Networks in the Aegean Islands. *Advances in Historical Studies*, *5*(2), 63–79. DOI: 10.4236/ahs.2016.52007

Boulanger, P. (1991), "L'Ile de Mytilène et le négoce français au XVIIIe siècle", in: *Les Villes dans l'Empire ottoman: activités et sociétés,* CNRS, Paris, 1991, t. 1, p. 273-298 (Greek translation in: *Λεσβιακά, Δελτίου της Εταιρείας Λεσβιακών Μελετών,* τ. XVI, Μυτιλήνη, 1996, σ. 176-200) (in French/Greek)

Conze, A. (1865). *Reise auf der Insel Lesbos*. Rümpler. (in German)

Cuinet V., (1892), *La Turquie d'Asie: géographie administrative, statistique, descriptive et raisonnée de chaque province de l'Asie-Mineure,* (4 vol.), Paris (in French)

Hellenic Statistical Authority (ESYE-ELSTAT), *Population Censuses:* 1920, 1928, 1940, 1951, 1961, 1971, 1981, 1991, 2001, 2011, 2021 [ΕΣΥΕ-ΕΛΣΤΑΤ, *Απογραφές Πληθυσμού:* 1920, 1928, 1940, 1951, 1961, 1971, 1981, 1991, 2001, 2011, 2021] (in Greek)

Kaitatzidis M., (2009), *Aegean Steam Navigation P.M. Courtgis & Co, 1883-1911,* Patras (bilingual edition, Greek/English)

Kampouris, Z. P. (1962), "The last years of the Ottoman period in Lesvos", *Lesviaka*, vol. 4, Mytilene, (pp. 102-188) [Καμπούρης Ζ., (1962), «Τα τελευταία χρόνια της Τουρκοκρατίας στη Λέσβο», *Λεσβιακά* τ.4, Μυτιλήνη, σ. 102-188] (in Greek)

Kampouris Z.P., (1978), *Natural disasters in Lesvos in the 19th century,* Mytilene [Καμπούρης Ζ., (1978), *Θεομηνίες στη Λέσβο του 19ο αιώνα,* Μυτιλήνη] (in Greek)

Karpat K., (1985), *Ottoman population 1850-1914, Demographic and Social Characteristics*, London

Madison Kolodny E., (1974), *La population des îles de la Grèce. Essai de géographie insulaire en Méditerranée Orientale*, 3 vol., Aix-en-Provence (in French)

Kolodny, E., & Darques, R. (2004), "Turcs, Grecs et réfugiés dans l'île de Lesbos au XXe siècle", *Méditerranée* 103 (pp. 65-74) (in French)

Launay L. de, (1897), *Chez les Grecs de Turquie*, Paris (in French) Mantran R., (1989), *Histoire de l'Empire Ottoman*, Paris (in French)

Mantzaris, A. (1928). *Our tobaccos*. Pyrsos. [Μάντζαρης Α., (1928), *Τα καπνά μας*, Πυρσός, Αθήνα] (in Greek)

Newton, C. T. (1865), *Travels and discoveries in the Levant*, London

Paraskevaidis P., (2013), *The refugees of the second persecution in Lesvos*, (2nd edition)

Tartaruga, Mytilene [Παρασκευαϊδης Π., (2013), *Η προσφυγιά του β' διωγμού στη Λέσβο*, Ταρταρούγα, Μυτιλήνη] (in Greek)

Rougon F., (1892), *Smyrne*, Paris (in French)

Sifneou, E. (1996a), *Lesvos. Economic and Social History*, 2nd Edition, 2016, Aeolida, Mytilene [Σιφναίου Ε., (1996), *Λέσβος. Οικονομική και Κοινωνική Ιστορία* 2η έκδοση, 2016, Αιολίδα, Μυτιλήνη] (in Greek)

Sifneou, E. (1996b), "The industrialization of Lesvos during the late Ottoman period", *Lesviaka, Bulletin of the Society of Lesvian Studies*, vol. XVI, Mytilene, p. 454-478 [Σιφναίου Ε., (1996), «Η εκβιομηχάνιση της Λέσβου κατά την ύστερη Τουρκοκρατία», *Λεσβιακά, Δελτίον της Εταιρείας Λεσβιακών Μελετών,* τ. ΙΣΤ' Μυτιλήνη, σ. 454-478] (in Greek)

Sifneou, E. (Ed.). (2002), Soapmaking in Lesvos. A memento, Livanis, Athens (bilingual edition, Greek/English)

Sifneou, E. (2007). Lesvos-Ayvalik-Adramytti: "Expansionism" or equal exchange in the trade and manufacture of olive-oil? In Kitromilides, P., & Michaelaris, P. (Eds.), *Mytilene and Ayvalik: A bilateral historical relationship in the north-eastern Aegean* (pp. 249–260). [Σιφναίου Ε., (2007), "Λέσβος-Κυδωνίες-Αδραμύτι: Οικονομικός επεκτατισμός ή ισότιμη συναλλαγή στην εμπορία και τη μεταποίηση του ελαιολάδου;" στο: Κιτρομηλίδης Π., Μιχαηλάρης Π., (Επιμ.), *Μυτιλήνη και Αϊβαλί (Κυδωνίες): Μια αμφίδρομη σχέση στο βορειοανατολικό Αιγαίο*, (σελ. 249-260), Αθήνα] (in Greek)

Strauss, W., & Howe, N. (1991). *Generations: The history of America's future, 1584 to 2069*. William Morrow.

Strohmeier, M. (2010), "Economy and society in the Aegean province of the Ottoman empire, 1840-1912", *Turkish Historical Review*, 1.2, (pp. 164-195) DOI: 10.1163/187754610X538825

Vafeas, N., & Mandamadiotou, M. (2023), *The islands of North-Eastern Aegean. From the Ottoman Empire to the Greek State,* Ed. Alexandria, Athens [Βαφέας N., Μανδαμαδιώτου M., (2023), *Τα νησιά του βορειοανατολικού Αιγαίου. Από την Οθωμανική Αυτοκρατορία στο Ελληνικό Κράτος*, Εκδ. Αλεξάνδρεια, Αθήνα] (in Greek)

Vogiatzis, A. (Ed.). (2012). *Lesvos in 1912. Detailed depiction of the state of the island by the first Greek Administration*. Aeolida. [Βογιατζής Α. (Επιμ.), (2012), *Η Λέσβος το 1912. Λεπτομερής αποτύπωση της εικόνας του νησιού από την πρώτη Ελληνική Διοίκηση*, Αιολίδα, Μυτιλήνη] (in Greek)

von Scherzer, K. (1873), *Smyrna*, Vienna (Greek Edition, 1995, Istoritis, Athens) [Scherzer K., *Σμύρνη,* Ιστορητής, Αθήνα] (in Greek)

Chapter 2
Analyzing the Determinants of Regional Unemployment in Turkey:
Evidence from Spatial Panel Data

Merve Baş
https://orcid.org/0000-0001-9856-2447
Independent Researcher, Turkey

Coşkun Akdeniz
https://orcid.org/0000-0002-3973-754X
Tekirdag Namık Kemal University, Turkey

ABSTRACT

Unemployment is one of the most important macroeconomic indicators for both developed and developing countries, serving as a key indicator of economic performance and social well-being. Economic crises, structural problems within economies, and regional disparities may increase unemployment. This study uses spatial econometric techniques to examine the determinants of regional unemployment in Turkey. Based on the NUTS-2 classification, 26 subregions were selected for empirical analysis for 2009-2021. The study considers income levels, general prices, education levels, youth population growth, and out-migration variables at the regional level as explanatory variables. The findings show that unemployment rates in Turkey are spatially dependent, while the random effect spatial error model results indicate that higher income levels and out-migration reduce unemployment. Conversely, general price levels, education levels, and a growing youth population increase regional unemployment rates.

DOI: 10.4018/979-8-3373-1554-6.ch002

1. INTRODUCTION

Unemployment is one of the main economic problems of all countries. Besides harming economic growth, it also distorts income distribution. Beyond the damages on macroeconomic parameters, unemployment directly affects social order (Gayawan et al., 2025). Due to its negative socio-economic impacts, unemployment remains a core issue that both developed and developing countries strive to minimize (Özsabuncuoğlu & Direkçi, 2012).

Similarly, Turkey has been struggling with unemployment for many years. As shown in Figure 1, the unemployment problem has been gradually aggravating. During the 2008 global financial crisis, unemployment climbed as a natural consequence of the economic slowdown (Kesebir, 2018). The moderate recovery in the early 2010s, marked by accelerating capital inflows, has been repeatedly interrupted by internal and external shocks (İlhan & Akdeniz, 2024), and unemployment reached a peak of 13.73% in 2019. The slowdown in global production due to lockdowns and stringency measures during the COVID-19 pandemic has made reducing unemployment even more difficult for Turkey.

Figure 1. Turkey's Unemployment Rate (1980-2022)

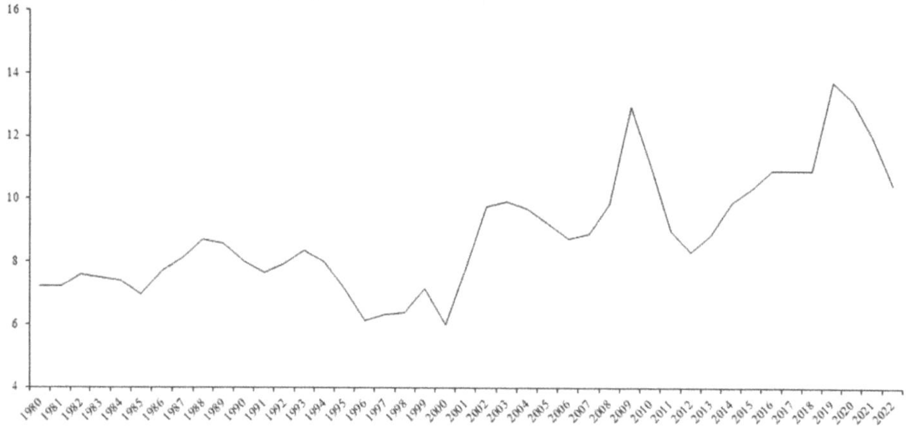

Note: *International Monetary Fund (IMF, 2024)*

Beyond domestic and global shocks, structural factors also push the unemployment rate in Turkey to high levels. Pre-deindustrialization, surplus generated by rapid growth in the youth population, uncontrolled rural-to-urban migration, and the limited propensity of educated individuals to enter labor markets make it difficult to increase employment (Uyar Bozdağlıoğlu, 2008). Moreover, inter-regional disparities within Turkey lead to increasing unemployment levels. These regional differences in economic development also exacerbate the distorting effect of unemployment on income distribution (Eser & Terzi, 2008). Therefore, this study aims to identify the determinants of unemployment across Turkey's 26 Nomenclature of Territorial Units for Statistics (NUTS) 2 subregions and reveal any potential spatial dependence in regional unemployment with spatial panel-data techniques for the 2009-2021 period.

The remainder of the paper proceeds as follows. The next section reviews selected empirical studies from the relevant literature. The third section describes the data set and methodological framework. Empirical results are presented and discussed in the fourth section. The last part of the study concludes with a summary of the main findings.

2. LITERATURE REVIEW

Although the national unemployment rate tends to dominate analyses of the causes of unemployment and the formulation of policies to combat it, regional disparities in economic development within countries must also be taken into account (Badinger & Url, 2002). This is because such regional economic differences can cause losses nationally and undermine balanced and sustainable development processes (Danova & Vozarov, 2020). Accordingly, the following literature review includes both national and regional unemployment studies. The general trend in the literature reveals a greater focus on the national level, often examining specific countries or groups of countries, whereas fewer studies address unemployment at the regional scale.

Among studies examining specific economies at the national level, Eita and Ashipala (2010) investigated the determinants of unemployment in Namibia focusing on macroeconomic factors. Analyzing data from 1971 to 2007 using cointegration analysis, the researchers found a negative relationship between unemployment and inflation, while unemployment was reduced by increases in aggregate demand and investment. Based on an autoregressive distributed lag (ARDL) framework, Subramaniam and Baharumshah (2011a) examined the determinants of unemployment in the Philippines over the period 1974-2003. According to the results, neither inflation nor foreign direct investment (FDI) had a statistically significant impact on unemployment. On the other hand, economic growth and open vacancy rates are found to increase unemployment, whereas public expenditures lead to decreased unemployment. Among the limited number of studies examining the Pakistani economy, Maqbool et al. (2013) identified an inverse relationship between unemployment and inflation in both the short and long run. Chowdhury and Hossain (2014) applied regression analysis to investigate the impact of macroeconomic variables on Bangladesh unemployment rates between 2000 and 2011. The empirical findings suggest a positive relationship between unemployment and inflation, conversely higher economic activity and exchange rate movements led to lower unemployment rates.

Oniore et al. (2015) examined the effects of macroeconomic factors on the unemployment rate in Nigeria between 1981 and 2014. According to the results, there are significant short-run relationships between unemployment and inflation, economic activity, trade openness, and private domestic investment. Additionally, these indicators have a negative impact on unemployment in the long-term. Similarly, Ogbeide et al. (2016) investigated the drivers of unemployment by utilizing ordinary least squares (OLS) and ECM approaches covering the period 1981 to 2013. It was found that there was no statistically significant relationship between inflation and unemployment, either in the short or long term. In another study focusing on the Nigerian economy, Raifu (2017) used cointegration analysis to explore the impact of the current account balance and trade openness on unemployment for 1981-2014. Based on the empirical results, the deteriorating current account balance leads to an increase in unemployment in the short term; however, it decreases unemployment in the long term. In addition to this, trade openness has a positive impact on unemployment in both terms. Alrabba (2017) demonstrated using a vector autoregression (VAR) model that private investments in Jordan decrease unemployment while inflation and population growth increase it between 1992 and 2015.

There have been several studies conducted on the Turkish economy. For example, Peker and Bölükbaş (2012) found that an increase in the real sector confidence index decreased unemployment between 2000Q1 and 2011Q1, whereas a rise in collective labor agreements increased it. Based on the ARDL bounds approach, Günaydın and Çetin (2015) examined the determinants of youth unemployment in Turkey between 1988 and 2013. According to the results, there is a negative relationship between youth unemployment and per capita income, FDI, and trade openness. Sever and İğdeli (2018) investigated the drivers of youth unemployment in Turkey for the period from 1988 to 2006. The study concluded that youth unemployment has a negative long-run relationship with per capita income. By employing the vector error correction model (VECM), Apaydın (2019) investigated unemployment dynamics in Turkey between 2005Q1 and 2018Q4. The analysis reveals that technological shocks, labor demand, and real wage growth contributed to a decline in the unemployment rate. In contrast, an increase in labor supply led to a rise in unemployment. Sağlam and Kaya (2020) used ARDL bounds testing to conclude that information and communication technology activities initially reduced unemployment but ultimately increased it.

Among the studies that examined country groups, Baccaro and Rei (2007) analyzed selected OECD members between 1960 and 1998 to identify the components of unemployment. Research has indicated that contractionary monetary policies by central banks have a greater impact on unemployment rates than institutional factors. Subramaniam and Baharumshah (2011b) investigated the relationship between unemployment rates and macroeconomic factors in Malaysia, Singapore, and the Philippines. According to the empirical findings, exports and foreign direct investments were the most significant determinants of unemployment in Malaysia. In the Philippines, government spending and export growth contributed to a reduction in unemployment, while in Singapore, exports alone played a significant role in determining unemployment rates. Caporale and Gil-Alana (2014) investigated the macroeconomic drivers of youth employment in European countries between 1980 and 2005 using autoregressive and cointegration approaches. Results from the study revealed a long-term relationship between economic activity, inflation, and unemployment.

Folawewo and Adeboje (2017) analyze the components of unemployment in the Economic Community of West African States for 1991-2014 and find that inflation significantly increased the unemployment rate. Notably, increased labor productivity and population growth were associated with an increase in unemployment, whereas both FDI and external debt resulted in a decrease in unemployment. Bayrak and Tatlı (2018) used panel data analysis to investigate the determinants of youth unemployment in OECD countries between 2000 and 2015. The empirical findings indicated that economic activity, inflation, and savings reduced youth unemployment, while higher labor productivity increased it. Azolibe et al. (2022) employed cointegration analysis and VAR models to analyze the impact of banking system credits on unemployment rates in Nigeria and South Africa from 1991 to 2018. Based on the findings, banking system credits were more effective in reducing unemployment in South Africa, whereas foreign direct investment was more important in Nigeria. Additionally, higher public expenditures led to a reduction in the unemployment rate in both countries.

Among the studies examining unemployment rates at the regional level, Filiztekin (2009) employed both parametric and non-parametric spatial analysis methods to investigate the Turkish economy from 1980 to 2000, using NUTS-3 and NUTS-2 regions. According to the findings, unemployment rates remained higher in the provinces and regional disparities widened during periods of increased national employment. Among NUTS-3 regions, the highest clustering of unemployment was found in the Southeastern Anatolia Region, which has human capital deficiencies and a lack of demand for labor. Trimurti and Komalasari (2014) examined the effect of economic activity, inflation, and minimum wage on unemployment rates across seven provinces in Indonesia between 2004 and 2012. The regression analysis revealed that economic activity and the minimum wage did not have a statistically significant effect on unemployment, whereas inflation had a significant and positive impact.

Based on spatial econometric analysis, Karaalp-Orhan and Gülel (2016) compared 26 Turkish subregions classified at the NUTS-2 level to identify the determinants of unemployment both before and after the 2008 global financial crisis. According to the analysis, regional unemployment rates were significantly affected by net migration, population growth rate, and employment in the agricultural sector during the crisis period. Following the crisis, migration rates and the share of employment in the agricultural sector remained associated with higher unemployment rates. A similar study by Aral and Aytaç (2018) examined regional unemployment dynamics at the NUTS-3 level in Turkey using data for 2011 to identify determinants of unemployment and assess regional dependence. The findings revealed a significant negative relationship between the female labor force participation rate and unemployment. Conversely, a significant positive relationship was observed between the youth population, the educated population ratio, and unemployment. The study also confirmed the presence of spatial dependence, indicating that unemployment in each region is influenced by the rates in neighboring regions.

3. DATA SET AND METHODOLOGY

In this study, the determinants of regional unemployment rates within the NUTS-2 classification of the Turkish economy were analyzed using spatial econometric methods. The study used annual data covering 26 subregions the period from 2009 to 2021, based on data availability. Appendix Table 6 presents the 26 NUTS-2 subregions of Turkey along with their respective codes.

The dependent variable was the unemployment rate for individuals aged 15 and over (un_{it}). The five macroeconomic explanatory variables are income per capita in US dollars ($income_{it}$), the inflation rate (inf_{it}), education level which is represented as the proportion of individuals aged 15 and over who have graduated from high school or an equivalent level (edu_{it}), the population growth rate of the 15-24 age group ($young_{it}$), and out-migration from each region (mig_{it}). Income level, education level, and migration were included in the analysis using their logarithmic values. In the absence of spatial dependence, the model estimated in panel data format is expressed as follows:

$$un_{it} = \alpha_i + \beta_1 income_{it} + \beta_2 inf_{it} + \beta_3 edu_{it} + \beta_4 young_{it} + \beta_5 mig_{it} + u_{it} \quad (1)$$

Spatial econometrics is a subfield of econometrics that focuses on incorporating spatial locations and interactions into models, and specifying and estimating classical regression models within this spatial framework (Anselin et al., 2013). In all fields concerned with regional science, it is essential to model variables in a way that captures the interaction of social, strategic, and spatial positions in order to achieve more accurate and meaningful results (Akerlof, 1997: 1005). Therefore, spatial data analysis, which is becoming preferred over classical regression analysis methods, includes geographical locations and spatial dependency in the model, based on the proposition that the values of variables in neighboring locations tend to be more similar or related than those in distant locations (Fischer & Wang, 2011). Consequently, the values of a variable observed at a specific location tend to resemble those in nearby locations, which consequently leads to spatial clustering (Anselin, 1992).

It is necessary to construct a spatial weight matrix (w) in order to take into account for the dependency among spatial units that influence each other. The weight matrix, defined as an nxn matrix, is constructed using contiguity and distance criteria to represent spatial relationships among geographical units (Cook et al., 2015; Shi & Goulias, 2024). Accordingly, n corresponds to the number of spatial units represented by the rows and columns, while w_{ij} refers to the entry in the i-th row, representing a specific location, and the j-th column, representing another location, of the matrix (Anselin et al., 2008). The spatial weight matrix w, designed to exhibit the mutual interactions among spatial units, is constructed in different ways depending on contiguity and distance.

The spatial weight matrix based on contiguity was calculated for NUTS-2 regions in this study. The neighborhood relationships between spatial units are represented by 0 and 1, where the matrix element is 1 (w_{ij}=1) if the units are connected by a common border (Anselin, 1988).

In order to determine whether there is a spatial dependence between geographic units, Moran's I statistic is applied to the regression residuals and expressed as follows (Anselin & Hudak, 1992; LeSage, 1999):

$$I = \frac{N}{S_0}\left(\frac{e'We}{e'e}\right) \tag{2}$$

Here, I represents the regression residuals, N denotes the number of observations, and e is the residual vector. S_0 refers to the standardized sum of all elements in the spatial weight matrix ($S_0 = \sum_i \sum_j w_{ij}$). $e'e$ indicates the sum of the squared residuals. Moran's I test determines whether an observation at spatial location i exhibits positive or negative spatial dependence with observations at other spatial locations j; however, it does not indicate what type of spatial dependence exists (Zeren, 2010).

The Lagrange Multiplier (LM) test, developed to determine the type of spatial dependence, is used to select the optimal model based on the maximum likelihood method (Abban et al., 2025: 15). For the spatial error model (SEM), the null and alternative hypotheses of the LM test are as follows: the null hypothesis assumes no spatial error (λ=0), while the alternative hypothesis assumes the presence of spatial error ($\lambda \neq$0). The LM test statistic is expressed as follows:

$$LM_{\lambda} = \{e'We / \sigma^2\}^2 / \, tr \, [W'W + W^2] \tag{3}$$

Here, $\sigma^2 = e'e/N$ and tr represent the error variance and the sum of the diagonal elements of the related matrix, respectively (Fischer & Wang, 2011). For the spatial lag model (SAR), the null and alternative hypotheses of the LM test are defined as follows: the null hypothesis assumes no spatial lag (ρ=0), while the alternative hypothesis assumes the presence of spatial lag ($\rho \neq$0). Accordingly, the LM test for the spatial lag model is defined as follows:

$$LM_{\rho} = \{e'Wy/\sigma^2\}^2 \; / \; \{((WXb)' MWXb)/\sigma^2) \; + tr \; [W'W + W^2]\} \tag{4}$$

Under the equality $M = I - X(X'X)^{-1}X'$, the term b represents the regression coefficients (Anselin & Hudak, 1992). If spatial autocorrelation is detected, then LM statistics are calculated to select the appropriate model, namely the one (either spatial lag or spatial error) that rejects the null hypothesis. If the null hypothesis is rejected for both models, then robust versions of the LM tests should be employed to determine the correct model (Anselin, 2005).

Another test used for identifying the appropriate model is the Likelihood Ratio (LR) test. For this test to be applied, spatial error or spatial lag models must first be estimated (Elhorst, 2014: 49). Additionally, since it allows for unbiased parameter estimation, the LR test evaluates the Spatial Durbin Model (SDM), which simultaneously includes spatially lagged dependent and independent variables, under the null hypothesis H_0: θ=0 (Elhorst, 2010; Abban et al., 2025). The classification of spatial regression models according to types of spatial dependence is presented in Figure 3 in the appendix.

Incorporating the time dimension into spatial regression analysis makes these models spatial panel data models. Spatial panel data, which combine both time series and cross-sectional data, are categorized as fixed effects and random effects models. The fixed effects model is appropriate when unobserved spatial and temporal effects are included in the model intercept and correlate with the independent variables. Conversely, the random effects model applies if the unobserved effects are part of the error term and are uncorrelated with the independent variables (Yılgör, 2019).

The Hausman test is employed to distinguish between fixed effects and random effects panel models. The null (H_0: $E(u_{it}/X_{IT})=0$) and alternative hypotheses (H_1: $E(u_{it}/X_{it}) \neq 0$) of this test correspond to the random effects and fixed effects models being valid, respectively (Baltagi, 2005).

4. EMPIRICAL FINDINGS

Spatial data analysis aims to reveal potential interactions between variables observed in a specific geographic region and those in other geographic locations. Accordingly, regional distributions must first be defined, while spatial clustering patterns should be visually represented (Fischer & Wang, 2011). Accordingly, the following analysis section reports the spatial distribution of unemployment across NUTS-2 regions in the Turkish economy. Figure 2 displays the spatial distributions of regional unemployment data at the NUTS-2 level for 2021 specifically.

In Figure 2, darker and lighter colors represent regions with higher and lower unemployment rates, respectively. The figure reveals a decreasing trend in unemployment from eastern and southeastern regions toward western regions, with clusters. Therefore, the distribution patterns of unemployment rates provide evidence of spatial dependence.

The seven regions marked in dark blue, indicating the highest unemployment rates, have values ranging approximately between 13.4 and 29.8 percent, whereas the six regions shown in lighter colors, representing the lowest unemployment rates, have values ranging between 5.8 and 9.39 percent. According to the NUTS-2 regional classification of the data for 2021, the highest unemployment rate of 29.8 percent is in TRC3, which includes Mardin, Batman, Şırnak, and Siirt. This region is also notable for having the second-lowest per capita income levels.

Figure 2. Spatial Distribution of NUTS-2 Level Unemployment Rates in Turkey

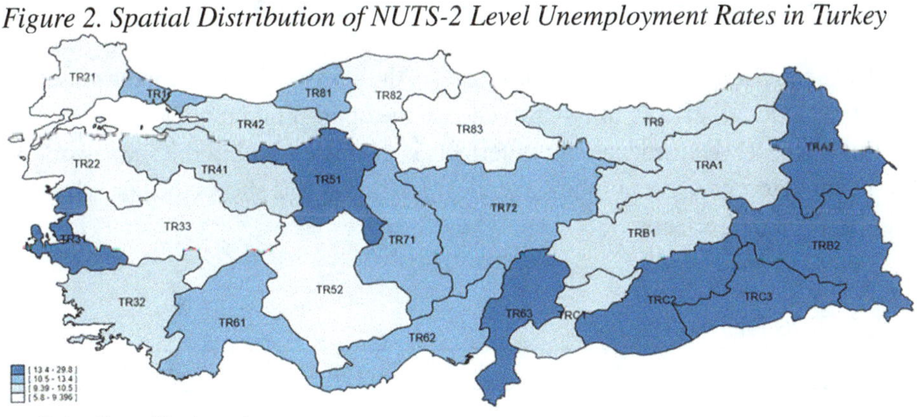

Note: *Created by the authors.*

One of the most important statistical measures for testing spatial dependence is Moran's I. Table 1, which presents the Moran's I statistics calculated for the unemployment rate for 2009-2021, provides evidence of its potential spatial interaction. A positive Z-score implies that an increase in the variable's value within one geographical region exerts a stimulating effect on neighboring regions with which it shares spatial dependence. Based on the Moran's I statistic p-values, the null hypothesis is rejected, generally at the 1% significance level, for every year except 2010, 2011, and 2012. In other words, regional unemployment rates in Turkey exhibit spatial dependence.

Table 1. Moran's I Statistic for the Unemployment Rate in Turkey (2009-2021)

Year	I	E(I)	Sd(I)	Z	Probability
2009	0.3123	-0.0400	0.1302	2.7054	0.0068
2010	0.0997	-0.0400	0.1299	1.0757	0.2820
2011	0.0850	-0.0400	0.1305	0.9579	0.3381
2012	-0.0292	-0.0400	0.1123	0.0958	0.9237
2013	0.2203	-0.0400	0.1221	2.1307	0.0331
2014	0.2860	-0.0400	0.1210	2.6940	0.0071
2015	0.2214	-0.0400	0.1199	2.1793	0.0293
2016	0.2282	-0.0400	0.1145	2.3415	0.0192
2017	0.1869	-0.0400	0.1157	2.3415	0.0499
2018	0.3781	-0.0400	0.1239	3.3759	0.0007
2019	0.4278	-0.0400	0.1210	3.8653	0.0001
2020	0.3985	-0.0400	0.1130	3.8811	0.0001
2021	0.3073	-0.0400	0.1130	3.1174	0.0018

Once spatial interaction has been confirmed, the next step is to test for the presence of unit (individual) effects in order to choose the appropriate model specification. For this purpose, the presence of unit effects is examined within the framework of the spatial autoregressive combined model (SAC). Table 2 shows the test statistics calculated for unit effects.

Table 2. Tests for Unit Effects

Tests	SAC	Probability
Breusch-Pagan LM-Two Side	705.2722	0.0000
Breusch-Pagan ALM-Two Side	492.6390	0.0000
Sosa-Escudero-Yoon LM-One Side	26.5570	0.0000
Sosa-Escudero-Yoon ALM-One Side	22.1955	0.0000
Baltagi-Li LM Autocorrelation Test	245.0552	0.0000

continued on following page

Table 2. Continued

Tests	SAC	Probability
Baltagi-Li ALM Autocorrelation Test	32.4219	0.0000
Baltagi-Li LM AR(1) Joint Test	737.6941	0.0000

The Baltagi-Li LM Joint test was employed to examine unit effects. The null hypothesis of this test $\left(\lambda = \sigma_\mu^2 = 0\right)$ is that there is neither autocorrelation nor unit effects, whereas the alternative hypothesis is that there are both autocorrelation and unit effects. As Table 2 shows, the null hypothesis was rejected at the 1% significance level, indicating that both autocorrelation and unit effects are present in the model. Based on the results of the Baltagi-Li LM autocorrelation test and its unit-effect-robust counterpart, the Baltagi-Li Adjusted LM (ALM) autocorrelation test, the null hypothesis (λ=0) was rejected at the 1% significance level, indicating that autocorrelation is indeed present in the model. The results of the Breusch-Pagan ALM-Two Side test, which examines the presence of unit effects and is robust to autocorrelation, also lead to rejection of the null hypothesis. Accordingly, the alternative hypothesis $\left(\sigma_\mu^2 \neq 0\right)$ that there were unit effects was accepted. The final test used to examine the presence of unit effects is the Sosa-Escudero-Yoon ALM test. Based on the test results, the null hypothesis ($\sigma_\mu^2 = 0$), which assumes no unit effects, was again rejected and the alternative hypothesis $\left(\sigma_\mu^2 > 0\right)$, which indicates the presence of unit effects, was accepted.

Overall, therefore, the test results establish the presence of unit effects in the model, making it necessary to determine whether the spatial fixed effects or random effects panel data model is appropriate. For this purpose, the Hausman test, which examines the random effects model, was conducted using the SAR model. The relevant test statistics are presented in Table 3. The null hypothesis of the Hausman test assumes that the random effects model is valid. According to the results, the null hypothesis was accepted at the 5% significance level, leading to the conclusion that the random effects model is appropriate.

Table 3. Hausman Test

Hausman Test	SAR	Probability
	367.651	0.5969

After establishing the presence of spatial dependence and confirming the validity of the random effects model, Global Moran's I, LM, and robust LM tests were conducted to identify the type of spatial dependence through spatial error and spatial lag models. The LM test results based on the SAR model are presented in

Table 4. According to the Global Moran's I test, the null hypothesis that there is no spatial autocorrelation was rejected at the 1% significance level and the alternative hypothesis of the presence of spatial autocorrelation was accepted. Since the Global Moran's I test does not provide information about the type of spatial dependence, LM tests were conducted to examine the structure of the spatial model.

Table 4. Spatial Panel Autocorrelation Tests

Tests	Test Statistics	Probability
Global Moran I	-0.1430	0.0001
LM Error (LM_{λ})	11.9795	0.0005
Robust LM Error (LM_{λ}^{*})	14.4980	0.0001
LM Lag (LM_{ρ})	0.0047	0.9451
Robust LM Lag (LM_{ρ}^{*})	2.5233	0.1122
LM SAC ($LM_{\rho\lambda}$)	14.5028	0.0007

The results of the LM SAC test, which jointly evaluate spatial error and spatial lag, rejected the null hypothesis of no spatial error and no spatial lag. In other words, both forms of spatial dependence were present. Examining the separate LM statistics for each specification, the null hypothesis that there is no spatial lag was accepted at the 5% significance level, whereas the LM test result for the spatial-error specification rejected the null hypothesis at the 1% level, thereby confirming the presence of spatial error in the model.

Based on the LM statistics, the appropriate specification incorporates spatial error dependence. Consequently, the final model to be estimated must belong to the class that captures spatial error, namely, the spatial error model (SEM) or the spatial Durbin error model (SDEM). The previous results indicated that the random-effects specification was valid, which implies that the unobserved unit and time effects are uncorrelated with the explanatory variables. Given this, we employed SEM rather than SDEM, which augments the model with spatial lags of the regressors. In this case u_{it} in equation (1) could be written for the SEM model in terms of the spatial error coefficient (λ) and spatial weight matrix (w_{ij}) in below:

$$u_{it} = \lambda \sum_{j=1}^{N} w_{ij} u_{jt} + \varepsilon_{it} \quad (5)$$

Table 5 reports the estimates obtained from the random effects SEM for the determinants of the regional unemployment rate for Turkey.

Table 5. Random Effect SEM Model

Variables	Coefficient	Standard Error	Z-Value	Probability
$income_{it}$	-0.5877***	0.0919	-6.39	0.000
inf_{it}	0.0593**	0.0260	2.27	0.023
edu_{it}	0.2143***	0.0661	3.24	0.001
$young_{it}$	0.9326***	0.1635	5.70	0.000
mig_{it}	-0.4008***	0.1265	-3.17	0.002
α_i	0.0013****	0.1265	0.01	0.992
λ	0.3512***	0.0641	5.47	0.000

Note: ***, **, * indicate statistical significance of the corresponding coefficient at the 1 %, 5 %, and 10 % levels, respectively.

The spatial error coefficient was positive and statistically significant, confirming the presence of spatial dependence in the disturbance term across neighboring regions. All other explanatory variables were likewise statistically significant. More specifically, per capita income and outward migration reduced regional unemployment, whereas inflation, the proportion of upper-secondary graduates, and the growth rate of the youth cohort exerted positive effects. The variable that increased unemployment the most was the growth rate of the youth cohort, whereas the general price level increased unemployment the least. Among the variables that reduced unemployment, per capita income growth had the largest impact.

5. CONCLUSION

The unemployment rate is one of the key macroeconomic indicators that determine the effectiveness of national economic policy. The reduction of unemployment is therefore a key priority for both developed and developing economies. Despite macroeconomic instability, structural problems, and economic crises being well-known drivers of unemployment, regional differences also play an important role.

Using spatial econometric techniques, this study investigates unemployment determinants across 26 NUTS-2 subregions in Turkey. To do this, it estimates a regional unemployment equation including variables such as per capita income, inflation, educational attainment, youth population growth, and outward migration. The analysis is based on annual data from 2009 to 2021, depending on the availability of data. The findings of the spatial dependence test supported the use of the SEM model among the alternative spatial specifications.

According to the estimation results, all variables had a statistically significant effect on regional unemployment. The result indicates that an increase in the per capita income leads to a decrease in unemployment. Conversely, the inflation rate had an adverse impact on the regional unemployment rate. This can be adhered to the fact that increases in prices due to cost inflation might end up with higher operating expenses for firms, leading to increased unemployment (Irewole, 2019: 13).

Furthermore, the findings demonstrated a positive impact of education on unemployment. This indicates that regions with a higher education level tend to experience higher unemployment rates. Wage expectations and job preferences of individuals may rise with increasing levels of education. Because of this, employment opportunities that are compatible with their skills might be difficult. On the other hand, employers do not prefer hiring workers with higher education as they demand higher wages and social benefits, leading to an increase the labor costs. This dynamic can result in a mismatch between the qualifications sought by employers and those held by job seekers (Kabaklarlı & Gür, 2011).

Another key finding of the study was that a growing youth population leads to increased unemployment levels, in accordance with the results of Karaalp-Orhan and Gülel (2016) and Aral and Aytaç (2018). Finally, out-migration has resulted in a decrease in unemployment rates. In other words, the reduction in regional unemployment occurs as a result of increased out-migration, which decreases the population density in the respective region.

REFERENCES

Abban, O. J., Rajaguru, G., & Acheampong, A. (2025). The Spillover Effect of Economic Institutions on The Environment: A Global Evidence from Spatial Econometric Analysis. *Journal of Environmental Management, 373*, 123645. DOI: 10.1016/j.jenvman.2024.123645 PMID: 39689530

Akerlof, G. A. (1997). Social Distance and Social Decisions. *Econometrica, 65*(5), 1005–1027. DOI: 10.2307/2171877

Alrabba, M. I. (2017). The Determinants of Unemployment Rate in Jordan: A Multivariate Approach. *International Journal of Economics and Finance*, *9*(11), 109–117. DOI: 10.5539/ijef.v9n11p109

Anselin, L. (1988). *Spatial Econometrics: Methods and Models*. Springer Science & Business Media. DOI: 10.1007/978-94-015-7799-1

Anselin, L. (1992). Spatial Data Analysis With GIS: An Introductıon to Application. In *The Social Sciences*. National Center for Geographic Information and Analysis.

Anselin, L. (2005). *Exploring Spatial Data with GeoDa TM: A Workbook*. Center for Spatially Integrated Social Science.

Anselin, L., Florax, R., & Rey, S. J. (Eds.). (2013). *Advances in Spatial Econometrics: Methodology, Tools and Applications*. Springer Science & Business Media.

Anselin, L., Gallo, J. L., & Jayet, H. (2008). Spatial Panel Econometrics. *In The Econometrics of Panel Data: Fundamentals and Recent Developments in Theory and Practice*, 625-660.

Anselin, L., & Hudak, S. (1992). Spatial Econometrics in Practice: A Review of Software Options. *Regional Science and Urban Economics, 22*(3), 509–536. DOI: 10.1016/0166-0462(92)90042-Y

Apaydın, Ş. (2019). Türkiye'de İşsizliğin Kaynakları: Bir Yapısal Vektör Hata Düzeltme Modeli. *Manas Sosyal Araştırmalar Dergisi*, *8*(1), 1002–1016. DOI: 10.33206/mjss.464857

Aral, N., & Aytaç, M. (2018). Türkiye'de İşsizliğin Mekansal Analizi. *Marmara Üniversitesi Öneri Dergisi*, *13*(49), 1–20.

Azolibe, C. B., Dimnwobi, S. K., & Uzochukwu-Obi, C. P. (2022). The Determinants of Unemployment Rate in Developing Economies: Does Banking System Credit Matter? *Journal of Economic and Administrative Sciences*.

Baccaro, L., & Rei, D. (2007). Institutional Determinants of Unemployment in OECD Countries: Does the Deregulatory View Hold Water? *International Organization, 61*(3), 527–569. DOI: 10.1017/S0020818307070221

Badinger, H., & Url, T. (2002). Determinants of Regional Unemployment: Some Evidence from Austria. *Regional Studies, 36*(9), 977–988. DOI: 10.1080/003434002200022558

Baltagi, B. H. (2005). *Econometric Analysis of Panel Data.* John Wiley & Sons.

Bayrak, R., & Tatlı, H. (2018). The Determinants of Youth Unemployment: A Panel Data Analysis of OECD Countries. *The European Journal of Comparative Economics, 15*(2), 231–248.

Caporale, G. M., & Gil-Alana, L. (2014). Youth Unemployment in Europe: Persistence and Macroeconomic Determinants. *Comparative Economic Studies, 59*(4), 581–591. DOI: 10.1057/ces.2014.29

Chowdhury, M. S., & Hossain, M. (2014). Determinants of Unemployment in Bangladesh: A Case Study. *Developing Country Studies,* 4(3).

Cook, S. J., Hays, J., & Franzese, R. (2015). Model Specification and Spatial Interdependence. *Research Methods in Political Science and International Relations*, 1-30.

Danova, M., & Vozarov, I. K. (2020). Regional Disparities in the Relationship Between Economic Growth and Unemployment. *Scientific Papers of the University of Pardubice,* 28(2).

Eita, J. H., & Ashipala, J. (2010). Determinants of Unemployment in Namibia. *International Journal of Business and Management, 5*(10). Advance online publication. DOI: 10.5539/ijbm.v5n10p92

Elhorst, J. P. (2010). Applied Spatial Econometrics: Raising The Bar. *Spatial Economic Analysis, 5*(1), 9–28. DOI: 10.1080/17421770903541772

Elhorst, J. P. (2014). *Spatial Econometrics: From Cross-Sectional Data to Spatial Panels, 479.* Springer. DOI: 10.1007/978-3-642-40340-8

Eser, B. Y., & Terzi, H. (2008). Türkiye'de İşsizlik Sorunu ve Avrupa İstihdam Stratejisi. *Erciyes Üniversitesi İktisadi ve İdari Bilimler Dergisi*, (30), 229–250.

Filiztekin, A. (2009). Regional Unemployment in Turkey. *Papers in Regional Science, 88*(4), 863–878. DOI: 10.1111/j.1435-5957.2009.00237.x

Fischer, M. M., & Wang, J. (2011). *Spatial Data Analysis: Models, Methods and Techniques*. Springer Science & Business Media. DOI: 10.1007/978-3-642-21720-3

Folawewo, A. O., & Adeboje, O. M. (2017). Macroeconomic Determinants of Unemployment: Empirical Evidence from Economic Community of West African States. *African Development Review*, *29*(2), 197–210. DOI: 10.1111/1467-8268.12250

Gayawan, E., Mpandiko, A. D., & Ojo, O. (2025). Modelling the Spatio-Temporal Dynamics of Youth Unemployment in Africa: A Bayesian Approach. *Regional Science Policy & Practice*, *17*(1), 100149. DOI: 10.1016/j.rspp.2024.100149

Günaydın, D., & Çetin, M. (2015). Genç İşsizliğin Temel Makroekonomik Belirleyicileri: Ampirik Bir Analiz. *Pamukkale Üniversitesi Sosyal Bilimler Enstitüsü Dergisi*, (22), 17–34.

İlhan, A., & Akdeniz, C. (2024). The Effect of Real Wages on Employment after the Global Financial Crisis: The Case of the Turkish Manufacturing Industry. *Sosyoekonomi*, *32*(61), 147–166. DOI: 10.17233/sosyoekonomi.2024.03.08

IMF (2024). World Economic Outlook Databases.

Irewole, O. E. (2019). Inflatıon and Unemployment Relationshıp: A Dynamic Report Of Nıgeria and Mexico in The Perspective of Phillips Curve From 1991-2016. *International Journal of New Economics and Social Sciences*, *2*(10), 11–30. DOI: 10.5604/01.3001.0013.8085

Kabaklarlı, E., & Gür, M. (2011). *Türkiye'de Genç İşsizlik Sorunu ve Ekonomik Belirleyicilerinin Uzun Dönem Eş-Bütünleşme Analizi. 14.* Uluslararası İktisat Öğrencileri Kongresi.

Karaalp-Orhan, H. S., & Gülel, F. E. (2016). The Impact of The Global Financial Crises on Regional Unemployment in Turkey. *International Journal of Economic Perspectives*, *10*(1), 151–160.

Kesebir, M. (2018). 2008 Küresel Finansal Kriz ve Bu Krizin Türkiye'deki Ekonomik Göstergelere Etkisi. *Anemon Muş Alparslan Üniversitesi Sosyal Bilimler Dergisi*, *6*(4), 531–538. DOI: 10.18506/anemon.378086

LeSage, J. P. (1999). The Theory and Practice of Spatial Econometrics. *University of Toledo*, *28*(11), 1–39.

Maqbool, M. S., Mahmood, T., Sattar, A., & Bhalli, M. (2013). Determinants of Unemployment: Empirical Evidences From Pakistan. *Pakistan Economic and Social Review*, *51*(2), 191–208.

Ogbeide, F. I., Kanwanye, H., & Kadiri, S. (2016). Revisiting the Determinants of Unemployment in Nigeria: Do Resource Dependence and Financial Development Matter? *African Development Review*, *28*(4), 430–443. DOI: 10.1111/1467-8268.12222

Oniore, J. O., Bernard, A., & Gyang, E. (2015). Macroeconomic determinants of unemployment in Nigeria. *International Journal of Economics. Commerce & Management, 3*(10), 215–230.

Özsabuncuoğlu, İ. H., & Direkçi, T. (2012). Şeçilmiş Bazı Makroekonomik Değişkenlerin Türkiye'de Bölgesel İstihdama Etkisi. *Turkish Economic Association*, Discussion Paper, 2012/86.

Peker, O., & Bölükbaş, M. (2012). Türkiye'de İşsizliği Belirleyen Faktörlerin Ampirik Bir Analizi. *Atatürk Üniversitesi İktisadi ve İdari Bilimler Dergisi, 26*(3-4), 33–43.

Raifu, I. A. (2017). On The Determinants of Unemployment in Nigeria: What Are The Roles of Trade Openness and Current Account Balance? *Review of Innovation and Competitiveness: A Journal of Economic and Social Research,* 3(4), 5-30.

Sağlam, C., & Kaya, H. F. (2020). Bilgi ve İletişim İktisadi Faaliyetinin İşsizlik Oranı Üzerindeki Kısa ve Uzun Dönem Etkileri. *Uluslararası Ekonomi ve Yenilik Dergisi, 6*(2), 261–275.

Sever, E., & İğdeli, A. (2018). The Determining Factors of Youth Unemployment in Developing Countries: The Case of Turkey. *International Journal of Social and Economic Sciences, 8*(1), 75–83.

Shi, H., & Goulias, K. (2024). Understanding The Zero-Emission Vehicle Market Spatial Diffusion and Its Determinants from 2019 to 2022 Using Spatial Econometric Models. *Energy, 313*, 133607. DOI: 10.1016/j.energy.2024.133607

Subramaniam, T., & Baharumshah, A. Z. (2011a). Determinants of Unemployment in the Philippines. *The Empirical Economics Letters, 10*(12), 1248–1257.

Subramaniam, T., & Baharumshah, A. Z. (2011b). Unemployment and Speed of Adjustment in ASEAN-3 Economies: A Cointegration Analysis. *The Singapore Economic Review, 56*(03), 327–347. DOI: 10.1142/S0217590811004353

Trimurti, C. P., & Komalasari, Y. (2014). Determinants of Unemployment: Empirical Evidences From 7 Province in Indonesia. [SCIRJ]. *Scientific Research Journal, 2*(8), 5–9.

TURKSTAT. (2024). Bölgesel İstatistikler Veri Tabanı. Access Date: 12.05.2024, Access Link: https://biruni.tuik.gov.tr/bolgeselistatistik/anaSayfa.do?dil=tr

Uyar Bozdağlıoğlu, E. Y. (2008). Türkiye'de İşsizliğin Özellikleri ve İşsizlikle Mücadele Politikalari. *Manas Üniversitesi Sosyal Bilimler Dergisi, 10*(20), 45–65.

Vega, S. H., & Elhorst, J. (2015). The SLX model. *Journal of Regional Science, 55*(3), 1–25.

Yılgör, M. (2019). *Geoda, ArcGis, Matlab Uygulamalı Mekansal Ekonometri*. Der Yayınevi.

Zeren, F. (2010). Mekansal Etkileşim Analizi. *İstanbul Üniversitesi İktisat Fakültesi Ekonometri ve İstatistik Dergisi*, (12), 18-39.

ENDNOTES

[1] This study is derived from the dissertation of the first author completed at Tekirdağ Namık Kemal University, Institute of Social Sciences.

APPENDIX

Table 6. Subregions and Codes within Turkey's NUTS-2 Classification

Region Code	Region Name
TRA1	Erzurum, Erzincan, Bayburt
TRA2	Ağrı, Kars, Iğdır, Ardahan
TRB1	Malatya, Elazığ, Bingöl, Tunceli
TRB2	Van, Muş, Bitlis, Hakkari
TRC1	Gaziantep, Adıyaman, Kilis
TRC2	Şanlıurfa, Diyarbakır
TRC3	Mardin, Batman, Şırnak, Siirt
TR10	İstanbul
TR21	Tekirdağ, Edirne, Kırklareli
TR22	Balıkesir, Çanakkale
TR31	İzmir
TR32	Aydın, Denizli, Muğla
TR33	Manisa, Afyon, Kütahya, Uşak
TR41	Bursa, Eskişehir, Bilecik
TR42	Kocaeli, Sakarya, Düzce, Bolu, Yalova
TR51	Ankara
TR52	Konya, Karaman
TR61	Antalya, Isparta, Burdur
TR62	Adana, Mersin
TR63	Hatay, Kahramanmaraş, Osmaniye
TR71	Kırıkkale, Aksaray, Niğde, Nevşehir, Kırşehir
TR72	Kayseri, Sivas, Yozgat
TR81	Zonguldak, Karabük, Bartın
TR82	Kastamonu, Çankırı, Sinop
TR83	Samsun, Tokat, Çorum, Amasya
TR90	Trabzon, Ordu, Giresun, Rize, Artvin, Gümüşhane

Note: Turkish Statistical Institute (TURKSTAT, 2024)

Figure 3. Spatial Regression Models

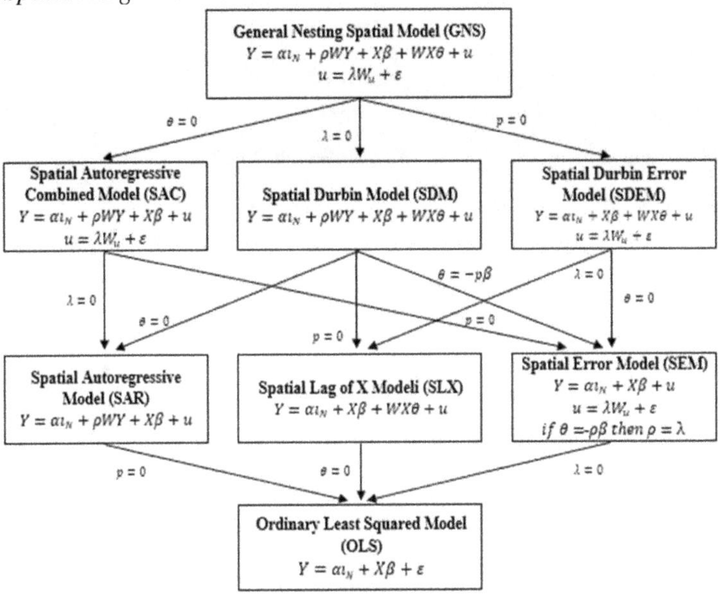

Note: *This figure is obtained from Vega and Elhorst (2015).*

Chapter 3
Sustainable Local Development and Local Administrations

Aylin Tepecik
https://orcid.org/0000-0002-3886-4301
Erciyes Üniversitesi, Turkey

ABSTRACT

Sustainable local development is important for the success of sustainable development. Local governments have become important actors in ensuring local development. As a basic service provider, local governments are the closest units to citizens in solving local problems and are easily accessible by citizens. This section aims to analyze the roles of local governments in sustainable local development. In this study, what are the roles and strategies for local governments in sustainable local development? In which areas are they successful or unsuccessful? What should local governments do to eliminate these failures? Answers were sought to the research questions. In order to find answers to the research questions, systematic literature review from qualitative research methods was used. This research examines the roles of local governments, one of the actors of sustainable local development, in local development and the strategies expected to be implemented by local governments.

INTRODUCTION

This section aims to analyze the roles of local governments in sustainable local development. In this study, what are the roles and strategies for local governments in sustainable local development? In which areas are they successful or unsuccessful? What should local governments do to eliminate these failures? Answers were

DOI: 10.4018/979-8-3373-1554-6.ch003

sought to the research questions. In order to find answers to the research questions, systematic literature review from qualitative research methods was used. For this purpose, a search was conducted in the databases Web of Science, Scopus, Science Direct, DergiPark, Researchgate, Google Scholar with the keywords ''local governments'' and ''sustainable local development''. The contribution of the study to the literature is to reveal the roles of local governments in sustainable local development, strategies for local governments and to offer suggestions for future studies. This research examines the roles of local governments, one of the actors of sustainable local development, in local development and the strategies expected to be implemented by local governments.

What is the Role of Local Governments in Sustainable Local Development?

The concept of sustainable local development is defined as increasing the quality of life so that the society can have a sustainable life through the sustainable use of factors such as people, environment, natural resources, and income growth (Kutluay Tutar et al., 2022). In sustainable local development, it is aimed to ensure that global and local actors act together and to ensure continuity in the development of local communities (Balkaya et al., 2022). Local actors affecting local development policies include units such as local governments, public institutions, private sector, regional development agencies, cooperatives, non-governmental organizations, enterprises, microfinance resources, local employment offices, education and training institutions, technoparks, and incubation centers (Atmaca, 2020; Balkaya et al., 2022; Kadiroğlu and Türko, 2024). In development, it is necessary for the relevant actors to exhibit a common management approach as well as active participation in management. It is important for local authorities and central authorities to act together in order to achieve sustainable local development goals. However, the development of the local economy, socially and humanly at the local scale is realized with a participatory approach (Eroğlu, 2010). In this direction, the local development process is a process in which all actors affecting development must act in cooperation and solidarity towards a common goal.

Local governments are institutions that promote economic development and welfare as they meet the needs of citizens for a wide range of public services such as health, education, housing, transportation services, cultural and entertainment facilities (Perugini, 2024). These governments are responsible for managing the administrative affairs of a particular region locally and ensuring public participation (Türk, 2024). Local governments, which are established as public legal entities that are closest to the local and address issues of primary concern to the society, are very important units that will facilitate local development (Çelebi and Tek, 2022; Türk,

2024). Local governments responsible for the management of cities are charged with important duties and responsibilities such as providing basic services to the people within their borders at the local level, ensuring social welfare at the local level, and implementing social and economic development policies. Nel (2001) argues that the primary responsibility of local governments is to find sustainable ways to meet the needs and improve the quality of life and to work with local communities to use these ways.

Local governments are given tasks that will contribute to the economic and social development of the region within their administrative borders (Zengin et al. 2014). The tasks of local governments identified as a result of the systematic literature review are given in Table 1.

Table 1. Important duties and responsibilities of local governments in terms of contributing to local development

References	Local Government Roles and Responsibilities
(Nel, 2001)	• To develop policies to reduce income inequality • To produce projects that will meet the needs of the local
(Sezer and Vural, 2009)	• To provide education, health, housing, transportation, infrastructure, development, water and sewage, environment and environmental health, solid waste, social and cultural services for the local people
(Şahin and Güven, 2011)	• Using local resources according to local needs • Determining and implementing social policies for the development of the region • Ensuring the welfare and happiness of the local people • Fighting against poverty
(Babaoğlu and Kocaoğlu, 2017)	• Making urban and environmental arrangements • Meeting the housing need • Providing social assistance • Providing vocational and skill-building activities to the local people in order to reduce unemployment • Opening shelters for women and children in need of protection
(Gündoğdu and Boyalı, 2018)	• Planning and implementation of activities for regional development • Leadership role in regional entrepreneurial activities • Providing the central administration with the necessary information for development plans and programs
(Tuğaç, 2019)	• Identifying urban-level hazards, risks and vulnerabilities to ensure urban resilience • Developing key strategies and practices to ensure urban resilience • Including capacity building and awareness projects that will raise public awareness about disaster preparedness
(Çelebi and Tek, 2022)	• It serves to make cities livable and sustainable places by providing infrastructure services such as roads, water and electricity
(Yıldız and Gümüş, 2023)	• Today, with the development of information and communication technology, it is possible to provide appropriate solutions for various service needs that encourage urban life and to carry out planned and programmed activities

continued on following page

Table 1. Continued

References	Local Government Roles and Responsibilities
(Türk, 2024)	• Managing the affairs of certain regions, developing solutions to local problems, ensuring that the needs of the people are met in order to increase social welfare
(Perugini, 2024)	• Supervising various public institutions in the areas of social welfare services, regional development, local transportation, child education, sports and cultural facilities, local security services, water distribution, waste disposal, and infrastructure expenditures
(Boz and Cay, 2024)	• Encouraging the preparation of smart city strategies and supporting local production on smart cities
(Du et al. 2024)	• Implementation and implementation of green policies • Addressing unsustainable issues • Developing strategic goals that include reducing greenhouse gas emissions • Developing and implementing green development strategies
(Lane et al. 2024)	• Developing new initiatives to reduce the amount of waste by local governments responsible for waste management. • Examine existing initiatives to separate and reduce household waste and develop household practices

Source: Drawn up by author.

When the studies identified as a result of the screening are examined, the central issues addressed in the studies in question are; preventing poverty, determining and implementing social policies that will meet local needs, activities aimed at reducing unemployment, making local areas livable and sustainable places, planning, approval and implementation duties of local governments in natural disasters and environmental protection. Current trends in the studies are seen as; adaptation to climate change, greenhouse gas reduction, use of smart technologies, disaster risk reduction, reduction of waste generation, environmental management, sustainability and resilience awareness raising.

The contribution of the duties and responsibilities of local governments to sustainable local development; development of compact city models and practices to prevent urban sprawl with zoning authority, open and green space system practices, implementation of practices to reduce urban heat island formation, promotion of renewable energy sources to reduce fossil fuel use and carbon emissions, promotion of energy efficient designs to reduce external dependency in energy can be addressed in cooperation with relevant institutions and organizations.

WHAT ARE THE STRATEGIES FOR LOCAL GOVERNMENTS IN PROVIDING SUSTAINABLE LOCAL DEVELOPMENT?

Local governments have an increasing responsibility in terms of economic development, social and public services. At this point, it is imperative for local governments to determine priorities, define strategies and then take action in order to be successful. Local governments have determined strategies to ensure sustainable development. These strategies can be listed as follows:

Poverty Reduction Strategies

Local governments are in a strategic position to determine the factors that cause poverty and strategies to reduce poverty. Strategies that local governments can follow in preventing poverty; determining the legal framework of urban poverty, improving slums, adopting gender-focused approaches, ensuring urban coordination and integration, making arrangements for street children, protecting local people against crime and natural disasters, implementing vocational training programs for the illiterate, providing microcredit to informal workers, developing entrepreneurial skills, creating new employment opportunities, providing vocational training for entering the labor market, providing minimum income support for school attendance, and conditional cash transfers (Wegelin and Borgman, 1995; Yeşildal, 2020a).

Environmental Management Strategies

In order to ensure sustainable management, local governments are responsible for reducing environmental problems in their regions and improving the quality of life. Strategies recommended for local governments to increase environmental performance; building a sustainable culture, environmental learning, reducing costs and increasing income, providing waste management, resource efficiency and recycling services, and providing sustainable environmental management (Yeşildal, 2020b).

Strategies to Combat Climate Change

Local governments play important roles in ensuring urban sustainability and resilience against disasters and risks. Strategies expected to be implemented by local governments in terms of adaptation to climate change; analyzing the unique conditions and problems of cities, addressing land use, urban planning and design with climate-sensitive approaches, ensuring urban integrated water management, conducting impact and vulnerability analyses within the scope of disaster risk management, preparing early warning systems for hazards and risks, developing nature-based approaches in cities, addressing green infrastructure systems as hybrid systems together with gray and blue infrastructures, evaluating the possible effects of climate change in terms of health infrastructure and disaster risks, addressing them in the context of population projections, reducing greenhouse gas emissions from tourism activities and targeting net zero emissions, and making the necessary plans in this direction (Tuğaç, 2021).

Strategies expected to be implemented by local governments for greenhouse gas emission reduction; preventing urban sprawl by taking a compact approach in reducing fossil fuel consumption, popularizing the green area approach at the neighborhood unit level, making sustainable transportation planning, developing different types of transportation, encouraging the use of alternative fuels other than fossil fuels in transportation vehicles, including mixed uses in cities, revitalizing public spaces with centers and sub-centers, reducing vehicle use, determining greenhouse gas emission sources, preparing greenhouse gas inventories, adopting a zero waste approach; reducing, separating and reusing waste at the source (Tuğaç, 2021).

Local governments can contribute to combating climate change through zoning plans. Issues that local governments should take into consideration are; taking environmental and climate conditions into consideration when preparing zoning plans, considering passive design strategies in zoning planning processes, giving importance to and encouraging ecological architecture, creating planned settlement areas, developing smart city applications, preventing construction in flood and floodplains, preventing landslide or landslide potential areas from being converted into settlements, and developing housing projects that are resilient to natural disasters (Uysal, 2022).

Urban Open and Green Space Management Strategies

Local governments play important roles in protecting, planning and developing management strategies for green areas and improving these strategies. Hassamancıoğlu (2021) determined management strategies associated with green areas in his study and examined these strategies by dividing them into local urbanism, environment and social categories. These strategies are; using ecological functions, protecting, improving and increasing green areas, ensuring access to parks and gardens for all segments of society, increasing the amount of green areas per capita, developing strategies for green areas in urban transformation activities, planning green areas in urban planning studies and providing details about planned projects, determining the amount of green areas per capita; and including information on the number of parks and recreation areas.

Energy Efficiency Strategies

Local governments play important roles in energy production, consumption and saving. It is evaluated that the use of local renewable energy types in the fight against climate change will also encourage low-carbon development and thus reduce greenhouse gas emissions (Tuğaç, 2021). The issues that local government units should pay attention to regarding energy efficiency, which is among the most basic strategies of countries; increasing alternative and renewable energy sources instead of traditional energy production methods with high carbon emissions, designing lighting and insulation options in newly constructed buildings in a way that will cause the least emissions in order to increase energy efficiency, encouraging efficiency-enhancing applications in existing and newly constructed buildings, using and encouraging renewable energy sources in urban transformation and large-scale housing projects, using low-energy technologies in traffic signs and street lighting systems, and developing projects that encourage energy saving by establishing an energy efficiency unit (Uysal, 2022).

Parking Strategies

Local governments have duties in the management and planning of parking lots. These duties are mainly to find solutions to problems in determining, planning and managing parking lots. Parking strategies, which are one of the methods used by local governments, play an important role in reducing traffic congestion and density, finding solutions to parking problems and increasing livability. Parking management contributes to making transportation systems more sustainable, improving traffic flow and facilitating daily life. Parking strategies of local governments include; parking regulations, more flexible parking standards, parking fees, financial incentives, car sharing, separation of parking and apartment sales, and limiting parking supply by changing regional regulations.

Smart City Strategies

Smart and sustainable city services have become more important for countries and local governments in terms of managing resources and using them efficiently and effectively. The goals and strategies for building smart and sustainable cities vary according to the national development programs of the countries. In Turkey, the 2024-2030 National Smart Cities Strategy and Action Plan and the Twelfth Development Plan were considered as reference documents regarding smart cities, and the smart city strategies of local governments were analyzed.

The policies that can be associated with the smart city initiatives of local governments in the Twelfth Development Plan include; implementing smart city applications within the framework of certain standards according to local needs, expanding the use of domestic products in smart city applications, expanding smart agricultural applications with digitalization, artificial intelligence and data-based business models, strengthening the smart grid infrastructure, expanding the use of structures that increase transportation efficiency with technological solutions such as smart transportation systems, smart roads, connected and autonomous driving systems, and expanding smart building designs that provide energy efficiency in buildings.

In the 2024-2030 National Smart Cities Strategy and Action Plan, the strategic objectives that can be associated with the smart city initiatives of local governments are; developing the smart city ecosystem, developing transparent and informative smart city applications, ensuring smart city transformation in urban services, developing monitoring, evaluation and reporting mechanisms in smart city activities. The actions expected to be carried out by local governments are; organizing training activities for the development of smart city applications, preparing development plans in cities to ensure smart city transformation, encouraging the use of low-carbon and sustainable transportation vehicles for transportation, increasing the use of multi-modal transportation systems, popularizing micromobility vehicles, planning activities related to waste management.

Urban Transformation Strategies

The duty of local governments during urban transformation is to make project selection, produce zoned lands excluding areas that need to be protected according to special laws and agricultural lands, build, sell, rent and purchase land for this purpose, make expropriation and cooperate with other institutions for this purpose and complete the project construction phase together when necessary (Özden, 2006). In urban transformation projects, local governments have a better understanding of the structure of the area in terms of analyzing and planning the area and can implement projects that are suitable for the region. However, local governments experience problems in undertaking urban transformation projects alone, finding financial resources and entitlement issues.

In their study where Çelebi and Alas (2024) proposed urban transformation strategies in local governments, they aimed to provide urban transformation in the whole city by creating a sustainable urban transformation cycle in the whole settlement. With the strategy proposed in the study, the regions of the city with value creation potential were determined, and it was revealed that the income obtained from these regions could finance urban transformation in other regions. In this context, the issues expected by local governments are; collection and analysis of data on settlements, analysis of data on population, preparation of strategy and action plans both in the whole city and at sub-scales based on the analyzes made in the whole city, determination of priority areas by zoning within the settlement.

Branding Strategies

Local governments need branding to strengthen local identity, increase tourism and tourism-related investments, gain competitive advantage, and ensure social development. The duties and responsibilities of local governments gain importance in the branding process. Local governments play a role in maximizing community benefits, making planning, infrastructure and economic arrangements, and providing other facilities (Özden et al., 2021). The actions expected to be carried out by local governments in the creation of branding strategies should contribute to organizing meetings, being a conciliator, ensuring that decisions are made effectively and quickly, transferring accurate information, providing technical assistance, and developing capacity (Özden et al., 2021).

The analyses made from the strategies expected to be implemented by local governments are as follows:

Local governments; Although they are an effective actor in preventing poverty at the local level, they fail to fight poverty alone. (Yeşildal, 2020a) stated that it is an absolute necessity for many actors to fight poverty. On the other hand, their successful aspects are the determination of the causes of poverty at the local level, their ability to make decisions on the local scale regarding reduction targets and what needs to be done.

Important strategies of local governments regarding environmental management are to take necessary measures in issues such as environmental protection and waste management. Local governments have responsibilities in informing the public about environmental awareness and managing the impact on the environment. However, the attitudes of local governments towards environmental problems, deficiencies in implementation, and inadequacies in the lack of qualified personnel have led to an increase in the damage given to the environment. It has not been possible to clearly reduce the impact of people on the environment and environmental damage with the actions developed in this direction. In this respect, the security measures taken are not completely successful.

Hassamancıoğlu (2021) examined the policies of local governments regarding green areas in his study and stated that the practices in the protection, planning and management of open green areas were inadequate and that there were deficiencies in management strategies. Local governments have successful aspects in terms of including green area planning and project development activities in their units, putting forward strategies regarding open green areas and setting targets in the design of these areas. The fact that more than one unit is authorized in carrying out these activities and the disruptions seen in the reflection of central policies to the local reveal their unsuccessful aspects.

Local governments with zoning authority offer significant opportunities. It has been stated that it is possible to manage the negative effects of climate change, to move from energy consumer to producer, to create green area systems, to reduce the urban heat island effect. In terms of preventing climate change; they make significant contributions in making decisions regarding building layout and form, implementing and encouraging sustainable design strategies, planning settlement areas considering climate conditions, saving energy resources, reducing urban traffic, supporting local development, increasing employment, and reducing greenhouse gas emissions. They fail in the face of setbacks such as insufficient financial resources of local governments, deficiencies in legal and administrative regulations, inadequacy of local infrastructure and technology, and uncertainties in authority.

Local governments play an active role in reducing traffic congestion, providing parking spaces, planning and designing parking areas. Parking strategies can be used as a method to provide solutions to traffic congestion and parking problems in urban planning and regulating mobility, saving on parking costs, and increasing environmental sustainability and livability. National scale documents emphasize the need to use smart transportation systems for smart city management and encourage the use of low-carbon and sustainable transportation vehicles. While solving problems such as traffic, environmental pollution, and climate change with the help of technology provides benefits to local governments, it is considered an opportunity for local governments to produce strategic solutions with smart city designs.

It is thought that by using information and communication technologies in local governments; citizens will participate in the administration, increase the quality of the service provided to citizens, direct the behavior and consumption of citizens with smart city designs, increase the efficiency and safety in transportation with smart transportation system, contribute to entrepreneurship and production capacity, and increase employment (Erdem and Bilgili, 2023). In order to reduce traffic and fossil fuel use in transportation, ensuring the use of low-carbon vehicles, encouraging bicycle and pedestrian transportation, and developing alternative transportation types are actions that will produce positive results.

The applications made under the name of urban transformation also include rent projects. Structures that are disconnected from the integrity of the city and contrary to the urban identity emerge. The proposed strategy allows urban rent to be used for the benefit of society and for the city. A strategic planning approach is important to achieve success in urban transformation processes. In order to develop a good strategy, the analysis, research, study, synthesis and planning decisions to be made by local governments are the most basic requirements. The main goal of local governments should be to approach urban transformation with a focus on housing, to eliminate uninhabitable, risky and obsolete structures, and to ensure that people live under healthier and safer conditions (Çelebi and Alas, 2024).

Özden et al. (2021) stated that the role of local governments in determining branding strategies varies according to the understanding of centralism in that country. In this direction, since the decisions to be taken or the investments to be made are determined by the central government, local governments remain in the background.

CONCLUSİON

There are deficiencies in the literature in the evaluation of the roles and strategies of local governments in sustainable local development. This study provides an idea about the roles of local governments in the literature and the strategies that can be implemented for local governments.

The responsibilities of local governments have increased within the new order brought by the globalization and localization process. The analyses conducted show that local governments are not only responsible for the provision of basic goods and services, but also, when current trends in studies are taken into consideration; they are responsible for activities such as adaptation to climate change, greenhouse gas reduction, use of smart technologies, disaster risk reduction, waste generation reduction, environmental management, sustainability and resilience awareness raising. These trends emphasize developing solutions to unsustainable problems at the local level, developing and implementing green policy, green development/smart city strategies, raising public awareness on disaster preparedness, and implementing initiatives and practices on reducing waste. In this context, the strategies expected to be implemented by local governments are poverty reduction, environmental management, combating climate change, management of open and green areas, ensuring energy efficiency, urban transformation, branding, parking lot management and smart city strategies.

Another research question of the study is the evaluations made about the issues where local governments are unsuccessful; the problems experienced in transferring resources from the central government to local governments. In order for sustainable local development to be successfully implemented, other actors need to support local governments in creating the infrastructure, technology, labor force and financing that local units need. Financial and organizational interventions in urban transformation projects should be carried out with the cooperation of central and local governments. The public should also have a say in urban interventions.

When evaluated in general, local government units play an encouraging role in implementing the strategies and actions specified in national documents for the transition to smart cities; they contribute to directing the public to environmentally friendly means of transport, implementing alternative means of transport, reducing carbon dioxide emissions, providing solutions for waste management, and encouraging the public to adopt energy efficiency. In this context, the innovations brought by the information age will be applied to local communities and increase the competitiveness of the local. Another contribution of the study to the literature is that the changing social structure and conditions have expanded the duties and responsibilities of local governments and revealed that they are more effective in management.

In line with the information obtained within the scope of the research, suggestions were made for the strategies and action plans to be developed regarding the roles of local governments in sustainable local development:

- Local governments should develop strategies for disaster risk management and water resources management against environmental shocks at the local level. In this direction, disaster emergency action plans, early warning systems, hazard and risk maps should be prepared at the local level and integrated with smart city technologies, waste management plans should be created, the number of environmentally friendly buildings should be increased, energy consumption should be reduced, and investments should be made in green infrastructure systems.
- The rate of elderly population is increasing all over the world. States are focusing on social and economic policies that support active aging. There is no study in the literature that clearly demonstrates active aging strategies for the elderly population for local governments. For future research; active aging strategies for the elderly population can be addressed by local governments.
- The organization of local government units regarding smart cities should be expanded. The capacities of these units regarding sustainable planning and policy making should be increased. Sustainable development policies should be strengthened at the local level. Technical and financial support should be provided to enable local governments to implement smart city applications at the national level.

When a general assessment is made, the irresponsible use of available resources has caused poverty to become more widespread, climates to change, natural resources to be destroyed, and environmental damage to occur. The solution to such problems is environmentally friendly smart city applications. Local governments are the most important actors that will help implement smart city applications at the local level, from urbanism to transportation, from combating climate change to waste management, from sustainability to resilience. By local governments; creating open and green space systems in urban planning, reducing greenhouse gas emissions, ensuring the use of environmentally friendly vehicles in transportation vehicles, ensuring public transportation-oriented development, establishing regional heating systems, ensuring the reduction of energy consumption with smart energy, establishing smart city infrastructures, minimizing losses in emergency and disaster situations by creating smart structures, and meeting energy needs from renewable energy sources at the local level are important applications that local governments can implement. Within the framework of the findings, it is seen that the duties and authorities defined for local governments set out conditions that require and encourage work on many issues in order to ensure sustainable development at the global and local levels.

REFERENCES

Atmaca, Y. (2020). An Evaluation on the Contribution of Development Agencies to Local Development Policies''. *Journal of Public Administration and Policy, 1*(2), 66–81.

Babaoğlu, C., & Kocaoğlu, M. (2017). The İssue of Urban İmmigrants and Local Governments. *Turkish Administration Magazine, 485*, 497–516.

Balkaya, D., Kutluay, T. F., & Tutar, N. F. (2022). Think Global Move Local: Micro Based Sustainable Local Development''. *International Academic Social Resources Journal, 7*(35), 299–309. DOI: 10.29228/ASRJOURNAL.57789

Boz, Y., & Cay, T. (2024). How Smart and Sustainable are the Cities in Turkiye? -National Policies and the Enthusiasm Level of the Local Governments. *Heliyon, 10*(4), e26002. DOI: 10.1016/j.heliyon.2024.e26002 PMID: 38390120

Çelebi, H. İ., & Tek, M. (2022). The İmportance of Municipal Cooperative Relations for Local Development: The Case of Hatay Metropolitian Municipality''. *Journal of Social Sciences, 58*(58), 190–209. DOI: 10.29228/SOBIDER.62659

Çelebi, İ., & Alas, B. (2024). Urban Transformation Strategy Proposition in Local Governments - Şişli Example. *JENAS Journal of Environmental and Natural Studies, 6*(1), 37–51.

Du, J., Zhu, X., Li, X., & Ünal, E. (2024). What Drives the Green development Behavior of Local Governments? A Perspective of Grounded Theory''. *Heliyon, 10*(6), 1–18. DOI: 10.1016/j.heliyon.2024.e27744 PMID: 38509992

Erdem, A. D., & Bilgili, A. (2023). Climate Change Combat Tools in Turkey: National Smart Cities Strategy and Action Plan. *Journal of Contemporary Local Governments, 32*(1), 51–78.

Eroğlu, H. T. (2010). Sustainable Local Development and City Councils. *Dumlupınar University Journal of Social Sciences, 26*(2), 44–55.

Gündoğdu, S., & Boyalı, H. (2018). ''*The Role of Local Governments in Local Development*'', Talas, M. and Hilal, E., (eds), Civil Society, Political System and Political Governance, İksad Publishing House, Turkey, 60-80.

Hassamancıoğlu, V. M. (2021). An Evaluation of Urban Green Space Management Strategies of Local Governments in Turkey. *OPUS İnternational Journal of Society Researches, 18*(39), 627–656.

Kadiroğlu, A., & Türko, E. S. (2024). *"Actors of Local Economic Development in Türkiye"*. Turkish Studies – Economics, Finance. *Politics*, *19*(1), 207–224.

Kutluay Tutar, F., Canbulut, S., & Tutar, N. F. (2022). Smart cities for a sustainable life in local economic''. *International Journal of Disciplines Economics & Administrative Scienves Studies*, *8*(43), 472–495.

Lane, R., Kronsell, A., Reynolds, D., Raven, R., & Lindsay, J. (2024). Role of Local Governments and Households in Low-Waste City Transitions''. *Environmental Innovation and Societal Transitions*, *52*, 100879. DOI: 10.1016/j.eist.2024.100879

Ministry of Environment and Urbanization Climate Change. (2011). 2024-2030 National Smart Cities Strategy and Action Plan (2023). Retrieved from, https://www.akillisehirler.gov.tr/wp-content/uploads/2023/12/Taslak-Eylem_Plani.pdf

Nel, H. (2001). A Project Management Approach to the Implementation of Development Programmes within the Local Government Sphere: An Emprical Analysis. *Development Southern Africa*, *18*(5), 605–624. DOI: 10.1080/03768350120097450

Özden, M., Kantar, G., & Çınar, S. (2021). The Importance of Local Governments in the Urban Branding Process. *Journal of Yasar University*, *16*(62), 1005–1021.

Özden, P. P. (2006). ''Proposals on Applicability of Urban Regeneration in Turkey''. Istanbul University Faculty of Political Sciences Journal, 35, 215-233.

Perugini, F. (2024). Local Government Efficiency and Economic Growth: The İtalian Case. *Socio-Economic Planning Sciences*, *91*, 101775. DOI: 10.1016/j.seps.2023.101775

Şahin, D., & Güven, A. (2011). ''Local Development and The Fight Against Poverty of Local Governments''. 26. *Turkish Finance Symposium*, 281-304.

Sezer, Ö., & Vural, T. (2009). ''*Within the Framework of Globalization-localization İnteraction, Development-oriented Local Social Policies in Local Governments*''. National Development and Local Governments: Proceedings of the 4th National Local Governments Symposium, 2.

Strategy and Budget Presidency. (2018). 2024-2028 Twelfth Development Plan (2023). Retrieved from, https://www.sbb.gov.tr/kalkinma-planlari/

Tuğaç, Ç. (2021). *"Duties and Responsibilities of Local Governments from Urban Resilience Perspective*". İdealkent. *Journal of Urban Studies*, *28*(3), 984–1019.

Türk, S. M. (2024). Local Governments and Democracy. *Conservative Thought Magazine*, *20*(66), 430–444.

Uysal, Y. (2022). The role of local governments in combating climate change and global warming: Findings and recommendations''. *The Journal of Kesit Academy*, *8*(30), 324–354. DOI: 10.29228/kesit.57382

Wegelin, E. A., & Borgman, K. (1995). Options for Municipal İnterventions in Urban Poverty Alleviation''. *Environment and Urbanization*, *7*(2), 131–151. DOI: 10.1177/095624789500700219

Yeşildal, A. (2020a). Local Government in the Struggle against Poverty: New Trends from Global to Local. *Journal of Social Policy Conferences*, *78*, 455–481.

Yeşildal, A. (2020b). Political Economy of Environment and Sustainable Development; Local Governments and Multilevel Governance. *Kocaeli University Journal of Social Sciences*, *1*(39), 189–208.

Yıldız, M. and Gümüş, E. (2023). ''*Digital Transformation in Local Governments: E-Municipality Applications*''. Journal of Economics, Business, Politics and International Relations, 9(2-1), 201-221.

Zengin, E., Başkurt, M., & Es, M. (2014). Local goverments and local development''. *Manas Social Journal of Research*, *3*(2), 95–124.

Chapter 4
Insularity and Sustainable Development Through Smart Strategies on Small and Medium-Sized European Islands

Despina Dimelli
https://orcid.org/0000-0002-0234-1293
Technical University of Crete, Chania, Greece

ABSTRACT

The development of smart strategies in small island territories contributes to environmental protection and the improvement of characteristics related to the permanent phenomenon of geographical discontinuity, aiming at sustainable development and eco friendly tourism. This is achieved by reducing environmental, social, and economic burdens through an integrated, holistic approach. The main environmental and socio-economic benefits rely on the implementation of innovative projects focused on energy and natural resource management, waste management, sustainable mobility, and energy-efficient building upgrades. Furthermore, expanding research in this area can contribute to the further development and application of solutions not only on small and medium-sized European islands but across all island regions, making the sector more efficient and resilient.

DOI: 10.4018/979-8-3373-1554-6.ch004

INTRODUCTION

The purpose of this research is to study and analyze smart strategies that have been implemented on small and medium-sized European islands (Table 1), contributing to infrastructure improvements, the preservation of natural resources, and the potential extension of the tourist season. These strategies help mitigate the intense fluctuations in economic activity caused by the seasonal nature of employment. Moreover, this study examines how these strategies address problems related to insularity, creating favorable living conditions for both permanent residents and tourists throughout the year.

Table 1. Classification of European islands by population and size (2011)

Category	Population size (permanent residents)	Number of states	Area sq.km	Number of islands
Very large islands	>500.000	5	>1.000	37
Large islands	50.000-500.000	10	500-1.000	32
Medium sized islands	5.000-50.000	44	100-500	141
Small islands	50-5.000	303	50-100	95
Very small islands	<50	228	<50	285

Islands have fascinated humans as both geographical entities and symbolic places. Often celebrated as "unique destinations," they offer distinctiveness that attracts interest across cultures and time. However, insularity is a complex concept, leading to various interpretations of island life. Efforts have been made to define insular regions, focusing on their socio-economic, environmental, and cultural contexts.

In socio-economic studies, insularity is defined by measurable characteristics like land area, population density, economic development, remoteness, and isolation. Beyond these factors, insularity is shaped by the natural environment and the cultural identity of many islands. These elements contribute to their "distinctiveness." While this uniqueness can offer advantages, it also creates challenges, as isolation can be both a strength and a vulnerability for island communities.

European island regions, in particular, face challenges that are inherent to their insular nature. These challenges are not merely incidental but structural, arising from the geographical discontinuity that defines islands. The isolation caused by being surrounded by water results in several vulnerabilities for island communities, including limited economic opportunities, environmental fragility, and demographic challenges. Due to their geographic separation from mainland areas, islands typically have fewer economic opportunities, which can stymie their growth and development.

Furthermore, island communities often struggle with the higher costs associated with transportation, energy provision, and other essential services, all of which are exacerbated by their isolation (Diagram 1).

Figure 1. Characteristics of the Island System

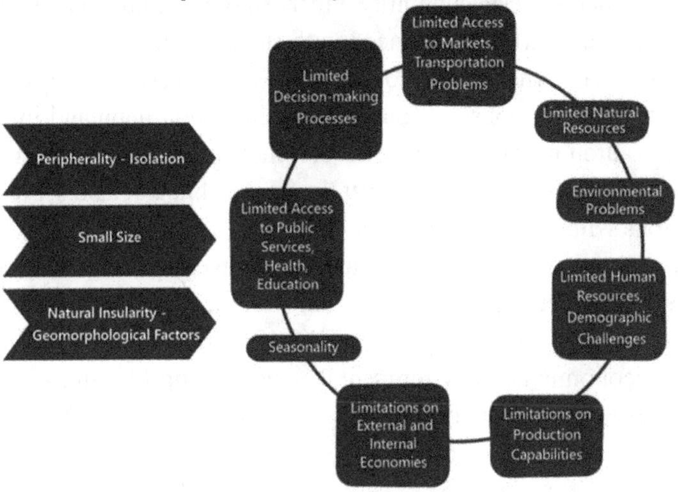

The most significant consequence of this isolation is its impact on the sustainability of island economies. While islands possess unique attributes—such as close-knit communities and the potential for self-sufficiency, these advantages are frequently overshadowed by the persistent challenges posed by their remoteness and limited access to broader markets. The traditional appeal of islands, characterized by their natural beauty, cultural heritage, and recreational opportunities, is particularly vulnerable during the off-season, usually in the winter months. This seasonal dip in tourism can contribute to economic instability, as reliance on tourism often leads to fluctuating incomes and employment levels, which in turn results in higher living and production costs during the off-season.

Tourism, often considered a potential solution to this economic vulnerability, brings its own set of challenges. In many Mediterranean islands, the rise of mass tourism has led to significant negative consequences, such as environmental degradation, overuse of natural resources, and the erosion of local cultures. Additionally, economic dependence on tourism makes islands vulnerable to global economic shifts, which can reduce demand for leisure travel, especially during economic downturns. As such, the benefits of tourism must be carefully managed to avoid overwhelming the island's infrastructure or disrupting the delicate environmental balance.

Despite these challenges, islands possess distinct advantages over mainland regions. The physical isolation of islands often fosters strong community engagement, as the need for cooperation and the sharing of resources brings people together in ways that may not be as evident in larger, more urbanized areas. The sense of community is a defining feature of island life, fostering a collective responsibility for local development and sustainability. Furthermore, isolation can be an asset in achieving energy self-sufficiency. Faced with the challenges of remote locations and dependence on external energy sources, many island communities have turned to renewable energy solutions to reduce their reliance on mainland imports. This drive for energy autonomy has led to the development of innovative technologies and sustainable practices that not only benefit the islands but can also serve as models for other regions with similar challenges.

In conclusion, while islands face unique challenges due to their geographical isolation, they also offer opportunities for innovation, particularly in areas such as sustainability, community development, and self-sufficiency. Islands' capacity to address socio-economic and environmental issues through localized and creative solutions is one of their most defining strengths. However, it is essential that islands strike a balance in their development efforts, ensuring that they avoid the pitfalls of mass tourism and overexploitation of natural resources. By focusing on sustainable growth, islands can create thriving, resilient communities that benefit both residents and visitors alike.

The following sections present projects and strategies of "smart" European islands and examine their relationship with sustainable tourism development and the improvement of problems arising from insularity.

METHODOLOGY

In this research, a literature review was conducted to investigate the relationship between insularity and sustainable development through smart strategies. Small and medium-sized European islands were examined using the case study methodology, utilizing secondary data. The aim is to explore the subject from different perspectives, group together results from various studies and highlight this issue in all its dimensions.

CASE STUDY ANALYSIS

According to comprehensive studies conducted in 2011 by the European Union, it has been revealed that approximately 88% of all islands within European territories remain entirely uninhabited, lacking any form of permanent settlement or residential infrastructure. Furthermore, of the islands that are indeed inhabited, it has been found that 39% of them accommodate fewer than 50 permanent residents, illustrating the widespread demographic challenge of extreme population sparsity across most European islands.

In this context, larger islands, when evaluated in relation to their population size and density, possess a significantly greater market capacity and exhibit stronger local demand for essential goods, products, and services. This is in stark contrast to smaller islands, which confront pronounced economic disadvantages. These disadvantages arise primarily from the increased operational costs faced by local businesses and private households, which are largely attributed to their inability to achieve economies of scale or benefit from economic concentration effects that are common in larger population centers.

Moreover, smaller islands frequently suffer from the additional constraint of limited natural resources, a consequence of their geographically restricted size and the absence of a hinterland that could otherwise supplement local production. This situation renders environmental protection, sustainable resource management, and tourism development considerably more complex and challenging. As a result, this reality accentuates the imperative need for the implementation of comprehensive and effective environmental management strategies to preserve ecological integrity while enabling socio-economic development. (Marmaras, 2019)

Given that smaller islands, particularly those with very few inhabitants, are disproportionately exposed to greater structural, environmental, and economic difficulties, this research strategically focuses on the concept and application of "smart" islands within this specific category. The research aims to showcase exemplary projects and innovative initiatives that have the potential to strengthen the attractiveness, viability, and long-term sustainability of these small-scale island communities.

Originally, the term "smart" emerged and was predominantly utilized within specialized fields such as materials science, advanced technologies, and infrastructure facilities. However, over time, this terminology underwent a conceptual expansion, extending its relevance to the macroeconomic and urban development domain, giving rise to the concept of the "smart city," which historically preceded the more recent notion of the "smart island." (Grydehøj & Kelman, 2016)

In contemporary discourse, a "smart" island is understood as an island that adopts a strategic set of practices and policies aimed at achieving sustainable development objectives and promoting the protection and preservation of the natural environment. These practices include the safeguarding of cultural heritage, approached through a holistic and integrated framework. Such islands combine proactive efforts for mitigating and adapting to the adverse impacts of climate change while simultaneously striving to foster sustainable local economic development. The ultimate objective is to improve the quality of life for both permanent residents and visiting tourists alike by applying innovative, intelligent, and integrated solutions for the management of infrastructure systems, public services, and natural resources. (Pantazis et al., 2017)

The European islands examined within the framework of this research study (Map 1) are categorized according to their population size and geographical area, distinguishing them as either small or medium-sized islands. These islands predominantly rely on tourism as their principal economic activity, a sector that is both vital and vulnerable. There is an urgent necessity to promote the sustainable expansion of tourism to avoid critical challenges such as over-tourism, environmental degradation, and the risk of unsustainable mass tourism practices. Furthermore, improved management of local infrastructure, financial resources, and natural assets is required to ensure balanced development.

Additionally, it is crucial to implement strategies aimed at retaining the permanent resident population, which can be facilitated by extending the duration of seasonal employment and creating diversified job opportunities throughout the year. The strategies examined in this research are operationalized through the analysis of specific projects that focus on key thematic areas such as environmental conservation, improvement of living conditions, enhancement of mobility and transport systems, economic development, human capital, and participatory governance mechanisms. (Clean Energy for EU Islands, 2019; Efthymiopoulos, 2016).

Figure 2. Map location of the European islands under study

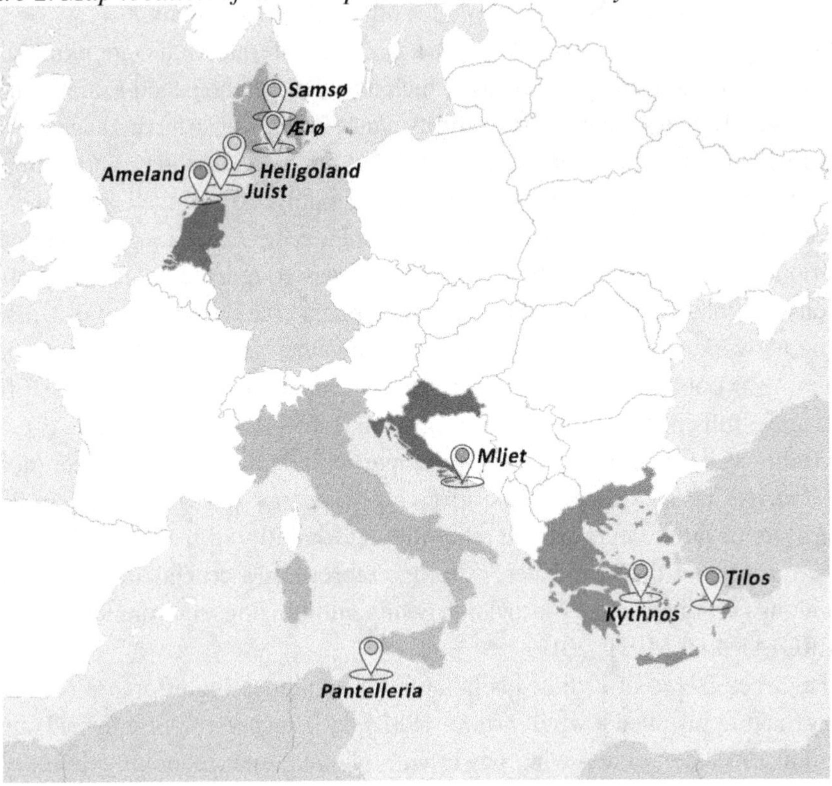

The first efforts to promote renewable energy production and energy self-sufficiency on small and medium-sized European islands began several decades ago, laying the foundations for the development of innovative and sustainable energy solutions adapted to the challenges and characteristics of insular territories. A pioneering example is found on the Danish island of Ærø, which emerged as a global leader in renewable energy projects as early as 1980. At that time, Ærø installed one of the largest wind farms in the world, demonstrating the island's early commitment to energy transition and environmental protection.

Subsequent developments have further reinforced the strategic direction toward sustainability in island territories. In 1996, the Danish island of Ærø implemented its first district heating plant, using a combination of solar energy and biomass to meet the heating needs of its residents. Initially, the plant covered 40% of the island's residential heating demand. The success of this initiative led to its expansion in 2013, with the addition of two more district heating facilities. These efforts are part of Ærø's broader goal to achieve net-zero carbon emissions by 2030 (Diagram 2). As a direct result of these integrated strategies, Ærø has successfully reduced its annual CO_2 emissions by 8.5%, making a significant contribution to combating climate change and setting an example for other island communities.

Similarly, the Danish island of Samsø has gained international recognition for its ambitious renewable energy projects, which have helped the island achieve energy autonomy and environmental sustainability. Since 2013, Samsø has managed to meet all its residents' electricity needs through wind power and exports approximately 70% of the electricity produced annually. Additionally, 70% of the island's heating requirements are covered by a combination of solar energy and biomass. The energy transition process in Samsø began in 1993 and was completed by 2005, involving the construction of three solar-thermal power plants, the development of a district heating network, and the installation of both onshore and offshore wind turbines (Diagram 2). Looking ahead, Samsø aims to eliminate its reliance on fossil fuels and achieve full energy autonomy through renewable sources by 2030.

To address increasing energy demands, particularly during the summer months when tourism peaks, Samsø began investing in energy storage systems in 2017. These systems aim to improve the reliability and stability of the renewable energy supply throughout the year. Energy storage represents a crucial step in ensuring continuous, efficient energy provision while minimizing environmental impact (ENERGIAKADEMIET, 2019).

The Greek Island of Kythnos is notable for its pioneering efforts in renewable energy, having installed a wind farm in 1982 and launched the first hybrid energy system in 1983, combining wind power with photovoltaic technology. Initially intended to meet 25% of the island's energy needs, this system eventually exceeded expectations, providing more than 75% of the island's energy demand (WISEGRID, 2024). In 2001, Kythnos advanced further by creating a microgrid with two autonomous hybrid photovoltaic systems, specifically designed to power the remote coastal settlement of Gaidouromantra, consisting of 12 holiday homes (Diagram 2) (DAFNI – Network of Sustainable Greek Islands, 2021).

Another significant development in the Greek context is Tilos, which initiated a smart microgrid project in 2015. This grid combines wind and solar energy with advanced energy storage systems to ensure a stable and reliable electricity supply for the settlement of Livadia (Diagram 2). Tilos' achievement of energy autonomy has led to annual savings of approximately 62,000 liters of oil and a notable reduction in greenhouse gas emissions, highlighting the economic and environmental advantages of renewable energy integration (CNN, 2018; EUNICE, 2020; Koutsopoulou & Coccossis, 2012).

Figure 3. Timeline of energy management, natural resource management, and waste management projects

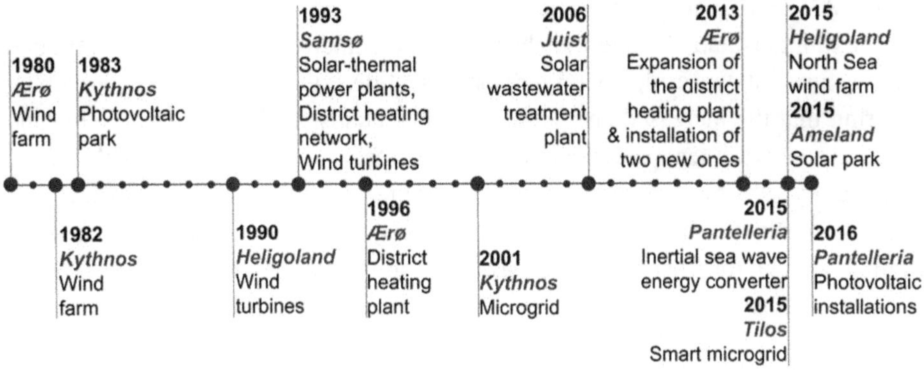

Similar renewable energy initiatives have been undertaken on the German island of Heligoland, where the installation of the first wind turbines began in 1990 with the goal of reducing the island's energy footprint by 2020. The completion of the North Sea offshore wind farm in 2015 marked a significant milestone, supplying electricity to approximately 320,000 homes and offsetting 850,000 tons of CO_2 emissions annually. This transition enabled the decommissioning of the island's old power plant, a pivotal step in sustainable energy production (BRITANNICA, 2019; Greening the Islands Foundation, 2019).

In addition to energy production, island territories face considerable challenges in managing wastewater and waste, particularly during peak tourist seasons. An innovative solution was implemented on the German island of Juist, where a solar-powered wastewater treatment plant was installed in 2006. This facility has not only improved wastewater treatment efficiency but also contributed to a reduction of 15 tons of CO_2 emissions annually (Diagram 2) (Meschede, 2015).

These examples demonstrate the growing trend of sustainability-focused initiatives on European islands, reflecting a commitment to energy autonomy, environmental stewardship, and innovative solutions to the challenges posed by insularity. By integrating renewable energy technologies, improving infrastructure, and fostering energy independence, these islands are not only addressing their own needs but also providing valuable models for other regions grappling with similar environmental and socio-economic challenges.

In the Netherlands, one of the most prominent renewable energy projects was carried out in 2015 on the island of Ameland, with the construction and operation of the country's largest solar park (Diagram 2). During periods of low tourist activity, the energy produced by this installation is sufficient to meet the entire energy demand of the island's permanent residents. The project has led to a significant reduction in the use of fossil fuels and has achieved a 70% decrease in carbon dioxide emissions, reinforcing Ameland's role as a model of sustainable insular energy management (Alfen, 2017; Ameland, 2018).

The Italian Island of Pantelleria has similarly implemented a series of integrated actions aimed at conserving natural resources while extending the tourist season by ensuring that the increased energy demand is met sustainably. In 2015, the island installed an inertial sea wave energy converter, a technology compatible with marine biodiversity, which currently supplies 42% of the island's electricity demand. This innovation prevents the emission of approximately 14,720 tons of greenhouse gases annually (Wave for Energy, 2022). Moreover, in the same year, a seawater reverse osmosis desalination system was introduced, reducing electricity consumption for water production by 50% compared to the island's previous system and achieving a 65% reduction in water-related costs. In 2016, Pantelleria further strengthened its commitment to renewable energy with the addition of photovoltaic systems to the island's existing thermoelectric power plant, significantly reducing its reliance on oil for energy production (Diagram 2) (SMART ISLAND, 2015; Ciriminna et al., 2016; Franzitta & Curto, 2017).

In addition to managing energy resources and protecting natural assets, improving connectivity with the mainland has become a priority for many European islands, especially those struggling with geographic isolation. Insularity creates separation that significantly impacts the socio-economic dynamics and demographics of island regions. This isolation often leads to population decline, particularly on smaller or more remote islands where access to essential services, education, healthcare, and employment is limited. To address these challenges, several islands have launched innovative sustainable mobility projects aimed at improving accessibility, ensuring year-round travel for residents and visitors, and boosting the local economy.

Sustainable mobility on islands extends beyond improving maritime connections, though these are essential. Efficient, quick, and environmentally friendly travel to and from the mainland is key to attracting tourists, retaining residents, and supporting local economies, especially those reliant on tourism. Consequently, islands have developed initiatives to modernize maritime transport infrastructure, with a focus on integrating clean energy technologies and reducing greenhouse gas emissions from transportation.

The Danish Island of Samsø has been a leader in this field, operating a vessel with "green technology" since 2007. This pioneering effort marked one of the first instances of environmentally friendly maritime transport in insular Europe. Following Samsø's example, the German island of Heligoland introduced similar sustainable maritime solutions in 2015, and the Croatian island of Mljet launched three solar-powered passenger ships in 2018. These solar-powered vessels not only reduce the carbon footprint of island transportation but also help to protect the delicate marine ecosystems that surround these islands.

One of the most notable examples of sustainable maritime transport comes from the island of Ærø in Denmark. Despite its small size and relatively limited population, Ærø became home to the world's largest fully electric ferry in 2019 (Diagram 3). This ferry serves as a groundbreaking model for sustainable island transport, contributing to the island's broader environmental goals while enhancing transportation efficiency. The introduction of electric and solar-powered vessels on these islands has led to reduced travel times, more frequent routes, and improved connectivity between the islands, their surrounding regions, and the mainland.

These projects reflect the growing recognition of the need for islands to balance environmental protection with practical, efficient transport solutions. By adopting electric and solar-powered vessels, these islands not only contribute to the reduction of emissions but also enhance their accessibility, resilience, and attractiveness as destinations for both residents and visitors. These sustainable mobility initiatives represent a critical step in ensuring that island communities can thrive in an era of environmental challenges, demographic shifts, and increasing mobility demands.

Ensuring adequate and sustainable connectivity throughout the year is not merely a matter of convenience but is considered one of the most decisive factors for the future viability of island communities. Enhanced accessibility fosters economic development by supporting the tourism sector, which for many islands represents their primary source of income, but it also improves residents' access to education, healthcare, and employment opportunities, thereby encouraging population retention and even potential population growth. In this context, the implementation of sustainable maritime transport systems complements broader strategies for the sustainable development of island regions.

In addition to investments in maritime transport, several islands have recognized the need to develop land-based sustainable mobility projects, further strengthening their environmental strategies and improving residents' quality of life. These initiatives contribute to the reduction of environmental impacts from local transport, promote energy savings, and often involve the provision of financial incentives by local authorities to encourage their adoption by permanent residents.

One of the earliest and most ambitious examples of sustainable land mobility projects was implemented on the Dutch island of Ameland in 2007, where public transport initially operated exclusively on natural gas. In 2018, this system was upgraded with the complete replacement of public transport vehicles with electric ones, marking a transition towards a fully electric mobility system. Similarly, on Samsø, a significant portion of the public vehicle fleet — amounting to 50% — was replaced by electric cars over the same period. These projects served as models for other islands, inspiring the adoption of similar measures elsewhere.

From 2015 onwards, the Croatian island of Mljet launched comparable initiatives, followed in 2016 by the Greek islands of Kythnos and Pantelleria in Italy, and later in 2019 by the Danish island of Ærø. These initiatives involved not only the acquisition of electric vehicles for public services but also the parallel development of the necessary charging infrastructure to support their widespread use (Diagram 3). Particularly notable are the incentive schemes implemented on Mljet, where financial subsidies were offered to permanent, residents wishing to purchase electric vehicles, covering between 40% and 60% of the total cost. Additionally, residents were granted free access to charging stations, further encouraging the adoption of sustainable mobility solutions and reducing overall transportation costs.

Overall, these comprehensive mobility strategies — encompassing both maritime and land-based transport — illustrate the holistic approach adopted by many European islands in their efforts to promote sustainability, improve residents' quality of life, and ensure the long-term viability of their communities.

Figure 4. Timeline of "smart" mobility projects

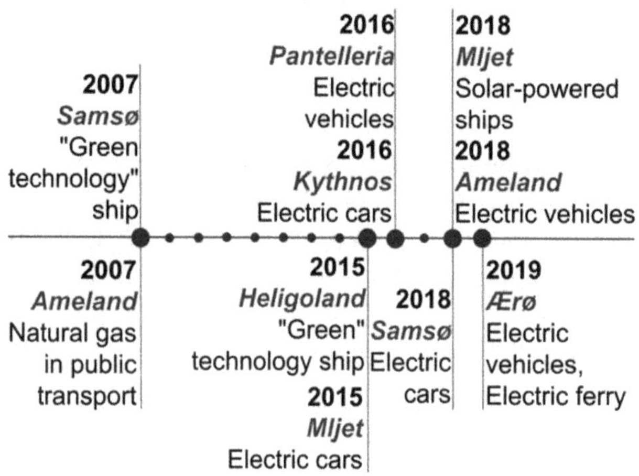

Alongside the development of sustainable mobility, many European islands have increasingly focused on "smart" street lighting projects as part of their environmental and sustainability strategies. These initiatives aim to modernize infrastructure while addressing critical challenges such as light pollution and energy consumption. Street lighting, often overlooked in sustainability discussions, has become a key area of innovation for island communities. By improving lighting systems, islands aim to balance human activity with the preservation of their fragile ecosystems and biodiversity.

Beginning in the early 2010s, several European islands started ambitious projects to replace traditional streetlamps with LED lighting. LEDs offer several advantages, including higher energy efficiency, longer lifespans, reduced maintenance costs, and lower carbon emissions. Furthermore, smart street lighting systems incorporate automated controls that adjust light intensity based on factors like pedestrian or vehicle presence, weather conditions, and time of night. This approach minimizes unnecessary energy use, reducing the ecological footprint of public lighting.

One of the earliest examples of this transition was on the island of Heligoland, which initiated the replacement of traditional streetlamps with LED technology as early as 2012. This pioneering effort was followed closely by similar projects on the islands of Juist and Ameland in 2013, both of which sought to modernize their public lighting networks in line with broader sustainability objectives. In 2021, the Greek island of Kythnos joined this trend, implementing its own smart street lighting system and further highlighting the growing recognition of smart lighting as a vital component of sustainable infrastructure on islands (Diagram 4).

Figure 5. Timeline of "smart" street lighting projects

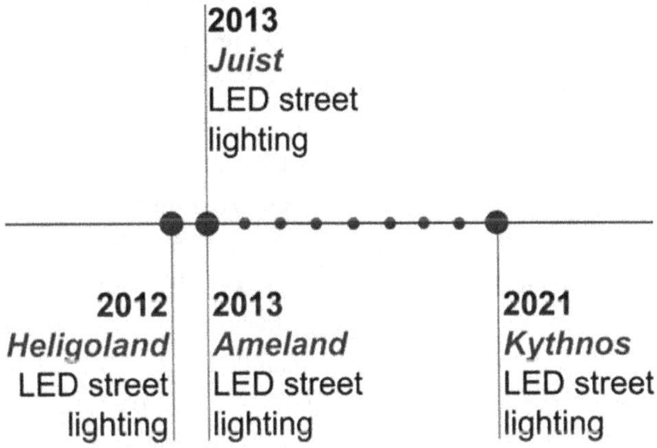

Among the case studies of sustainable development on European islands, some particularly innovative solutions stand out, especially in the design and implementation of street lighting systems on the islands of Juist and Ameland. These islands have not only embraced the latest technological advancements but have also integrated environmental considerations in ways that make their approaches especially noteworthy.

On Juist, for example, the island's unique climatic conditions—characterized by persistent, strong winds—have proven to be a significant asset in powering the island's street lighting systems. By harnessing the power of wind energy, Juist has managed to create a lighting system that operates entirely on renewable energy, ensuring that the public lighting network is not only energy-efficient but also energy-autonomous. This approach aligns seamlessly with the island's broader sustainability goals, which focus on maximizing the use of renewable energy sources and reducing the dependency on conventional fossil fuels. By tapping into the consistent and reliable wind resources available on the island, Juist can provide a sustainable, self-sufficient solution for its street lighting, effectively contributing to the island's environmental objectives. The decision to use wind energy for street lighting serves as a demonstration of how island communities can leverage their natural resources to create energy solutions that are both locally appropriate and environmentally responsible.

In addition to the use of renewable energy, Juist's wind-powered lighting system also plays a key role in reducing the island's carbon footprint. With the continuous need for street lighting in a community, transitioning from conventional energy sources to wind-powered solutions significantly reduce carbon dioxide emissions, contributing directly to the island's efforts to combat climate change. The implementation of wind-powered street lighting not only addresses local needs but also positions Juist as a leading example of how island territories can incorporate sustainable practices into their infrastructure development.

On Ameland, the approach to street lighting has been equally innovative, albeit with a different focus. Rather than primarily concentrating on energy autonomy, Ameland has placed considerable emphasis on reducing the ecological impact of its lighting system on the local wildlife, particularly migratory birds and nocturnal animals. Artificial lighting, particularly bright, uncontrolled lights, is known to disorient wildlife, especially during migration periods, leading to fatal collisions or behavioral disturbances. On Ameland, a particularly thoughtful solution was implemented: the installation of specially designed lamps that emit a customized light spectrum. These lamps are crafted to reduce the impact on wildlife, ensuring that they are less disruptive to the animals' natural behaviors. The modified light spectrum significantly decreases disorientation, helping migratory birds and nocturnal creatures navigate more safely through the island without experiencing the stress and danger often caused by artificial light.

This effort to minimize the ecological footprint of street lighting on Ameland underscores the island's commitment to biodiversity conservation. The island's ecosystem, like many island ecosystems, is particularly vulnerable due to its relatively small size and the concentration of unique species. Protecting these fragile habitats from disturbances caused by artificial lighting is crucial in ensuring that the island maintains its biodiversity. Ameland's innovative approach to street lighting reflects a deep understanding of the interdependence between human activities and the natural environment. By addressing the issue of light pollution, Ameland is not only ensuring the protection of local wildlife but is also setting a valuable precedent for other islands and communities that face similar challenges with preserving biodiversity in the face of urbanization.

Together, these two initiatives demonstrate how smart street lighting projects can fulfill a variety of strategic objectives simultaneously. On one hand, they serve to enhance energy efficiency by utilizing renewable energy sources such as wind and solar power. On the other hand, they contribute to the preservation of natural ecosystems by mitigating the negative impact of artificial lighting on wildlife. Beyond their environmental benefits, these projects also offer economic advantages to the islands by reducing operational costs for municipalities. The adoption of energy-efficient street lighting systems, particularly those powered by renewable energy, reduces the long-term financial burden on local governments. These savings, in turn, can be redirected toward further sustainability initiatives or other community development projects that contribute to the overall well-being of residents and visitors alike.

Moreover, these smart street lighting systems improve public safety, as they provide adequate lighting during the night while being intelligently managed to avoid excess illumination. By incorporating automated control mechanisms that adjust the intensity of light based on specific conditions—such as the presence of pedestrians, vehicles, or changing weather—the systems ensure that energy is used only when and where it is needed. This contributes to a safer public environment while minimizing unnecessary energy consumption and light pollution. The ability to dynamically adjust lighting according to real-time needs not only improves safety but also enhances the overall quality of life for residents and visitors.

More than just technological upgrades, these projects exemplify how islands can lead by example in the broader global effort to create more sustainable and resilient living environments. By combining cutting-edge technology with environmental stewardship, both Juist and Ameland are showcasing the potential of island regions to serve as laboratories for innovation. These communities, with their unique geographical and environmental constraints, are well-positioned to adopt forward-thinking solutions that can benefit both their local populations and the global community. The success of these projects not only contributes to achieving sustainability goals but also sets a powerful example for other islands and communities around the world that face similar challenges related to energy use, environmental preservation, and public infrastructure.

In conclusion, the smart street lighting projects on Juist and Ameland serve as excellent case studies of how islands can leverage their unique characteristics to create sustainable, innovative, and environmentally conscious solutions to modern challenges. These initiatives demonstrate the capacity of small island communities to adopt technologies that benefit both human and ecological health. By prioritizing renewable energy use, minimizing environmental impact, and improving public safety, these islands are advancing toward a more sustainable and resilient future. Their success underscores the importance of integrating environmental considerations into infrastructure design, and they provide a model for other communities looking to make meaningful strides in sustainability and innovation.

In this context, smart street lighting becomes more than just a tool for modernizing infrastructure; it serves as a model for integrating environmental concerns into everyday urban planning. These projects reveal that, even in the face of geographical isolation and limited resources, islands can implement innovative solutions that reduce their ecological footprint while enhancing the quality of life for both residents and visitors. By adopting such sustainable practices, islands are positioning themselves as key players in the global effort to transition to more sustainable, resilient, and environmentally responsible communities.

The final category of projects identified across the islands examined in this study pertains to the area of sustainable facilities management, which has increasingly emerged as a crucial and multifaceted component of integrated island sustainability strategies. Sustainable facilities management encompasses a wide range of interventions and practices, each designed to optimize the environmental performance of buildings and infrastructure while simultaneously enhancing the overall living experience for residents and visitors alike. This category is broad in scope and includes various approaches, ranging from the application of isolated energy-saving technologies within existing structures to more comprehensive and large-scale upgrades aimed at significantly improving the energy efficiency of entire buildings. Additionally, it also includes the construction of entirely new facilities that are designed from the outset according to eco-friendly, sustainable principles, incorporating modern green technologies and building practices.

These initiatives, which span a diverse array of actions, reflect a growing recognition among island communities of the vital role that buildings and facilities play in achieving broader sustainability goals. Not only do they aim to reduce energy consumption and minimize environmental impact, but they also seek to improve the quality of life for island residents, ensuring that local communities can thrive in an environmentally responsible manner. At the same time, these sustainable facility management projects enhance the attractiveness of these islands as destinations for tourists who are increasingly seeking eco-friendly accommodation and experiences. By adopting green building standards, islands can position themselves as leaders in sustainable tourism, appealing to a growing demographic of environmentally conscious travelers.

One of the primary objectives of sustainable facilities management projects is the reduction of energy consumption. This is often achieved through the implementation of energy-saving technologies and improvements in building design. For example, islands may retrofit existing buildings with energy-efficient windows, insulation materials, and smart heating, ventilation, and air conditioning (HVAC) systems that reduce energy demand. These upgrades are typically implemented in both public and private buildings, ranging from residential homes to government offices and tourist accommodation. By enhancing energy efficiency in this way, islands can achieve significant reductions in their overall carbon footprint, contributing to broader climate change mitigation efforts.

In addition to retrofitting existing buildings, many island communities are investing in the construction of new, sustainable facilities designed with eco-friendly materials and technologies from the very beginning. This involves integrating renewable energy systems such as solar panels, wind turbines, and geothermal heating into the design of new buildings. The emphasis is placed on creating structures that are energy-sufficient, capable of generating their own power through renewable resources while minimizing their reliance on external energy sources. These buildings are also often designed to be highly adaptable to changing environmental conditions, with features such as rainwater harvesting systems, green roofs, and energy-efficient lighting and appliances that further contribute to their sustainability.

A key aspect of these initiatives is the integration of sustainable practices into the design and operation of public buildings, which are particularly important for fostering a sense of environmental responsibility and community engagement. For example, schools, libraries, and community centers in island communities may be constructed with energy-saving technologies and sustainable building materials, setting a positive example for residents and visitors alike. The construction of such buildings provides a clear demonstration of the islands' commitment to sustainable development, offering a model for how eco-friendly principles can be incorporated into public infrastructure.

Sustainable facilities management is also closely linked to improving the resilience and adaptability of island communities to climate change. By designing buildings that are energy-efficient, resilient to extreme weather events, and able to function autonomously in case of disruptions to energy supply, islands are better equipped to withstand the challenges posed by climate change. These efforts not only enhance the physical resilience of island infrastructure but also promote social and economic resilience by reducing energy costs and supporting local economies. Furthermore, the integration of climate adaptation measures into the design of new buildings helps to mitigate the risks associated with rising sea levels, extreme weather events, and other climate-related hazards that disproportionately affect island communities.

In addition to the environmental and resilience benefits, sustainable facilities management initiatives also contribute to the economic sustainability of island communities. The reduction of energy costs associated with building upgrades or new energy-efficient constructions results in long-term savings for island governments, businesses, and residents. These savings can be reinvested into other sustainability initiatives or used to reduce the financial burden on local communities. For example, energy-efficient buildings may lead to lower utility bills for both residents and businesses, improving economic stability and enhancing the affordability of living and working on the island. In many cases, the initial investments made in sustainable facilities management are offset over time through energy savings, making these projects both environmentally and economically viable in the long run.

Furthermore, the construction of sustainable facilities and the adoption of energy-efficient technologies can have a positive impact on the attractiveness of islands as tourist destinations. Eco-friendly tourism is a rapidly growing market, and many visitors are now actively seeking out destinations that prioritize sustainability and environmental stewardship. By offering accommodation and facilities that align with these values, islands can attract a new demographic of eco-conscious tourists. This shift towards sustainable tourism can help to diversify local economies and reduce the reliance on traditional, often resource-intensive industries, such as mass tourism. By aligning their facilities management practices with sustainable tourism principles, islands can create new economic opportunities while simultaneously protecting their natural environments and cultural heritage.

In summary, the integration of sustainable facilities management practices across European islands represents a vital component of their broader sustainability strategies. These initiatives, which range from energy-saving upgrades to existing buildings to the construction of new, eco-friendly facilities, not only reduce energy consumption and minimize environmental impact but also improve the quality of life for residents and enhance the islands' appeal to environmentally conscious visitors. By investing in sustainable facilities management, islands are positioning themselves as leaders in the transition to more sustainable, resilient, and economically viable communities. These efforts underscore the importance of buildings and infrastructure in shaping the future of island sustainability and demonstrate how innovation in facilities management can drive progress toward broader environmental and social goals.

The initial efforts in this field were primarily focused on the implementation of standalone energy-saving solutions within building facilities. A notable pioneer in this regard was the island of Samsø, which launched its first projects in 2007. These projects introduced heating and cooling systems powered entirely by renewable energy sources, such as wind and solar energy, significantly reducing the dependence of residential and public buildings on conventional fossil fuels. These early interventions set an important precedent for other island communities, demonstrating the feasibility and tangible benefits of decentralized renewable energy systems at the building level.

Subsequently, more ambitious projects began to emerge, combining multiple energy technologies to achieve complete energy autonomy within specific facilities. A particularly significant case is that of Mljet, where, in 2012, an energy-autonomous tourist complex commenced operation. The energy needs of this complex are met entirely through renewable sources, utilizing solar collectors and a small-scale wind turbine installed on-site (Sekulo, 2018). This facility serves as an exemplary model of how energy autonomy can be integrated into tourism infrastructure, reducing environmental impact while simultaneously promoting sustainable tourism practices.

In 2012, the island of Heligoland undertook one of the most significant sustainable building initiatives observed in this study, marking a key milestone in its long-term strategy to address pressing demographic and environmental challenges. The project involved the construction of 68 eco-friendly residential units, designed primarily to provide free housing for workers and their families. This was a direct response to the persistent issue of population decline common problem faced by many isolated island communities. By providing high-quality, affordable housing, Heligoland aimed to not only attract new residents but also stabilize its population, which had been dwindling due to the lack of economic opportunities and the challenges associated with island isolation.

The design and construction of these residences emphasized sustainability at every stage. During the building phase, environmentally friendly materials and techniques were employed to minimize the carbon footprint. Additionally, the homes were equipped with energy-saving technologies and efficient building systems intended to reduce the environmental impact throughout their lifespan. This approach ensured that the residents would continue to contribute positively to environmental goals long after construction. The incorporation of energy-efficient systems meant that residents would benefit from lower energy consumption and reduced living costs, further incentivizing the decision to relocate to the island. By offering free housing and energy-efficient living, Heligoland not only sought to attract new residents but also aimed to foster long-term community stability and sustainability.

This project on Heligoland, along with the energy-autonomous developments on Mljet, exemplifies the importance of sustainable facilities management in island contexts. Both initiatives represent significant achievements in terms of environmental sustainability, technological innovation, and social policy. By addressing fundamental issues like housing and energy costs, these projects create tangible benefits for island residents and contribute to the long-term resilience of island communities. The integration of energy-saving technologies, such as solar panels and advanced insulation systems, reduces the carbon footprint of both individual homes and public infrastructure, ensuring that the environmental benefits are felt over time. Furthermore, these projects promote a sense of self-sufficiency in energy production, a crucial step in reducing dependence on external sources and increasing the island's resilience to potential energy supply disruptions.

In addition to their environmental impact, these projects also have significant socio-economic implications. Energy self-sufficiency in residential buildings directly lowers living costs, making the island a more attractive place to live and work. This, in turn, supports population retention and potentially encourages population growth, as new families and workers are incentivized by the prospect of affordable housing and reduced living expenses. Moreover, by extending the tourist season beyond the summer months, these projects help to diversify the island's economy, which is often heavily dependent on seasonal tourism. A more stable population, coupled with the potential for year-round tourism, creates a more sustainable economic model for these islands, making them less vulnerable to the fluctuations typical of the tourism industry.

In the years following these initiatives, other islands also began to embrace similar sustainable building strategies. For example, in 2013, Ameland initiated a project to upgrade the energy efficiency of its school buildings, improving thermal comfort for students and staff while simultaneously reducing energy costs for the local administration. This initiative was part of a broader effort to enhance the energy performance of public buildings, demonstrating a growing recognition of the role that building energy efficiency plays in local development. Similarly, in 2016, Pantelleria undertook a series of interventions aimed at improving the energy efficiency of its educational infrastructure, further reinforcing the island's commitment to sustainability and energy conservation (Diagram 5).

These projects not only contribute to the islands' environmental goals but also have significant educational value. By improving the energy efficiency of public buildings such as schools, the islands provide students and local communities with a firsthand understanding of the importance of sustainable practices. This awareness-building among younger generations is crucial for fostering a culture of sustainability and energy conservation, ensuring that future island residents and decision-makers will continue to prioritize environmental stewardship.

In conclusion, the examples of sustainable facilities management on islands like Heligoland, Mljet, Ameland, and Pantelleria highlight the critical role that building energy efficiency and sustainable infrastructure play in the broader framework of island sustainability strategies. These projects demonstrate that by combining technological innovation with social policy measures, island communities can achieve greater resilience and self-sufficiency. In doing so, they can effectively address contemporary challenges such as energy dependence, climate change, and population decline, while also creating more attractive and sustainable places to live and visit.

Figure 6. Timeline of sustainable building management

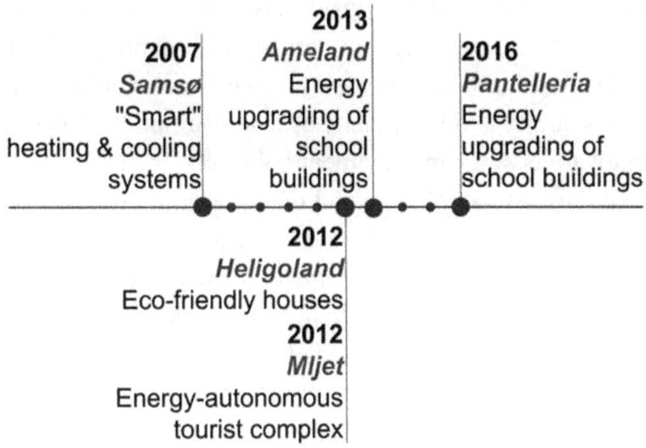

CONCLUSIONS

In 2007, European Union policy documents introduced the concept of "smart islands" and "smart cities," emphasizing the importance of six key characteristics or dimensions that define these terms. These dimensions, identified as the "smart environment," "smart living," "smart mobility," "smart economy," "smart people," and "smart governance," have since guided the development of sustainability initiatives across various regions. While most of the sustainable development projects implemented on the islands examined in this study fall primarily within the first three categories, the very nature of these projects has inevitably contributed to advancements in the remaining three dimensions as well. By attempting to classify these various initiatives into the corresponding categories (Table 2), a noticeable correlation was identified between the scope and scale of these projects and the population size of each island, alongside the extent of tourism activity. In general, larger islands with more substantial populations and marked increases in tourist arrivals exhibit a trend toward addressing all six dimensions, thereby fostering more integrated and holistic approaches to sustainable development.

The first of these key dimensions, the "smart environment," places a strong emphasis on improving environmental conditions and the sustainable management of natural resources, particularly energy and water. The majority of the islands under study demonstrate a clear commitment to environmental protection, and many have achieved significant success in reducing emissions through a variety of environmentally friendly and renewable energy projects. These initiatives are particularly prominent in energy-related developments, where the use of wind farms, photovoltaic systems, and microgrids has been instrumental in reducing the islands' carbon footprints and achieving greater sustainability. Notably, all of the islands have made substantial strides in managing water resources, addressing issues such as water scarcity and inefficiencies in consumption. Several islands have implemented innovative water management solutions, including the development of advanced desalination technologies and the integration of rainwater harvesting systems, significantly improving local water security.

In addition to energy and water management, waste and wastewater management systems have undergone transformative changes, with a focus on implementing more efficient, sustainable practices. The islands have introduced various cutting-edge waste recycling programs and wastewater treatment solutions that not only reduce environmental contamination but also contribute to the sustainable use of resources. Many of these systems rely on local, decentralized infrastructure, which aligns with the overall goals of reducing environmental impact and promoting self-sufficiency. The priority in these projects has consistently been the protection of local ecosystems, wildlife, and marine biodiversity, which are especially vulnerable to the pressures of tourism, development, and climate change. Through these coordinated efforts, the islands are working toward a sustainable and balanced coexistence between human development and environmental conservation.

The second dimension, “smart living,” addresses the quality of life for residents and visitors alike, focusing on creating more livable, resilient, and inclusive communities. While much of the emphasis in the projects discussed above is on environmental protection and energy efficiency, these initiatives also contribute directly to improving the living conditions on the islands. The integration of renewable energy sources, energy-efficient buildings, and sustainable mobility solutions all contribute to a higher quality of life by reducing utility costs, enhancing air quality, and fostering a sense of community engagement. Moreover, the incorporation of innovative technologies into daily life, such as smart grids and energy storage systems, empowers residents to take more control over their energy consumption, while also supporting local businesses and industries that rely on sustainable practices.

The “smart economy” dimension focuses on building a more sustainable and competitive local economy, which can be achieved through the effective use of technology, innovation, and entrepreneurship. On the islands studied, many of the energy-related projects discussed above also align with the goals of the “smart economy.” These projects not only address energy consumption and sustainability but also contribute to local economic development by creating new business opportunities, generating jobs, and enhancing the islands' appeal as destinations for tourists and investors. Notably, many of the islands have achieved energy surpluses through the development of renewable energy projects, such as wind farms and solar arrays, which produce more energy than is required for local consumption. This surplus is often stored in energy storage systems or exported to neighboring islands or the mainland, generating additional revenue streams for the islands. These economic benefits extend beyond energy production, as islands are also able to offset the costs associated with importing fossil fuels, creating a more resilient and self-sufficient economy.

Starting from 2015, the Croatian island of Mljet began to implement a series of groundbreaking initiatives aimed at enhancing sustainable mobility on the island. This move toward a more sustainable and eco-friendly transportation system was followed closely by other European islands that recognized the importance of addressing the environmental and logistical challenges posed by their insular nature. In 2016, both the Greek islands of Kythnos and Pantelleria in Italy joined the efforts by introducing similar initiatives. By 2019, the Danish island of Ærø also became part of this growing movement, demonstrating a commitment to sustainable practices that would benefit both residents and the environment. These islands did not just acquire electric vehicles (EVs) for public services; they also recognized the need to develop and establish the necessary charging infrastructure to ensure that these vehicles could be used efficiently and effectively across their communities.

The introduction of electric vehicles on these islands was part of a broader strategy to encourage the adoption of clean and renewable energy alternatives in daily life, not just limited to public transport. On Mljet, the local government further supported this transition by offering substantial financial incentives to residents. The incentive program provided subsidies that covered 40% to 60% of the purchase price of electric vehicles, depending on the applicant's circumstances. This initiative aimed to make electric vehicles more accessible to island residents, addressing concerns about the higher upfront cost of environmentally friendly vehicles compared to traditional gasoline or diesel options.

In addition to financial subsidies, Mljet's local government introduced free access to charging stations, which was crucial in making the transition to electric vehicles (EVs) more accessible and affordable for residents. By eliminating charging costs, the authorities removed a significant barrier to the widespread adoption of EVs, further promoting the use of cleaner transportation. This combination of financial incentives and infrastructure development helped reduce transportation costs and made it easier for islanders to embrace sustainable mobility.

The sustainability initiatives on Mljet were part of a broader, integrated approach to improving the island's sustainability and residents' quality of life. These efforts included the installation of electric vehicle charging stations across the island to support both public and private EVs. Strategically placed in key locations, these stations ensured easy access for all residents, promoting a reliable and sustainable transport system.

In parallel, these initiatives targeted both land-based and maritime transport. Since many islands rely heavily on tourism, ensuring efficient, environmentally friendly travel to and from the mainland is vital for economic success and environmental sustainability. To address this, several islands modernized their maritime transport infrastructure, integrating clean energy technologies to reduce the environmental impact of travel between the islands and the mainland. Electric and solar-powered vessels replaced conventional fuel-based ships, leading to significant emission reductions in the transportation sector.

The integration of electric vehicles into daily operations has proven transformative for island communities. By incorporating green energy solutions into public transport and charging infrastructure, islands like Mljet, Kythnos, Pantelleria, and Ærø set a strong example for other island territories and even mainland communities. These islands demonstrate that it is possible to maintain high mobility standards while minimizing environmental impact. In the long run, adopting sustainable transport options enables islands to reduce their carbon footprints and create more resilient, self-sufficient transportation networks that rely less on fossil fuels and external energy supplies.

These mobility strategies, integrating land and maritime transport, address the unique challenges of island communities. By adopting electric vehicles and green transport solutions, these islands are making significant progress toward their sustainability goals. This approach enhances residents' quality of life by providing efficient, affordable transportation year-round, while also ensuring the islands remain attractive to eco-conscious tourists.

In conclusion, integrating electric vehicles and clean energy into island mobility systems is crucial for sustainability. These initiatives, backed by community involvement and long-term planning, not only improve the islands' sustainability but also provide a model for other regions facing similar challenges.

Table 2. Evaluation of the islands by application field based on the development level of their projects

Islands	Area Km2	Population (residents)	Population density (residents/ km2)	Smart Economy	Smart Environment	Smart People	Smart Mobility	Smart Living	Smart governance
Samso	114	3.724	32,7	+	+++	+	+++	++	
Kythnos	99,43	1.456	14,6		++	+	++	++	+
Mljet	98	1.088	11,1		++	+	+++	++	
Ero	88	6.669	75,8	+	++	++	+++	+	
Panteleria	83	7.729	93,1	++	+++	+	++	+++	

continued on following page

Table 2. Continued

Islands	Area Km2	Population (residents)	Population density (residents/ km2)	Smart Economy	Smart Environment	Smart People	Smart Mobility	Smart Living	Smart governance
Tilos	61,19	829	13,5	+	++			+	+
Ameland	58,83	3.716	63,2		+++		++	++	
Juist	16,43	1.539	93,7	+	++		++	+	+
Heligoland	1,70	1.407	827,6	+	++		++	++	

The "smart people" category is significantly developed across all case studies through collaborations with research and educational institutions. These partnerships have facilitated the establishment of educational initiatives and seminars focused on promoting energy efficiency among residents. These activities, organized throughout the year, aim to raise awareness and provide valuable information on sustainable living practices, effectively contributing to the development of educational and scientific tourism. Such a form of tourism is characterized by its focus on intellectual engagement rather than mass consumption, thus supporting a more sustainable approach to tourism development. By integrating these educational initiatives into the local fabric, the islands not only enhance their residents' awareness of energy conservation but also foster a form of tourism that is less impactful on the environment and more enriching for both tourists and locals. This dual approach ensures that the islands remain attractive as both destinations for learning and as sustainable communities, contributing to long-term environmental goals and the sustainable development of local economies.

Another important category addressed by the islands under study is "smart living," which focuses on enhancing quality of life through better infrastructure, social care initiatives, and increased tourist appeal. The islands have prioritized sustainable building practices, ensuring that residential and public infrastructure align with energy efficiency and environmental sustainability. Many islands have implemented energy-saving solutions for buildings, and some have undertaken large-scale projects, such as constructing eco-friendly buildings and sustainable tourist accommodations. These efforts aim to reduce the environmental footprint while offering practical benefits to residents and tourists, such as lower energy costs and improved living standards.

Improving living conditions is particularly important for retaining a permanent population, as it directly influences the island's attractiveness as a place to live. Additionally, it enhances tourism experience by offering sustainable accommodation and services that meet eco-conscious expectations. Therefore, these smart living initiatives are essential for maintaining vibrant, sustainable communities while also boosting the islands' appeal as tourist destinations.

In addition to “smart living,” “smart mobility” is a key focus for most of the islands studied. This category includes projects aimed at enhancing transportation systems to reduce environmental impact and improve connectivity. A major emphasis is placed on developing ecological marine transport systems, which help protect the marine environment by reducing the carbon footprint of traditional ferries. These projects also enhance connectivity with the mainland and nearby regions, boosting the local economy and improving residents' and visitors' quality of life.

Another important aspect of smart mobility is the adoption of electric vehicles (EVs) for internal island transport. Many islands have integrated EVs into public and private transport systems, reducing emissions, improving air quality, and offering a more energy-efficient mode of transport. This is especially important for islands facing challenges like isolation and reliance on traditional transport systems with limited schedules and long travel times.

The weaknesses of traditional transportation, primarily characterized by long distances, limited schedules, and a dependence on public transport services, have often led to difficulties in maintaining year-round connectivity for both residents and tourists. By incorporating ecological marine transport systems, the islands can achieve multiple benefits, such as energy savings, increased frequency of transportation schedules, reduced operational costs, and minimized travel times. Furthermore, the monopolistic nature of many shipping companies can be alleviated through the development of alternative transport services, thus improving the island’s overall transportation infrastructure. This increased connectivity not only enhances the quality of life for the local population but also supports tourism by ensuring that visitors can access the islands more easily, irrespective of the season.

The analysis of these European islands leads to several key conclusions about the role of renewable energy sources in the islands' sustainable development. On an environmental level, the shift toward renewable energy is crucial for reducing emissions and minimizing the islands’ dependence on imported fossil fuels. However, beyond environmental considerations, the use of renewable energy sources also plays a significant socio-economic role, particularly in the creation of employment opportunities and the generation of local economic activity. Large-scale projects, such as the establishment of wind farms, photovoltaic parks, and other renewable energy systems, have proven to be a catalyst for increased demand for labor in various sectors, ranging from construction and operation to maintenance and management. These energy projects help to mitigate the challenges faced by island communities, particularly the issue of limited economic activities, by creating a more diverse and resilient economy. In this way, renewable energy projects help to sustain or even increase the population of the islands by offering stable employment opportunities, which are especially important in the context of population decline and youth migration.

The development of new economic activities has dual importance for improving the islands' socio-economic characteristics. It addresses both the challenge of providing sustainable employment and the issue of demographic decline, especially among young people and families who are typically more mobile and likely to leave islands in search of better job opportunities. The introduction of renewable energy projects provides a steady, year-round source of employment, which helps reduce the economic fluctuations that are often associated with seasonal work in tourism-dependent regions. In this sense, renewable energy plays a central role in stabilizing the local economy and improving the overall resilience of island communities.

Tourism is one of the most developed industries globally, and it plays a central role in the economy of the islands under study. However, despite its significant socio-economic benefits, tourism also contributes to a range of environmental challenges, including greenhouse gas emissions, waste management issues, and high resource consumption. Through the implementation of innovative projects in energy, waste management, and water conservation, these islands have been able to mitigate the negative externalities of tourism. By adopting sustainable practices, the islands can reduce the environmental impact of tourism while also fostering a more resilient and sustainable tourism industry. Moreover, these efforts help to maintain the permanent population, as they contribute to the overall improvement of living conditions, making the islands more attractive to both residents and tourists alike.

In conclusion, the sustainable development projects implemented across these European islands demonstrate the interconnection between renewable energy, socio-economic development, and environmental sustainability. By leveraging renewable energy technologies, the islands have not only reduced their environmental impact but also generated economic growth, job creation, and a more sustainable form of tourism. These initiatives, combined with efforts to improve quality of life, mobility, and governance, are paving the way for more resilient and self-sufficient island communities. Through these efforts, the islands are setting an example of how small, remote regions can address global challenges such as climate change, economic instability, and demographic decline, while fostering sustainable development for future generations.

REFERENCES

Alfen. (2017). *Alfen Annual Report.* Almere: Alfen Beheer B.V. Retrieved 10 15, 2024, from https://alfen.com/search?search_api_fulltext=ameland

Ameland. (2018). *Sustainability.* Retrieved 10 15, 2024, from https://www.vvvameland.com/practical/general/trivia/duurzaamheid

BRITANNICA. (2019, January 3). *Helgoland island, Germany.* Retrieved 10 15, 2024, from https://www.britannica.com/place/Helgoland

Ciriminna, R., Pagliaro, M., Meneguzzo, F., & Pecoraino, M. (2016, April 9). Solar energy for Sicily's remote islands: On the route from fossil to renewable energy. Retrieved 10 15, 2024, from https://www.sciencedirect.com/science/article/pii/S2212609015301254

Clean Energy for EU Islands. (2019). *EU ISLANDS IN THE ENERGY TRANSITION A CATALOGUE OF GOOD PRACTICES.* Clean Energy for EU Islands Secretariat.

CNN. (2018, August 21). *Τήλος, το πρώτο ενεργειακά αυτόνομο νησί στην Ελλάδα.* Retrieved 10 15, 2024, from https://www.cnn.gr/perivallon/story/143481/tilos-to-proto-energeiaka-aytonomo-nisi-stin-ellada

DAFNI –Network of Sustainable Greek Islands. (2021, July 20). *Sustainable Islands Prosperous people.* Retrieved 10 15, 2024, from https://dafninetwork.gr/ylopoiisi-ergou-ksi/

Efthymiopoulos, I. (2016). *Smart Islands Projects and Strategies.* Friedrich-Ebert-Stiftung.

ENERGIAKADEMIET. (2019). *Samsoe Version 2.0.* Retrieved 10 15, 2024, from https://energiakademiet.dk/en/uafhaengig-af-fossile-braendsler/

EUNICE. (2020). *Tilos Project.* Retrieved 10 15, 2024, from EUNICE: https://eunice-group.com/gr/projects/tilos-project/

Franzitta, V., & Curto, D. (2017). Sustainability of the Renewable Energy Extraction Close to the Mediterranean Islands. Retrieved 10 15, 2024, from https://www.mdpi.com/1996-1073/10/3/283/htm

Greening the islands foundation. (2019, May 1). *Germany's Helgoland island calls for legislative changes to support its pioneering role in sustainability and zero-emission ambition.* Retrieved 10 15, 2024, from https://greeningtheislands.org/germanys-helgoland-island-calls-for-legislative-changes-to-support-its-pioneering-role-in-sustainability-and-zero-emission-ambition/

Grydehøj, A., & Kelman, I. (2016). *Island Smart Eco-Cities: Innovation, Secessionary Enclaves, and the Selling of Sustainability*. Urban Island Studies.

Koutsopoulou, A., & Coccossis, C. (2012). *DYNAMICS OF ISLAND SYSTEMS: TOWARDS A STRATEDY FOR INTEGRATED DEVELOPMENT OF THE SMALL GREEK ISLANDS*. University of Thessaly, Department of Planning and Regional Development. (In Greek)

Marmaras, E. (2019). *Cycladic aspects of settlement development and new human geography in the islands*. Gutenberg. (In Greek)

Mergos, G., Papadaskalopoulos, A., & Christofakis, E. (2004). *Economic Characteristics and Development Strategy of Greek Islands*. Academy of Athens, Economic Studies Office. (In Greek)

Meschede, H. (2015, November). SMART ENERGY SYSTEMS FOR USING RES UP TO 100%-THE COMBINATION OF 100% RES AND TOURISM EXAMPLES FROM "LA GOMERA" AND "JUIST" ISLANDS. Colimbari, Crete: University of Kassel. DOI: 10.13140/RG.2.2.21071.18088

Pantazis, D., Moussas, V., Murgante, B., Daverona, A., Stratakis, P., Vlissidis, N., . . . Gadolou, E. (2017, October). SMART SUSTAINABLE ISLANDS VS SMART SUTAINABLE CITIES. Athens: ISPRS.

Sekulo, E. (2018, March). Sustainable Island Tourism: the Future of the Island. Dubrovnik: RIT Croatia Digital repository - Rochester Institute of Technology. Retrieved from https://repository.acmt.hr/islandora/object/acmt%3A33/-datastream/PDF/view

SMART ISLAND. (2015, January). Retrieved 10 15, 2024, from http://www .smartisland.eu/en/replicability/pantelleria.html

Wave for Energy. (2022, November 9). *Pantelleria*. Retrieved 10 15, 2024, from https://www.waveforenergy.com/news/portfolio-items/pantelleria/

WISEGRID. (2024). *Kythnos*. Retrieved 10 15, 2024, from https://www.wisegrid .eu/pilot-sites/kythnosgr

Chapter 5 International Organizations Policies for Inclusive Sustainable Development

Anastasios Sepetis
http://orcid.org/0000-0002-8882-7242
University of West Attica, Egaleo, Greece

ABSTRACT

The definition of the term inclusive growth can be found in different political and scientific concepts, as several scientific and institutional articles define inclusive growth as a combination of social views with economic development and equitable management of natural resources through policy approaches. Literature research on the policies of international organizations reveals a general and growing recognition that economic growth alone does not provide equal opportunities to different segments of society and it is not possible without the development of international policies to reduce social inequalities, optimal management of natural resources and intergenerational justice. Deepening that study of the definition of the concept of inclusive sustainable development by international organizations, we find that it is defined in a broader concept of development, including various non-financial aspects of life, such as health, employment prospects and environmental issues with a focus on local communities and the Sustainable Development Goals.

DOI: 10.4018/979-8-3373-1554-6.ch005

1. INTRODUCTION

Social exclusion, or even the perception of exclusion, can lead to certain groups of citizens being marginalized from participating in markets, accessing social services, and securing employment, with adverse consequences both for individuals and the broader local economy. Despite explicit commitments rooted in global treaties and operational initiatives, studies such as those by Stiglitz (2012) and Piketty (2014) highlight that, prior to the Paris Agreement, the Global Climate Change Treaty, and the adoption of the Sustainable Development Goals (SDGs) in 2015, the global trends of increasing social inequality and environmental degradation persisted. This occurred even in the context of inclusive development policies at international, national, and local levels. Global and multinational financial institutions, such as the United Nations (UN), Organisation for Economic Co, operation and Development (OECD), World Bank, International Monetary Fund (IMF), World Economic Forum, and Asian Development Bank (ADB) which play a pivotal role in defining conceptual frameworks, indicators, and policies aimed at promoting inclusive growth. Within these institutions, inclusive growth is framed as a multidimensional concept encompassing economic growth, social well, being, and the protection of natural ecosystems. To illustrate, several definitions of inclusive growth have been put forth. The International Policy Centre for Inclusive Growth (IPC, IG) [1] emphasizes the importance of social participation, advocating not only for the equitable distribution of growth benefits. Also, the active involvement of people in the processes of wealth creation and decision, making (Rainieri & Ramos, 2013). Similarly, the OECD describes inclusive growth as reducing the gap between rich and poor while ensuring equitable sharing of growth benefits to improve living standards and quality of life outcomes, including health, employment, skills, environmental quality, and community support (OECD, 2013).

The World Bank has defined inclusive growth since 2009 as economic growth that is robust enough to lift large populations out of poverty while encompassing the majority of the workforce in the development process (World Bank, 2009) [2]. The ADB adds that addressing the discrimination faced by the most marginalized social groups is integral to inclusive development, stating that these groups have historically lagged behind in poverty alleviation and economic progress. Facilitating their active participation and benefit from economic opportunities is central to achieving inclusive growth (ADB, 2011). Klasen (2010), in his study for the ADB, argues that inclusive growth is broader than pro, poor growth, as it seeks to benefit all levels of society. He also includes extremely poor households, those on the verge of poverty, middle, income groups, and even the wealthy and describes this as "disadvantage, reducing growth." McKinley's work (2010) expanded earlier interpretations of inclusive growth by highlighting two fundamental requirements. The first concerns the need for development strategies that not only support economic expansion but also ensure that such expansion is environmentally and socially sustainable over time. The second relates to the distribution of opportunities: economic progress must be paired with mechanisms that allow people from all parts of society to access these opportunities and share their benefits. Building on this perspective, the World Bank approaches inclusive growth through a practical framework that seeks to widen opportunities while confronting the structural barriers that restrict participation for many groups. Its policies center on strengthening the ability of individuals to take part in economic and social life, while also tackling forms of inequality that are embedded in institutions and systems.

Thus, a key element of this approach is the emphasis on citizen engagement, encouraging people to express their needs, influence policymaking, and seek accountability from public authorities. Through this lens, inclusive development becomes not only a matter of economic policy but also of democratic participation and social justice. Such policies are critical for fostering local development and alleviating poverty, not only in the present but also for future generations. Additionally, the World Bank often emphasizes development at the national and regional government levels, while paying special attention to local communities, particularly those with marginalized or vulnerable populations or those in remote areas.

It is important to note that within the discourse on inclusive growth, mechanisms of social exclusion described in inclusive growth models vary significantly. Some are market, based, while others pertain to broader issues of social inequality, environmental sustainability, and just transition. Certain mechanisms are specific to individual nation, states, while others are transnational. Among these cross, sectoral interdependencies lies the critical question: *"How can we collectively manage and utilize the planet's natural resources in a fair and sustainable manner?"* In scientific analyses and institutional frameworks, there is a gradual shift toward redefining economic growth and development within a broader paradigm of sustainable and inclusive growth. This redefinition prioritizes improvements in human well, being and considers the interactions between economic development and social changes that foster environmentally balanced development, with a strong emphasis on intergenerational equity (OECD, 2018; UN, 2020; World Bank, 2019; IMF, 2017).

2. THE HISTORY OF INTERNATIONAL ORGANIZATIONS AND THE BASIC GLOBAL CONDITIONS FOR INCLUSIVE SUSTAINABLE DEVELOPMENT

The earliest foundations were laid in the 1940s with the adoption of the Universal Declaration of Human Rights, which established dignity, equality, and freedom as core principles that development must uphold. During the 1950s, attention turned to the structural disadvantages facing countries with limited economic capacity, leading to international efforts aimed at mobilising investment and facilitating development in lagging regions. The 1960s expanded this framework further through a series of human rights treaties that formally linked political, economic, and social rights to global development goals[3,4].

By the 1970s, concerns about unemployment, deepening inequality, and persistent poverty brought social equity and income redistribution to the forefront. This period also marked the beginning of a global awareness of environmental degradation and the need for coordinated action to manage cross, border ecological pressures. These debates paved the way for the landmark 1987 report of the World Commission on Environment and Development (WCED), which introduced sustainable development as a core pillar of global policy, integrating environmental considerations directly into economic planning[5]. The early 1990s added another layer by reframing development through the lens of social inclusion and community participation. Agenda 21, adopted in 1992, underscored the importance of empowering marginalized groups, addressing social inequalities, supporting grassroots movements, and promoting gender equality, all within an environmentally responsible framework that recognized the challenges faced by developing countries[6,7].

In 2000, the Millennium Development Goals formalized a global commitment to combating extreme poverty, social injustice, and inadequate access to basic services. This period intensified international cooperation and mobilized political will around measurable social outcomes. By 2015, the scientific community's growing emphasis on climate change and inclusive development converged with global policymaking in the adoption of the Sustainable Development Goals. The SDGs represent the culmination of decades of evolving thought, integrating environmental sustainability, economic development, and social inclusion into a single comprehensive agenda.

These Global Treaties complemented a number of emblematic institutional and operational actions:

(1) The initiatives of global financial institutions for *"Sustainable Finance"* of the United Nations (UNEP FI, 2002[8])
(2) The World Bank's *"Inclusive Green Growth: The Pathway to Sustainable Development" (World Bank, 2012),*
(3) The institutional and operational actions for *"Sustainable Finance",*[9] *"Sustainable Taxonomy*[10]*", "Green Deal"*[11] and *"Just Transition",*[12] the *"European Pillar of Social Rights*[13]*"* of the European Union (EU, 2018) which in turn promote inclusive growth within environmental and social boundaries.

Recent academic work and empirical findings increasingly show that macroeconomic policy plays a decisive role in shaping the conditions for inclusive growth, both over short policy cycles and across longer periods of structural change. The interaction between macroeconomic choices and social inequality is far from straightforward. It runs in both directions, making it difficult for researchers to isolate clear causal pathways or to determine precisely how specific policies influence distributional outcomes.

This complexity was evident in the conclusions of a joint OECD–World Bank conference held in 2011, where participants emphasized that inclusive growth depends on a broad set of interconnected policy areas. Issues such as the development of financial systems, the design of social policy, the regulatory environment, and the overall political economy were identified as central themes. The discussions underscored that structural reforms can play a pivotal role in widening investment opportunities and enabling economies to diversify their sources of growth. At the same time, such reforms can strengthen social cohesion by ensuring that the gains from development are shared more evenly.

A recurring message from the conference was that no single policy domain can secure inclusive growth on its own. For instance, innovation does not flourish in isolation; it requires sustained investment in education and skills, along with regulatory frameworks that support competition and entrepreneurial activity. Innovation is also a central component of green growth, where traditional industries are transformed through new knowledge and technologies that generate environmentally sustainable jobs and contribute to wider social welfare (De Mello & Mark, 2012).

Global financial institutions, including the World Bank and the International Monetary Fund, note that the push for inclusive growth has become a common priority across governments. Social and political events such as the Arab Spring, the widening divide between financial markets and everyday economic realities in many advanced economies, and the emergence of a fragmented global economy operating at multiple speeds have all intensified the need to understand how growth can be achieved in a fairer way.

Despite this renewed attention, research on the determinants of inclusive growth remains incomplete. Fundamental questions continue to shape the debate: How should inclusive growth be conceptualized? In what ways can micro, level inequality and macroeconomic performance be integrated into a coherent analytical framework? Which forces have historically underpinned inclusive development in emerging markets? (De Mello & Mark, 2012). More recent scholarship suggests that new analytical models that incorporate diverse and heterogeneous factors may offer promising avenues for deepening understanding of these issues (Davoodi et al., 2021).

3. THE DEFINITION OF THE CONCEPT OF INCLUSIVE SUSTAINABLE DEVELOPMENT

According to Pouw and McGregor, and further developed in the work of Gupta and colleagues, the idea of inclusive growth can be interpreted through four broad thematic orientations that capture both its conceptual and practical dimensions. First, the term is often positioned as a response to the limitations of the dominant neoliberal model, challenging the assumption that market, driven expansion automatically translates into equitable development outcomes. Second, it draws on a wider set of related notions, such as an inclusive economy, human well, being, social justice, and fundamental rights, linking economic debates to ethical and rights, based considerations. Third, inclusive growth serves as an analytical lens for examining how poverty and inequality interact, shedding light on the structural processes that sustain deprivation. Finally, the concept bridges two complementary perspectives: a focus on social systems and human behaviour, and an ecological perspective that recognises environmental limits and the interdependence between people and ecosystems.

The diversity of the literature is also shaped by distinct disciplinary concerns. One line of research examines the institutional, legal, and regulatory mechanisms used to address poverty and inequality. Another strand focuses on the obligations that derive from international human rights law and the extent to which states must protect socio, economic entitlements. A further body of work considers social protection frameworks and the argument that marginalised populations require secure conditions to participate in society with dignity and safety.

Economic discussions contribute additional layers, highlighting the importance of intergenerational equity, the sustainable use of natural resources, and production systems that do not compromise future well, being. At the same time, political economy analyses point to how global economic structures can produce or reinforce patterns of poverty, not merely through neglect but through the normal functioning of powerful economic actors.

An emerging integrative perspective brings together questions of environmental stewardship, distributive justice, and democratic participation. This includes debates on how natural resources should be allocated, how welfare systems can promote fairness, and how decision, making processes can remain transparent and participatory. Contemporary references to ESG frameworks align closely with this approach, embedding environmental, social, and governance considerations into both public policy and corporate decision, making.

Based on the literature review, a working definition of Inclusive Sustainable Development can be proposed, incorporating the following key elements: 1. Broader Goals Beyond Income and GDP Growth: Inclusive Sustainable Development encompasses objectives that extend beyond mere economic growth, urging governments to adopt proactive measures to achieve these broader goals. It emphasizes that positive outcomes, such as human development and enhanced social well, being, cannot be assumed to naturally follow economic growth but must be deliberately pursued. 2. Equitable Distribution of Prosperity: The gains of development must be shared fairly across all segments of society, including the most marginalized groups, to ensure that no one is excluded from its benefits. 3. Addressing Poverty and Social Inequality: Poverty reduction and the mitigation of social inequalities must be integral components of Inclusive Sustainable Development policies. 4. Active Social Participation: Beyond focusing solely on distributional outcomes, Inclusive Sustainable Development prioritizes active participation by all citizens in the economy. This includes not only benefiting from economic growth but also contributing to and influencing the direction of economic development. 5. Environmental Sustainability and Climate Action: Inclusive sustainable development advocates for the responsible and sustainable use of natural resources and the protection of the environment, in line with climate change policies and the Sustainable Development Goals (SDGs).

4. THE CURRENT CHALLENGES OF INTERNATIONAL ORGANIZATIONS FOR INCLUSIVE SUSTAINABLE DEVELOPMENT

Evidence from World Bank research and operational practice shows consistently that promoting social mobility and social inclusion requires expanding real opportunities for groups that remain at the margins of society. This means enabling them to participate in markets, access essential public services, benefit from technological developments, and engage fully in social life as equal citizens. When discrimination and exclusion persist over time, the consequences are substantial. Individuals face restricted life chances, while communities and national economies absorb significant social and financial costs. For example, global estimates indicate that gender inequality has reduced human capital by approximately 160.2 trillion US dollars[14]. This figure illustrates the enormous economic losses associated with persistent structural disparities. A further indication of the scale of exclusion is the fact that around 90 percent of children with disabilities in developing countries are not enrolled in school[15]. Social protection systems can counteract some of these pressures and are estimated to prevent nearly 150 million people from falling into extreme poverty, according to World Bank assessments[16].

In Europe, the Roma population represents one of the clearest examples of long, standing marginalisation. The 10 to 12 million Roma living across the continent have experienced centuries of discrimination, which has generated entrenched cycles of limited opportunities, reduced aspirations, and chronic disadvantage. These dynamics have produced persistent socio, economic gaps between Roma and non, Roma. In several Eastern European countries, about seven out of ten Roma households live in severe poverty. Even in middle, income settings, many Roma communities experience forms of deprivation similar to those observed in some of the poorest regions globally. A World Bank study estimated that Romania alone loses roughly 887 million euros in productivity as a result of Roma exclusion, a finding that illustrates how social exclusion can also impede macroeconomic performance.[17]

In Europe, the Roma population represents one of the clearest examples of long, standing marginalisation. The 10 to 12 million Roma living across the continent have experienced centuries of discrimination, which has generated entrenched cycles of limited opportunities, reduced aspirations, and chronic disadvantage. These dynamics have produced persistent socio, economic gaps between Roma and non, Roma. In several Eastern European countries, about seven out of ten Roma households live in severe poverty. Even in middle, income settings, many Roma communities experience forms of deprivation similar to those observed in some of the poorest regions globally. A World Bank study estimated that Romania alone loses roughly 887 million euros in productivity as a result of Roma exclusion, a finding that illustrates how social exclusion can also impede macroeconomic performance.[18]

As a result, the global extreme poverty rate was expected to rise from 8.23 percent in 2019 to a range between 8.82 and 9.18 percent in 2020. This would mark the first global increase since 1998 and would erase the gains achieved since 2017. Even under favourable assumptions, analysts anticipated that any subsequent decline in poverty would be slow and uneven. In parallel, IMF projections suggested that forty, five developing countries, with a combined population of roughly one billion people, would experience slower per capita growth than advanced economies. Consequently, existing income gaps were expected to widen rather than narrow.[19]

Public concern about inequality intensified during the pandemic. A large international survey carried out by Glocalities in early and late 2020, involving more than 8,700 respondents from 24 countries, revealed a marked rise in the proportion of people who believe that income disparities should be reduced. The economic shock associated with COVID, 19 made many respondents more aware of the importance of secure employment for quality of life. Ronald Inglehart observed that the crisis served as a collective awakening, exposing long, standing vulnerabilities and reinforcing the need for more active government involvement. He argued that strategic investments in well, paid jobs, particularly in childcare, environmental protection, and infrastructure, could help address growing public frustration with widening income inequality.[20]

Young adults expressed the greatest concern. Among respondents aged 18 to 34, one, third identified income inequality as their most pressing issue at the end of

2020. This represents an increase from 29 percent at the beginning of the same year. Martin Lambert noted that feelings of confusion, fear, and frustration had intensified, making it essential for governments to design interventions capable of directing these emotions in a constructive way. Inglehart highlighted similar concerns based on his teaching experiences at the University of Michigan. Even high, performing students were finding positions that did not correspond to their expectations, while many others struggled to secure employment at all. These patterns suggest that the labour market has become increasingly precarious for young people, even in advanced economies.[21]

5. THE UN INSTITUTIONAL AND OPERATIONAL FRAMEWORK FOR ACTION FOR INCLUSIVE SUSTAINABLE DEVELOPMENT

For the United Nations, social protection is not simply an economic tool; it is treated first and foremost as a matter of human rights.[22] When systems of social protection are designed so that they interact meaningfully with labour market policies and employment programmes, they can play an important role in reducing inequality and supporting a more inclusive pattern of growth. Ensuring that no one is left behind, however, requires more than fine, tuning existing welfare schemes. It calls for policies that reach those who face the highest risk of exclusion and give them genuine access to the resources, opportunities and public services needed to participate in economic, social, cultural and political life. In practice, this means addressing the deeper causes of exclusion, social attitudes and biases, gaps in legislation, institutional practices, and the way public authorities and partner organisations structure and deliver services.

Within the UN system, the Department of Economic and Social Affairs (UN DESA)[23] is the main body pushing for stronger international cooperation on issues linked to social inclusion and inequality. Its work centres on improving the effectiveness of inclusive development policies while fostering sustained global discussion on the social dimensions of development.

A significant part of the Directorate's mandate is to coordinate, monitor and support the implementation of the major global commitments made in the field of social development. These include:

1. **The World Summit on Social Development** and its Copenhagen Declaration and Programme of Action.
2. **The World Programme of Action for Youth.**
3. **The UN Declaration on the Rights of Indigenous Peoples** and the outcome document of the World Conference on Indigenous Peoples.

4. **The Convention on the Rights of Persons with Disabilities**, together with the outcomes of the High, Level Meeting on Disability and Development, and the Madrid International Plan of Action on Ageing.
5. **The objectives linked to the International Year of the Family** and the International Year of Cooperatives.
6. **The Third UN Decade for the Eradication of Poverty.**

Importantly, UN DESA approaches its work through a multidimensional lens. Its analyses look simultaneously at social, economic and environmental dimensions, and at the ways these intersect with rights, based frameworks and peacebuilding priorities. This allows the organisation to offer integrated policy advice that captures the complexity of contemporary development challenges rather than treating them in isolation.

5.1. UN Political and Operational Actions for Inclusive Sustainable Development

Within the broader agenda of inclusive and sustainable development, UNDP[24] places strong emphasis on ensuring that individuals not only contribute to economic growth but also share in its benefits, while keeping the environmental footprint as low as possible. The organisation articulates this clearly in its mandate, noting that the creation of inclusive growth conditions helps narrow economic, social, and ecological divides and forms a cornerstone of sustainable development. Central to this vision is the meaningful participation of people who are typically excluded from mainstream decision, making processes.

The UN's policy work on social protection and social inclusion runs throughout the 2030 Agenda and is reflected in several areas of action. UNDP assists countries in building social protection systems that are appropriate to national contexts and that extend reliable coverage to poor and vulnerable groups. It also promotes universal health coverage by encouraging policies that reduce financial risk, expand access to essential health services, and ensure the availability of affordable medicines and vaccines. Recognising the importance of unpaid care and domestic work, the organisation advocates for investments in public services, infrastructure, and social protection measures that allow such work to be more equitably shared. In the labour domain, UNDP emphasizes full and productive employment for all, grounded in safe working conditions and respect for labour rights. This approach aligns with broader efforts to enhance social, economic, and political inclusion, which also involve carefully designed tax, wage, and social protection measures aimed at reducing inequalities. A further priority is the strengthening of institutions so that they operate transparently, responsibly, and in ways that enable inclusive and representative decision, making across all levels of governance.

5.2. The UN Reports on Inclusive Sustainable Development

As part of its broader commitment to inclusive and sustainable development, the United Nations uses its evidence, driven analyses to help ensure that progress is shared across society, particularly among groups that face a heightened risk of being left out. This includes ethnic minorities, people with disabilities, internally displaced populations, rural communities, as well as women and young people. The UN's approach makes clear that inclusive development is not only about extending economic benefits but also about creating genuine opportunities for participation in social and economic life. Achieving this requires safeguarding natural resources and maintaining a long, term balance between human needs and environmental limits, so that development does not undermine the ecological foundations on which future well, being depends.

5.2.1. The 2014 UN Human Development Report *"Sustaining Human Progress: Reducing Vulnerabilities and Building Resilience"*

The report notes that, although human development indicators have generally improved worldwide, people across all age groups continue to confront serious risks to their well, being, ranging from natural hazards to man, made crises. It highlights that vulnerability rarely stems from a single source; rather, it arises from multiple pressures that operate simultaneously at both global and local scales, making coordinated action at every level essential. According to the UNDP (2014), the sluggish pace of improvement in living standards, combined with widening social inequalities, has intensified political divisions and weakened social cohesion in many regions of the world.

5.2.2. The United Nations Strategy (UNDP) *"For Inclusive and Sustainable Development"*

According to this strategic document, the agency outlines a set of services designed to assist Programme Countries as they design and implement national and regional policies aimed at advancing inclusive and sustainable growth, alongside the creation of full and productive employment. The strategy identifies three overarching areas that are considered essential for making progress toward this objective (UNDP, 2017).

5.3. UN Measurement Methodologies for Inclusive Sustainable Development

Around three decades ago, the United Nations Development Programme introduced a fundamentally different approach for assessing development and societal progress. Moving away from the traditional reliance on GDP growth as the primary indicator of advancement, the UNDP proposed a framework that evaluates countries on the basis of human development. This approach places emphasis on whether individuals have the freedom, capabilities, and real opportunities to pursue the kinds of lives they value, rather than simply measuring economic output.

5.3.1 The Human Development Index (HDI)

The Human Development Index was introduced to shift attention toward people themselves, their capabilities, choices, and potential, as the primary benchmark for assessing national progress, rather than relying solely on indicators of economic expansion.

5.3.2. The 2020 Human Development Report[25]

The UNDP's 2020 report puts forward the view that meaningful participation and the empowerment of citizens in public decision, making are essential if societies are to take the kinds of collective actions required to live more sustainably and more fairly. The report stresses that humanity has entered a period without historical precedent, in which human activity has become the dominant force influencing the planet's systems.

6. THE OECD'S INSTITUTIONAL AND OPERATIONAL FRAMEWORK FOR ACTION ON INCLUSIVE SUSTAINABLE DEVELOPMENT

The OECD has developed a framework for institutional and operational action to *"inclusive growth leaving no one behind towards a more robust and secure future"*.[26] With this institutional and operational framework for action, the OECD proposes *"moving away from the growth first and distribution later" approach and fully integrating social fairness into policy, making as a driver of growth and productivity"* (OECDa, 2018). In line with the institutional and operational framework for action of the OECD, *"inclusive growth and its upward trajectory in the global economic outlook creates opportunities to ensure that growth benefits all"* (OECDb, 2018).

The OECD emphasises several points in outlining its position on inclusive growth. It acknowledges that, although some countries have recorded progress, economic expansion has not consistently translated into broader improvements in people's living conditions, indicating that further effort is required. It also notes that firms still have room to strengthen their productive capacity and that wage developments should correspond more closely with productivity gains. Rising disparities in income and access to social opportunities across many OECD members and partner countries are presented as an urgent economic concern. Such inequalities can weaken public confidence in markets and trade, and over time may undermine both economic stability and long, term growth prospects. The OECD argues that directing productivity, oriented policies together with measures that reinforce social cohesion can produce outcomes that are more meaningful for citizens. At the same time, the organisation warns that failing to act on inclusive growth carries considerable risks, both socially and economically.

Its operational and institutional framework of action recommends investing in people and spaces, supporting entrepreneurial dynamism and the creation of more inclusive labour markets, which will help rebuild trust and strengthen social cohesion.

6.1. OECD Business Reports and Actions on Inclusive Sustainable Development

The report "*Productivity, Inclusiveness Nexus*" according to the relevant report "*Productivity, Inclusiveness Nexus*" *of*[27] the OECD. Under the OECD's Operational Action Plan, governments and local communities can do more to empower citizens and local communities to help them shape policies for citizens, thriving local businesses, shared prosperity and a sustainable planet. The key prompts and principles according to the report on inclusive local development are (OECD, 2018):

1. Invest in people and places left behind

According to the OECD, any policy agenda that genuinely aims to foster inclusive development must begin by widening the opportunities available to people from an early age. Unequal access to early education, healthcare or stable employment can constrain an individual's prospects long before they reach adulthood, and for this reason the framework places strong emphasis on measures that expand social mobility.

2. Supporting the dynamism of local businesses and inclusive labour markets.

The OECD argues that an effective starting point for advancing inclusive growth is to ensure that both workers and businesses operate in environments that allow them to be productive, competitive and adaptable to future changes. This requires coordinated action from policymakers and the private sector, since neither group can address these challenges alone.

3. Governments to regain the trust of citizens with effective and responsive governments in their institutional and social role.

For people to feel that they have a meaningful stake in the direction of economic development, their perspectives must be considered at every stage of policymaking. Citizens need to see themselves not as passive observers, but as contributors who can influence how change unfolds. The OECD's work suggests that this sense of agency can be strengthened in several ways.

6.2. Centre on Well, Being, Inclusion, Sustainability and Equal Opportunity, WISE[28]

WISE's operational agenda seeks to ensure that efforts to improve people's lives, both in the present and in the longer term, remain central to the strategies adopted for responding to and recovering from the COVID, 19 crisis. During a period marked by economic contraction and heightened uncertainty, WISE has placed particular emphasis on strengthening how societies measure well, being, inequality, social inclusion, and sustainability.

6.4. Business for Inclusive Growth, B4IG[29]

B4IG is a partnership between the OECD and a global coalition of companies fighting income and equal opportunity inequalities. More and more people in all societies are left behind, and cannot benefit from the enormous wealth created by globalization and technology. In the context of the COVID, 19 crisis, this fight against inequality is more necessary than ever.

6.5. The Champion Mayors Initiative for Inclusive Growth in March 2016

As socioeconomic divides continued to widen within cities and local communities, the OECD introduced the initiative *Leading Mayors for Inclusive Growth*. The goal was to bring together a group of mayors prepared to take coordinated action against inequality and to strengthen the foundations for more inclusive development at the urban level.

6.6. OECD Development Cooperation Report 2018: Joining Forces to Leave no one Behind[30]

In its assessment of current international development practices, the OECD raises a series of fundamental questions concerning the capacity of existing cooperation frameworks to support inclusive and sustainable development. The report argues that the international community cannot realistically meet the collective commitment embodied in the Sustainable Development Goals, particularly the pledges to "leave no one behind" and to prioritise those furthest from opportunity, without a substantial recalibration of how development cooperation is conceived and implemented (OECDc, 2018).

6.7. OECD Measurement Methodologies for Inclusive Growth

The degree to which governments and local authorities can advance inclusive development objectives is closely tied to their capacity to measure progress accurately, assess trade, offs, and collaborate effectively with a wide range of stakeholders in addressing challenges that extend beyond the reach of national or international policymaking alone. Recognising these constraints, the OECD has positioned itself to strengthen multilateral cooperation by developing clearer and more robust guidance.

6.7.1. Better Life Initiative[31]

Within this broader approach to assessing inclusive growth, a range of specialised software tools has been developed, drawing on the technical expertise of the OECD's statistical services. These tools align with the organisation's multidimensional framework for evaluating social well, being, which is applied across European, national, and regional contexts. An illustrative example is the methodology used to assess social well, being across different regions of Greece, demonstrating how these measurement systems operate in practice.[32]

6.7.2. The OECD report *"How is Life? Measuring Well, Being in 2020"*

The report offers an assessment of whether living conditions have improved for populations across 37 OECD member states and four partner countries, examining developments at both national and regional scales. The analysis evaluates how individual countries and their subnational regions perform across these dimensions (OECD, 2020).[33]

7. THE WORLD BANK'S INSTITUTIONAL AND OPERATIONAL FRAMEWORK FOR INCLUSIVE SUSTAINABLE DEVELOPMENT

Inclusive growth and social inclusion in line with its institutional framework for Environmental and Social Policy is vital and contributes to achieving the World Bank's twin goals of ending extreme poverty and enhancing shared prosperity[34]. The World Bank's environmental and social framework, which applies to all financing of investment projects, stresses that *"social inclusion is critical to all World Bank development interventions and to achieving inclusive and sustainable growth"* (World Bank, 2017).

7.1. World Bank Operational Flagship Actions for Inclusive Sustainable Development

The World Bank's policies for inclusive growth and social inclusion argue that if the exclusion of disadvantaged social groups is not addressed, then the outcome can be costly for states and local communities. When exclusion is examined from the perspective of individuals themselves, several recurring consequences tend to appear. People who experience discrimination often face lower earnings over their lifetimes, fewer opportunities to develop their abilities, and weaker educational and employment outcomes. These forms of marginalisation frequently manifest in health as well, with racism and other discriminatory practices contributing to both physical strain and psychological distress.

7.1.1. The World Bank's Operational Action for People With Disabilities

The World Bank places the inclusion of persons with disabilities, together with the expansion of fair opportunities, at the centre of its broader effort to foster sustainable and socially inclusive communities. This priority is directly aligned with the institution's overarching aims of eradicating extreme poverty and advancing shared prosperity.[35]

7.1.2. The Operational Action of the World Bank for the Social Inclusion of Roma

For more than twenty years, the World Bank has repeatedly drawn attention to a persistent and troubling fact: many marginalized groups continue to face significant barriers in securing equal opportunities and accessing mechanisms of accountability.

Early work in this area concentrated largely on improving the evidence base, particularly by documenting the socio, economic conditions of Roma communities[36].

7.1.3. The Management of Environmental and Social Risks of Investment Projects Through the Institutional Framework of Environmental and Social Policy of the World Bank

The World Bank's Environmental and Social Policy framework embeds explicit commitments to prevent discrimination and to safeguard the rights of individuals and groups who are disadvantaged or otherwise vulnerable. This orientation is closely aligned with the Bank's overarching objective of fostering shared prosperity, which directs attention to the living conditions and opportunities of the poorest 40 percent of the population in each country. Within this segment, marginalized racial and ethnic communities frequently constitute a substantial share.

7.1.4. Jobs Diagnostics, JDs

In the context of inclusive growth and employment expansion, the World Bank helps countries identify labour challenges by analysing macroeconomics, labour market and private sector development, compared to international benchmarks[37]. Jobs Diagnostics (JDs) [38,39] point to market and policy failures that hamper private investment and job creation and suggest ways to correct them.

7.1.5. The Operational Action of the World Bank's Citizen Engagement Program

The World Bank's Citizen Engagement programme adopts a broad, community, wide approach that goes beyond simply strengthening local groups. It also attends closely to the *quality* of their interactions with public authorities and private actors. In practical terms, this strategy encourages the formation of producer associations, partnerships with private sector stakeholders, and, where feasible, more formalised structures for sustained civic participation.

7.1.6. The Strategic Framework for Integrating the Participation of Citizens of Local Communities in the Operations of the World Bank

The World Bank's approach to inclusive local development is grounded in a broader institutional effort to listen more carefully to citizens and to strengthen the role of communities in shaping development outcomes. The underlying conviction is that citizen involvement, when it is meaningful and sustained, can reinforce transparency, bolster accountability, and widen the scope for social inclusion, thereby improving the effectiveness and legitimacy of development interventions at the local level. The framework defines citizen engagement as a reciprocal exchange between citizens and public authorities, or between citizens and private actors, carried out within the scope of Bank, supported activities (World Bank, 2014).

7.1.7. Community Driven Development Programs, CDD

Within the broader agenda of inclusive local development, the World Bank has also advanced a long, standing set of Community, Driven Development (CDD) initiatives. [40]. These programmes rest on four core principles, transparency, participation, accountability, and strengthened local capacity, and have become a central mechanism through which the Bank engages communities directly in development decision, making.

7.1.8. The World Bank's Operational Action for Inclusive Cities

The World Bank’s dual mandate, to eradicate extreme poverty and foster shared prosperity, places sustainable, inclusive local development at the centre of its operational agenda”[41]. This commitment aligns closely with Sustainable Development Goal 11, which calls for cities that are “inclusive, safe, resilient, and sustainable.” Within this framework, the Bank conceptualises *inclusive cities* as complex systems shaped by interdependent spatial, social, and economic forces. Advancing inclusion at the urban level therefore requires attention to several interconnected domains. [42]

8. THE INTERNATIONAL MONETARY FUND'S INSTITUTIONAL AND OPERATIONAL FRAMEWORK FOR INCLUSIVE SUSTAINABLE DEVELOPMENT

In recent years, the International Monetary Fund (IMF) has paid increased attention to inclusive growth and the detrimental macroeconomic consequences of social inequality with a focus on local communities. Loungani 2017 report outlines the International Monetary Fund's inclusive growth policies that seek to promote social inclusion in local communities. According to the report, these issues have, over time, become increasingly central to the IMF’s mission, as they have a direct impact on economic performance and stability in many countries. Within this context, Loungani presents the main findings from several initiatives that have sought to identify IMF policies in the three core areas of inclusive growth:

(1) The reduction of unemployment because, according to IMF policies, job growth is the basis for citizens to feel socially integrated and have a sense of dignity.
(2) Social inequality because according to IMF policies large sections of the population should have the opportunity to participate in the prosperity of a country.
(3) Environmental sustainability because, according to IMF policies, development must be shared not only between this generation but also with future generations.

8.1. IMF Reports and Operational Actions on Inclusive Sustainable Development

The IMF's 2013 study, Inclusive Growth: Measurement and Determinants, introduced a comprehensive methodological framework designed to capture inclusive growth in a more coherent and analytically consistent manner. In this report, the authors develop a unified model that they then apply to a wide set of emerging economies. The approach brings together information on economic performance and the distribution of income over roughly thirty years, combining national accounts data on GDP per capita with household survey evidence. Their empirical findings point consistently to three pillars as crucial for fostering inclusive growth: maintaining macroeconomic stability, investing in human capital, and implementing structural reforms capable of transforming the productive base of the economy (IMF, 2013[43]).

8.1.1. The IMF's 2017 report on inclusive growth

The IMF's 2017 annual report on inclusive growth highlights a paradoxical development: although income disparities between countries have gradually diminished, inequalities *within* countries, and even within individual local communities, grew steadily from the mid, 1980s through the mid, 2000s, a trend particularly pronounced in advanced economies. According to the Fund, a combination of structural and policy factors contributed to this shift. Rapid technological innovation tended to reward those already possessing capital or advanced skills, widening the gap with lower, skilled workers.

8.1.2. The IMF Report *"Inequality of Opportunity, Income Inequality and Economic Growth"*

In this report, the IMF argues that the crucial link in the relationship between inequality and growth is inequality of opportunity[44]. In societies where opportunities are unequal, including between generations, rising income inequality tends to become entrenched, limiting the potential and prospects of low, income earners and hampering long, term growth. The ability to increase social spending is also essential to achieve the Sustainable Development Goals (IMF, 2019).

8.1.3. The IMF Report on "Fiscal Policy and Growth: Human, Social and Natural Investment for the Sustainable Development Goals" [45]

In its 2019 assessment, the IMF stresses that the scale and nature of climate, related action differ markedly across countries, largely because their economic structures and development needs are not the same. The report notes that achieving the Sustainable Development Goals will require substantial annual increases in public spending, particularly in areas such as healthcare, education and essential infrastructure. For emerging market economies, the additional resources needed are projected to rise to roughly 4 percent of GDP by 2030.

8.1.4. The IMF Report Regional Disparities, Growth and Exclusion

This report analyses regional disparities on the basis of economic performance and living standards. Based on some key facts, the authors provide a conceptual framework to help analyze whether these regional disparities are the result of national policies or reflect market failures. They also examine whether political efforts to reduce regional disparities necessarily involve an exchange between social justice and economic efficiency (IMF, 2021)[46]

8.1.5. The IMF Report on Inclusive Local Development and World Trade

A recent IMF report examining the links between globalisation, international trade, and inclusive growth, particularly at the level of local economies, shows that the relationship is far more complex than is often assumed. On the one hand, deeper integration into global markets is consistently associated with higher per capita income, reductions in poverty, and expanded employment opportunities. On the other hand, these aggregate gains have frequently been accompanied by widening disparities within countries, as the benefits of trade are distributed unevenly across regions, sectors, and population groups. By reducing distortions in global markets and promoting more predictable trading conditions, international institutions, such as the World Trade Organization, can support a trading system that distributes the benefits of globalisation more broadly and contributes to genuinely inclusive local development (IMF–Bacchetta et al., 2021).

8.1.6. The IMF's Fiscal Monitor

The IMF argues that economic efficiency and social equity can and should go hand in hand. Policymakers have many options to achieve effective and fair outcomes. The IMF Fiscal Monitor[47] focuses on three policy discussions: progressive taxation, the guaranteed minimum income, and public spending on education and health. The IMF supports two key principles for social inequality: a) that fiscal policy can make a difference in tackling social inequality and b) that economic efficiency and social equity must go hand in hand. The use of digital tools in tax collection can also be part of a comprehensive strategy to boost domestic revenues. Reducing corruption can improve citizens' social participation and increase trust in government. Most importantly, these strategies can secure the necessary resources to invest in expanding opportunities for local communities and individuals who have been left behind.

8.1.7. The IMF Gender Budget

One particular aspect of inequality that the IMF has focused on is gender inequality and women's participation in the workforce. In many countries around the world, integrating women into the workforce can improve the match between skills and jobs and increase per capita income (Christiansen *et al.,* 2017).

8.1.8. The IMF's Strategic Commitment to Social Spending

Over the past decade, IMF efforts to address social inequality have been integrated into monitoring, lending, research and capacity building work, and this will continue over the next decade. The cornerstone of the IMF's approach to economic inclusion is its social spending strategy[18].

8.1.9. IMF Technical Assistance for the Sustainable Development Goals

The International Monetary Fund (IMF) provides technical assistance and training to help countries build effective financial institutions that can implement the right policies[49]. These capacity, building efforts help countries achieve their development goals and contribute significantly to their progress towards the Sustainable Development Goals (SDGs).

8.1.10. IMF Capacity Development, CD

The IMF's capacity building[50] Capacity building occupies a central place within the IMF's mandate. It is one of the institution's three core functions and absorbs a substantial share of its annual budget. The logic behind this emphasis is clear: without capable economic institutions, it is extremely difficult for any country to maintain macroeconomic stability, support inclusive patterns of growth, or create conditions in which employment can flourish.

Although income gaps between countries have narrowed over time, inequality within countries remains a persistent and often deepening concern. The patterns vary by region, income level and local context. IMF research has consistently shown that entrenched inequality is associated with weaker long, term growth prospects, reduced resilience to economic shocks and greater financial fragility. Addressing inequality is therefore not an add, on to the Fund's work; it is integral to its contribution to inclusive development. At the local level, effective measures include expanding access to good, quality education and health services for disadvantaged groups, investing in infrastructure that supports economic participation, strengthening social protection, and adopting policies that encourage higher female labour force participation. These interventions are regarded as essential pathways for countries seeking to pursue growth that is not only faster, but also fairer (IMF, 2019).

9. CONCLUSION

Concluding our concise yet in, depth research on inclusive sustainable development across the most prominent international organizations, this study highlights key institutional and operational initiatives undertaken by the OECD, the UN, the World Bank, and the International Monetary Fund. These organizations have emerged as leading global advocates for a multidimensional approach to human social well, being, articulated through a framework that defines the concept of inclusive growth.

This institutional and operational framework acknowledges the dynamic and complex interplay between processes and outcomes, emphasizing that the notion of inclusive sustainable development cannot be confined solely to economic dimensions or measured exclusively through traditional economic indicators such as GDP. It is widely recognized that outcomes such as employment, advancements in health and education, and the optimal management of natural resources are critical components of human well, being. These elements not only enhance the quality of life but also expand citizens' opportunities to actively participate in the economy.

All the aforementioned institutional initiatives and operational actions are rooted in the shared conviction that it is possible to establish a common agenda that acknowledges the necessity of economic growth while ensuring that its benefits are distributed in a manner that is socially equitable, environmentally sustainable, and inclusive. The pathway to achieving this reconciliation lies in examining the interconnection between economic growth, human social well, being, and environmental sustainability

REFERENCES

Asian Development Bank. (2011): *Framework of Inclusive Growth Indicators Key Indicators for Asia and the Pacific 2011* Special Supplement https://sustainabledevelopment.un.org/content/documents/739indicators.pdf

Asian Development Bank. (2021): *Inclusive growth. Retrieved,* https://www.adb.org/themes/social, development/ poverty, reduction/inclusive, growth

Bacchetta, M., Cerra, V., Piermartini, R., & Smeets, M. (2021). *Trade and Inclusive Growth.* International Monetary Fund.

Christiansen, L. E., Lin, J. M., Pereira, P., & Topalova, R. T. (2017): *Unlocking the Potential of Greater Female Employment in Europe, in: Intereconomics*, Review of European Economic Policy, 52(1), 2017, pp. 5, 16

Davoodi, R. H., Montiel, J. P., & Martirosyan, T. A. (2021): *Macroeconomic Stability and Inclusive Growth.* IMF https://www.imf.org/en/Publications/WP/Issues/2021/03/19/Macroeconomic, Stability, and, Inclusive, Growth, 50259

De Mello, D. L., Mark A. (2012): *Promoting Inclusive Growth: Challenges and Policies. Paris,* OECD and the World Bank. https://openknowledge.worldbank.org/handle/10986/16948 License: CC BY, NC, ND 3.0 IGO.

E.U. (2018): Towards Sustainable Development: Mainstreaming Environment and Climate Change into Development, E.U. SOCIAL PROTECTION pp 1, 16

Gupta, J., & Vegelin, C. (2016): "Sustainable development goals and inclusive development", *Int Environ Agreements* 16, pp. 433–448 https://link.springer.com/content/pdf/10.1007/s10784, 016, 9323, z.pdf

Gupta, J., Pouw, N., & Tonen, R. M. (2015). Towards an Elaborated Theory of Inclusive Development'. *European Journal of Development Research*, *27*(4), 541–559.

IMF. (2013): *Inclusive Growth: Measurement and Determinants*, IMF Working paper publ, WP/13/135

IMF. (2017): *Fiscal Monitor: Tackling Inequality,* IMF Working paper publ, https://www.imf.org/en/Publications/FM/Issues/2017/10/05/fiscal, monitor, october, 2017

IMF. (2017): *FOSTERING INCLUSIVE GROWTH G, 20 Leaders,* Summit, July 7, 8, Hamburg, Germany https://www.imf.org/external/np/g20/pdf/2017/062617.pdf

IMF. (2019): *Inequality of Opportunity, Inequality of Income and Economic Growth,* IMF Working paper WP/19/34 publ, https://www.imf.org/en/Publications/WP/Issues/ 2019/02/15/Inequality, of, Opportunity, Inequality, of, Income, and, Economic, Growth, 46566

IMF. (2019): *Fiscal Policy and Development: Human, Social, and Physical Investments for the SDGs,* Working paper SDN/19/03 https://www.imf.org/en/Publications/ Staff, *Discussion, Notes/Issues/2019/01/18/Fiscal, Policy, and, Development, Human, Social, and, Physical, Investments, for, the, SDGs, 46444*

IMF. (2021): *Regional Disparities, Growth, and Inclusiveness,* IMF Working paper WP/21/39 https://www.imf.org/en/Publications/WP/Issues/2021/02/13/Regional, Disparities, Growth, and, Inclusiveness, 50076

Klasen, S. (2010): Measuring and Monitoring inclusive growth: Multiple Definitions, Open Questions and some Constructive Proposals, Sustainable Development Working Paper Series, Asian Development Bank, Manila

Loungani, P. (2017). The Power of Two: Inclusive Growth and the IMF. *Review of European Economic Policy*, *52*(2), 92–99.

McKinley, T. (2010): *Inclusive Growth Criteria and Indicators: An Inclusive Growth Index for Diagnosis of Country Progress*, ADB Sustainable Development Working Paper Series, No. 14. Asian Development Bank, Mandaluyong City, Philippines

OECD. (2013): *Together We Stand: Inclusive Growth,* Speech by OECD Secretary General Angel Gurria, at OECD Workshop on inclusive growth, 3 April 2013, Paris

OECD. (2018): *Beyond GDP: Measuring What Counts for Economic and Social Performance,* OECD Publishing, Paris https://policydialogue.org/files/publications/ papers/FINAL, Beyond, GDP_, Measuring, What, Counts, for, Economic, and, Social, Performance.pdf

OECD. (2018): *Opportunities for all: a framework for policy action on inclusive growth*, OECD Publishing, Paris https://doi.org/, en.DOI: 10.1787/9789264301665

OECD. (2018): *The Framework For Policy Action On Inclusive Growth,* OECD Publishing, Paris https://www.oecd.org/mcm/2018/documents/C, MIN, 2018, 5, EN.pdf

OECD. (2018): *The Productivity, Inclusiveness Nexus*, OECD Publishing, Paris, https://doi.org/, en. https://www.oecd.org/publications/the, productivity, inclusiveness, nexus, 9789264292932, en.htmDOI: 10.1787/9789264292932

OECD. (2020): H*ow's Life? 2020 Measuring Well, being,* OECD Publishing, Paris, http://www.oecd.org/statistics/how, s, life, 23089679.htm

OECD. (2020): *Regional Policy for Greece Post-2020*, OECD Publishing, Paris, https://www.oecd, ilibrary.org/urban, rural, and, regional, development/regional, policy, for, greece, post, 2020_cedf09a5, en

OECD. (2018): *Joining Forces to Leave No One Behind Multilingual Summaries* Development Co, operation Report, OECD Publishing, Paris, https://www.oecd, ilibrary.org/sites/53d35549, el/index.html?itemId=/content/component/53d35549

el Piketty T. (2014): *Capital in the 21st century*, Cambridge, MA: Harvard University

Pouw, N., & McGregor, A. (2014): *An Economics of Wellbeing: What Would Economics Look Like if it were Focused on Human Wellbeing?* Institute of Development Studies Working Paper No. 436, London: IDS. Press

Ranieri, R., & Ramos, R. A. (2013): *After All, What is Inclusive Growth?* International Policy Centre for Inclusive Growth, Brasilia http://www.ipc, undp.org/pub/IPCOnePager188.pdf

Ranieri, R., & Ramos, R. A. (2013): *Inclusive growth: building up a concept*, IPC, IG Working Paper Number 104, International Policy Centre for Inclusive Growth, Brasilia

Ranieri, R., & Ramos, R. A. (2013): *Inclusive Growth: The Building up of a concept,* International Policy Centre for Inclusive Growth, Brasilia http://www.ipc, undp.org/pub/IPCWorkingPaper104.pdf

Stiglitz, J. (2012). *The price of inequality: how today's divided society endangers our future*. Penguin.

UNDP. (2014): *Human Development Report Sustaining Human Progress: Reducing Vulnerabilities and Building Resilience* https://www.undp.org/content/undp/en/home/librarypage/hdr/2014, human, development, report.html

UNDP. (2017): *Strategy for Inclusive and Sustainable Growth*, UNDP's Strategy for Inclusive and Sustainable Growth | UNDP

UNDP. (2020): *The 2020 Human Development Report The next frontier Human development and the Anthropocene* http://hdr.undp.org/sites/default/files/hdr2020.pdf

World Bank. (2009): *What is Inclusive Growth*? World Bank, Washington DC

World Bank. (2012): *Inclusive Green Growth: The Pathway to Sustainable Development*, World Bank Washington DC https://openknowledge.worldbank.org/handle/10986/6058

World Bank. (2014): *Strategic Framework For Mainstreaming Citizen Engagement In World Bank Group Operations,* World Bank Washington DC https://consultations .worldbank.org/sites/default/files/materials/consultation, template/engaging, citizens, improved, resultsopenconsultationtemplate/materials/finalstrategicframeworkforce_4.pdf

Word Bank. (2017): *Environmental and Social Framework,* International Bank for Reconstruction and Development, World Bank Washington DC https://pubdocs .worldbank.org/en/837721522762050108/Environmental, and, Social, Framework. pdf

World Bank. (2017): Portraits Of Labor Market Exclusion 2.0 Country Policy Paper (CPP) for Greece, International Bank for Reconstruction and Development, World Bank Washington DC http://documents1.worldbank.org/curated/en/ 118191522094488076/pdf/Country, Policy, Paper, for, Greece.pdf

World Bank. (2017): Promoting inclusive growth by creating opportunities for the urban poor (Philippines Urbanization Review Policy Note). Washington, DC: World Bank Group. http://documents.worldbank.org/curated/en/90447149580 8486974/ Promoting, inclusive, growth, by, creating opportunitiesfor, the, urban, poor

World Bank. (2019): *Breaking the Cycle of Roma Exclusion in the Western Balkans,* World Bank Washington DC https://www.worldbank.org/en/region/eca/publication/ breaking, cycle, of, roma, exclusion, in, western, balkans

ENDNOTES

1 See. About the website https://ipcig.org/ Date of access 15/01/2025

2 See. on the website, https://www.worldbank.org/en/topic/socialsustainab ilityDateaccessed15/01/2025

3 See. About on the website WCED (1987) Our Common Future, Oxford University Press, Oxford; New York, https://sustainabledevelopment.un.org/content/ documents/5987our-common-future.pdf Ημερομηνία πρόσβασης 15/01/2025

4 The Universal Declaration of Human Rights is a declaration adopted by United Nations General Assembly resolution 217 A(III) [REMOVED HYPERLINK FIELD] on 10 December 1948 (A/RES/217, 10 December 1948), outlining their view of basic human rights. The draft was written by John Peters Humphrey from Canada, with the help of Eleanor Roosevelt of the USA, René Cassin of France, P. S. Chang of China and others. Although not a legally binding

document, it became the basis for two legal treaties, the International Covenant on Civil and Political Rights and the International Covenant on Economic, Social and Cultural Rights. It continues to be cited by academics, jurists and constitutional courts. There is disagreement among international lawyers on which parts of it represent customary international law. Opinions vary between a few articles and the entire declaration.

5 See. About the website https://www.mfa.gr/missionsabroad/un/greece-in-organization/oikonomike-kai-koinonike-anaptuxe.html?page=1Dateaccessed15/01/2025

6 See. About on the website https://www.who.int/news-room/fact-sheets/detail/millennium-development-goals-(MDG) Date accessed: 15/01/2025

7 See. About the website https://sdgs.un.org/goalsDateaccessed15/01/2025

8 See. about the website https://www.unepfi.org/ Date accessed 15/01/2025

9 See. on the website, https://ec.europa.eu/info/business-economy-euro/banking-and-finance/sustainable-finance/overview-sustainable-finance_enDateaccessed15/01/2025

10 See. on the website, https://ec.europa.eu/info/business-economy-euro/banking-and-finance/sustainable-finance/eu-taxonomy-sustainable-activities_enDateaccessed15/01/2025

11 See. on the website, https://ec.europa.eu/info/strategy/priorities-2019-2024/european-green-deal_elDateaccessed15/01/2025

12 See. on the website, https://ec.europa.eu/info/strategy/priorities-2019-2024/european-green-deal/actions-being-taken-eu/just-transition-mechanism_elDateaccessed15/01/2025

13 Βλ. σχετικάστηνιστοσελίδα, https://ec.europa.eu/info/strategy/priorities-2019-2024/economy-works-people/jobs-growth-and-investment/european-pillar-social-rights/european-pillar-social-rights-20-principles_elDateaccessed15/01/2025

14 See relevant on the website, https://openknowledge.worldbank.org/handle/10986/33396Dateaccessed15/01/2025

15 See. on the website, https://openknowledge.worldbank.org/handle/10986/34237Dateaccessed15/01/2025

16 See. On the website, https://www.worldbank.org/en/news/feature/2020/07/13/everyone-equal-making-inclusive-growth-a-priority-for-ethnic-minoritiesDateaccessed15/01/2025

17 See. on the website, https://ec.europa.eu/info/policies/justice-and-fundamental-rights/combatting-discrimination/roma-eu/roma-equality-inclusion-and-participation-eu_en, Date accessed 15/01/2025

18 See. On the text Projected poverty impacts of COVID-19 (coronavirus) June 8, 2020 https://pubdocs.worl:dbank.org/en/461601591649316722/Projected-poverty-impacts-of-COVID-19.pdf Date accessed 15/01/2025

19 See. On the website, IMF report: October 2019 World Economic Outlook https://www.imf.org/en/Publications/WEO/weo-database/2019/OctoberDateaccessed15/01/2025

20 See. on the website, https://glocalities.com/latest/reports/valuestrendsDateaccessed15/01/2025

21 See. on the website, https://www.trust.org/inclusive-economies/ Date accessed 15/01/2025

22 See. On the website https://www.eurasia.undp.org/content/rbec/en/home/sustainable-development/Social-protection-and-inclusion.html Date accessed 15/01/2025

23 See. On the website, https://www.un.org/development/desa/en/about/desa-divisions/social-policy.html Date accessed Date accessed 15/01/2025

24 See. About the website https://www.undp.org/content/undp/en/home/2030-agenda-for-sustainable-development/prosperity/development-planning-and-inclusive-sustainable-growth.html Date accessed 15/01/2025

25 See. Detailed report UNDP (2020) The 2020 Human Development Report The next frontier Human development and the Anthropocene http://hdr.undp.org/sites/default/files/hdr2020.pdf Date accessed 15/01/2025

26 See. on https://www.oecd.org/inclusive-growth/ Date accessed 15/01/2025

27 See details on OECD (2018), *The Productivity-Inclusiveness Nexus*, OECD Publishing, Paris, https://doi.org/10.1787/9789264292932-en. https://www.oecd.org/publications/the-productivity-inclusiveness-nexus-9789264292932-en.htm Date accessed 15/01/2025

[28] See. About https://www.oecd.org/wise/, https://www.oecd.org/wise/Peoples-well-being-at-the-top-of-the-agenda-WISE-mission.pdf Date accessed 15/01/2025

[29] See. About https://www.oecd.org/inclusive-growth/businessforinclusivegrowth/ Date accessed 15/01/2025

[30] See OECD in detail . (2018): *Joining Forces to Leave No One Behind Multilingual Summaries* Development Co-operation Report https://www.oecd-ilibrary.org/sites/53d35549-el/index.html?itemId=/content/component/53d35549-elDateaccessed15/01/2025

[31] See. About on the website http://www.oecdbetterlifeindex.org/#/11111111111 Date accessed 15/01/2025

[32] See in detail the analysis of indicators per Greek Region according to the methodology for measuring well-being of the OECD https://www.oecdregionalwellbeing.org/EL64.html Date accessed 15/01/2025

[33] See. see OECD report (2020), How's Life? 2020: Measuring Well-being, OECD Publishing, Paris, https://doi.org/10.1787/9870c393-en. http://www.oecd.org/statistics/how-s-life-23089679.htm Date accessed 15/01/2025

[34] See. see the related Word Bank report. (2017): *Environmental and Social Framework*, International Bank for Reconstruction and Development, World Bank Washington DC https://pubdocs.worldbank.org/en/837721522762050108/Environmental-and-Social-Framework.pdf Date accessed 15/01/2025

[35] See. about website, https://www.worldbank.org/en/topic/disability#2 Date accessed 15/01/2025

[36] See. on the website, https://www.worldbank.org/en/region/eca/publication/breaking-cycle-of-roma-exclusion-in-western-balkansDateaccessed15/01/2025

[37] See. on the website, Date accessed 15/01/2025 https://www.worldbank.org/en/topic/jobsanddevelopment/overview#2

[38] See. about the website, https://www.worldbank.org/en/topic/jobsanddevelopment/publication/jobs-and-development-list-of-publicationsDateaccessed15/01/2025

[39] See. about the website, https://www.worldbank.org/en/topic/jobsanddevelopment/publication/jobs-and-development-list-of-publicationsDateaccessed15/01/2025

40 See. on the website, https://www.worldbank.org/en/topic/communitydrivendevelopmentDateaccessed15/01/2025

41 See. about the website, https://www.worldbank.org/en/topic/inclusive-cities #1 Date accessed 15/01/2025

42 See. about the website, https://www.worldbank.org/en/topic/inclusive-cities #1 Date accessed 15/01/2025

43 See. on the IMF report. (2013): *Inclusive Growth: Measurement and Determinants,* WP/13/135 Date accessed 15/01/2025

44 Date accessed 15/01/2025 https://www.imf.org/en/Publications/WP/Issues/2019/02/15/Inequality-of-Opportunity-Inequality-of-Income-and-Economic-Growth-46566Dateaccessed15/01/2025

45 See. about the website, https://www.imf.org/en/Publications/Staff-Discussion-Notes/Issues/2019/01/18/Fiscal-Policy-and-Development-Human-Social-and-Physical-Investments-for-the-SDGs-46444Dateaccessed15/01/2025

46 See IMF . (2021): *Regional Disparities, Growth, and Inclusiveness,* WP/21/39 Date accessed 15/01/2025

47 See. about the website, https://www.worldbank.org/en/topic/inclusive-cities #1 https://www.imf.org/en/publications/fm/issues/2017/10/05/fiscal-monitor-october-2017Dateaccessed15/01/2025

48 See. about the website, https://www.imf.org/en/Publications/Policy-Papers/Issues/2019/06/10/A-Strategy-for-IMF-Engagement-on-Social-Spending-46975Dateaccessed15/01/2025

49 See. on the website, https://www.imf.org/en/About/Factsheets/imf-capacity-developmentDateaccessed15/01/2025

50 See. on the website, https://www.imf.org/en/About/Factsheets/imf-capacity-developmentDateaccessed15/01/2025

Chapter 6
The Role of Spatial Planning in Balancing Development and Environmental Conservation

Charuka Deshan Jayalath
https://orcid.org/0009-0006-7259-4743
University of Moratuwa, Kadawatha, Sri Lanka

Udayangani Kulatunga
University of Moratuwa, Sri Lanka

Vijitha Disaratna
University of Moratuwa, Sri Lanka

ABSTRACT

This chapter examines how spatial planning contributes to the realization of sustainable development goals by balancing development demands and environmental conservation. It examines the impacts of spatial planning, highlighting its potential to address economic, social, and environmental challenges while ensuring value for money in development projects. This chapter draws insights to identify the concept of spatial planning and its importance in integrating it into city development planning in order to optimize resources, reduce life cycle costs, and enhance investment value. It emphasizes the need for environmental and economic assessments to mitigate risk associated with improper spatial arrangements. It also examines the historical significance of spatial planning in mitigating environmental impacts, increasing

DOI: 10.4018/979-8-3373-1554-6.ch006

resilience to natural hazards, and safeguarding investments. By providing theoretical foundations and practical approaches, this chapter serves as a comprehensive resource for policymakers interested in implementing effective spatial planning for sustainable development.

1. INTRODUCTION

In the beginning, prehistoric humans started to fulfill their basic needs by constructing the first freestanding building which is called home further, they have transformed or modified different types of environments to fulfill their needs, such as urban, peri-urban, rural, and agricultural (Vileniske & Arturas, 2012). However, in ancient times humans had achieved their changing needs knowing the limitations of nature. Despite this fact, the over-exploitation of environmental resources, vital resource depletion, and natural disasters are some of the reasons for the collapse of the highly developed ancient societies (Janssen & Scheffer, 2004). In this context, a great Socio-ecological system should have the ability to absorb changes in human needs without shifting to a new regime with a different set of processes and structures (Holling, 1973). In order to do that society should consist of unique knowledge, and according to Silva (2008), the social construction of knowledge is created within a given society and it is based on the general knowledge and social human behavior concerning their culture and mind.

According to the ecological footprint analysis, today we use the equivalent of 1.7 planets to provide the resources we use and to absorb our waste. If this trend continues, in the 2030s we will need the equivalent of two earths to support us. Thus, these experiences make us ponder anew the meaning of development. Nevertheless, already more than half the world's seven billion people live in cities, and by 2050, urban areas will be home to nearly 70% of the global population (Department of economic and social affairs, 2014). Consequently, planners also need to consider the ever-increasing demand, particularly for housing, and have them accommodated sustainably and be given priority (Vileniske & Arturas, 2012). However, in light of the foregoing, the construction industry has made a significant contribution to global environmental issues. Every human activity affects the environment and constraints are established by nature in any revolutionary operational activities (Lantitsou, 2017). Therefore, city planners should develop a thorough understanding of the surrounding environment and the homeostasis phenomena of nature, which are evident in occurrences such as "a steel piece becoming corroded" and "a landslide. This evolution phenomenon needs to be identified, modeled, and predicated during the planning stages of any kind of economic infrastructure development project on Earth in order to resolve the conflict between meeting our needs and protecting the environment (Jayalath & Jayalath, 2019). Thus, nowadays we have to use another dimension to measure the relationship of the development with the natural environment. Otherwise, on a nearby day, humans have to face the same collapse that ancient societies had to face.

Thus, humans should recognize the natural homeostasis phenomena, as well as social and cultural knowledge, and incorporate these into the spatial planning of the built environment to enable the selection of appropriate design approaches. In this context, spatial planning can be defined as the key instrument for development which is an important lever for promoting sustainable development and improving the Quality of life and it does balance demands for development with the need to protect the environment to achieve social and economic objectives (Economic Commission for Europe, 2008). Moreover, spatial planning can be identified as an instrument to coordinate socioeconomic development by preventing environmental problems and simultaneously protecting the natural environment and the cultural environment (Ministry of the Environment, 2001). According to Cilliers and Timmermans (2013), spatial planning is constantly faced with conflicts between pro-development approaches and pro-environment approaches. Hence, we have to deviate from this contradictory phenomenon, otherwise protecting the environment and meeting our needs will be two concepts that could not be attained simultaneously. Furthermore, spatial planning helps to enhance the value for money of the built environment by giving maximum benefit for the investment and the client can attest to value for money when they feel that they have received a service that was worth the price that they paid for it (Olatunji, Olawumi, & Awodele, 2017).

However, in the modern construction industry, although reorienting plans for place and space is widely supported, there is little common understanding of what "spatial planning" means and what concepts of space and location underpin spatial planning approaches (Davoudi & Strange, 2009). At the same time, due to the number of weak investments & rapid growth in unsustainable debt, Sri Lanka's economy was already showing signs of weakness before the COVID-19 pandemic (The world Bank, 2022). Also, Sri Lanka's environmental challenges are closely connected to the economic policies pursued by successive governments (Senaratne, 2019). Today, the dimensions that are used to measure sustainable development are mainly focused on Human, Social & Environmental well-being (Tiwari & Gupta, 2023). Especially, an economist would see development in terms of an increase in the Gross National Product (GNP) and per-capita income of individuals in the system (Kobani & Douglas, 2021). But nowadays Sri Lanka has to face environmental, social, and economic drawbacks in reference to city planning and development. Moreover, investors do not have the stability and confidence for investing to the economic development due to unsuccessful previous spatial planning, because if the spatial plan fails, the investment in the construction industry will be worthless. There are available some studies that were done on spatial planning and how it affects the social, environmental, and economic aspects. However, no study has been done yet related to the Sri Lankan context about how spatial planning affects the cost of construction and the value for money of the built environment. In this context, the main objective of this chapter is to identify the role of spatial planning in social, environmental, and economic aspects while identifying how spatial planning affects the cost of construction and the value for money of the built environment. Consequently, this chapter contributes to the global discourse on sustainable development by positioning spatial planning as a cornerstone of economic, social, and environmental resilience. It highlights the critical role of spatial planning in optimizing value for money, enhancing the quality of life, and fostering a sustainable future. By providing a robust theoretical foundation, practical insights, and actionable recommendations, the chapter serves as a valuable resource for academics, practitioners, and policymakers seeking to leverage spatial planning for sustainable and economically viable development.

2. LITERATURE REVIEW

The archaeological records confirm hundreds of circumstances where communities have formed strong relationships with their environments (Redman, 1999). But nowadays our needs are becoming more complex compared to the ancient period. Thus, we have transformed or modified different types of natural environments to fulfill our needs (Vileniske & Arturas, 2012). Accordingly, environmental discourses strongly concentrate on the relationship between humans and nature (Reid, Teamey, & Dillon, 2007). we have to think about what should be changed in the basic concepts of planning, designing & Construction. we need a viable solution while meeting demands for developments with the objectives to gain social and economic prosperity to mankind and to protect nature and spatial planning can be identified as an instrument to coordinate socioeconomic development while preserving the environmental concerns (Economic Commission for Europe, 2008). According to the Economic Commission for Europe (2008), when converting the space into a place that is more important to us using this planning tool we will be able to achieve the objectives successfully such as reducing environmental damage, enhancing and preserving natural and cultural resources, develop energy resources while ensuring security, Limit the impact of natural disasters and improve relations between the city and the countryside successfully. Considering this situation as a backdrop, before discussing the role of spatial planning in environmental and economic aspects, it is essential to explore the concept of spatial planning and its challenges.

2.1 Concept of Spatial Planning

Geographically space means that everything exists and continues in any direction or in an empty area to be used (Singh, 2019). Thus, in comparison to the word 'space,' which doesn't have significance unless we mean it, a place becomes meaningful if there are connections that are important to us (Wijesundara, 2015). Without a comprehensive exploration of this converting process, existing studies fail to provide actionable insights to mitigate environmental, and economic foremost impediments. In this context, spatial planning can be utilized as an instrument to transform space into a place that maximizes benefits across various aspects. As previously mentioned, spatial planning can be described as a key development instrument which is a significant lever in support of sustainable development and quality of life and balances development demands against environmental protection to achieve social and economic goals (Economic Commission for Europe, 2008). Spatial planning can be used as a tool to coordinate socioeconomic growth by avoiding environmental problems and protecting the natural and cultural environment at the same time (Ministry of the Environment, 2001). The scope of spatial planning differs greatly between countries, but most have a range of similarities. Terms like "land-use planning" (e.g., Ireland), "land planning" (Italy), "spatial planning" (Germany: Raumordnung), "town and country planning" (UK), others use "spatial development" (Poland), "regional development planning" (France) is used in a number of countries to imply the spatial planning (Fleischhauer, 2008). Spatial planning in all countries nearly focuses on establishing long and medium-term spatial goals, plans for land use, urban development, and the management of sectoral policies such as transport, agriculture, and environment (Koresawa & Konvitz, 2001).

In the planning world, spatial planning was the symbol of a modern trend; the profession used the term to point out modern goals and future paths in particular, in Europe (Natarajan, 2015). Moreover, the spatial planning system is considered an ensemble of arrangements for territorial governance that aim to form spatial development patterns, in particular in places (Nadin & Stead, 2008). Thus, the development of a spatial planning system has been identified as one of China's leading national agenda items, and several pilot cities have explored different plan integration mechanisms to provide practical support for forming a spatial scheme (Zhou, Lu, Lian, Chen, & Wu, 2017). Notably, although spatial planning has been utilized in various projects worldwide, its overall purpose can be concluded into specific objectives.

2.2 Purpose of Spatial Planning

Compared to the previous regulatory land use planning approaches, today spatial planning is used for promoting long-term strategic visions, ensuring spatial dimension to better align a variety of sectoral plans and activities, supporting "rational" approaches to sustainable development, and enhancing interaction with stakeholders and the public (Nadin, 2007). Amongst those purposes of spatial planning, spatial planners and governments have been trying to steer urbanization processes with the aim of developing sustainable cities and regions, can be identified as a significant role in spatial planning (Hersperger, et al., 2018). Using this planning system to face climate changes globally can be identified as another usage of spatial planning. As a result of driving shifts in ecological systems globally in terms of climate change, human beings face the challenge of adapting to changing environmental factors and a more unpredictable future (Adams, et al., 2017). Also, advocates view spatial planning as a way to put divergent voices together in order to develop and agree on shared goals for future spatial development (Allmendinger & Haughton, 2010). Spatial planning aims at achieving socioeconomic outcomes – stronger local economies, more sustainable growth, and a better quality of life. Further planners also need to consider the ever-increasing demand, particularly for housing and have them accommodated sustainably and be given priority (Vileniske & Arturas, 2012). However, proper spatial planning is a necessity in the current world, as there is a lack of stability and confidence in investing in economic development due to the failures of previous spatial planning efforts. When a spatial plan fails, investments in the construction industry risk becoming futile and worthless. Consequently, the reasons for implementing spatial planning can be summarized as follows:

- To safeguard people's rights. When people have access to land, all rights and responsibilities are effectively accomplished. It is necessary to manage change in such a way that those rights and obligations are respected.
- Natural systems to defend. Natural systems have their own operational requirements which must be respected in order to achieve sustainable human development as a long-term solution and to prevent, or at least mitigate, significant environmental destruction.
- To use resources efficiently. There is a shortage of resources such as land, water, electricity, finance, building materials, skills, etc. Those available must be wisely applied in all circumstances to ensure full benefit.
- In order for all areas of government to have a higher standard of service.
- Coordinating actions and planning in order to optimize the positive effects of resource investment

• Cooperation of various forms of actions and investments of public authority and enhanced cooperation between public and private acts.

2.3 Implementation of Spatial Planning

Although numerous factors advocate for the implementation of spatial planning, it is essential to explore the associated challenges and implementation strategies in depth. According to recent studies, spatial planning in all countries nearly focuses on establishing long and medium-term spatial goals, plans for land use, urban development, and the management of sectoral policies such as transport, agriculture, and environment (Koresawa & Konvitz, 2001). They are interlinked and interdependent in such a way that they affect each other. Changes in every item therefore bring about changes in the entire pattern of the region (Singh, 2019). Thus, spatial planning implementing conditions shall depend on the plan which shall determine for each spatial arrangement. According to Krajewska, Zrobek, & Kovac (2014), there are certain conditions including cultural heritage, nature conservation, preservation of the environment and natural assets, water management, the protection of human health, state security, and protection against natural and other disasters which can be generally found in spatial planning. However, even when spatial planning is developed with these conditions in mind, it must contend with numerous challenges arising from current global trends. According to Bruin, Goosen, Ierland, & Groeneveld (2013), these are some challenges which can be identified in spatial planning.

1. The demand for space: How does the government respond to the continuing urbanization? The demand for rural housing and high-quality housing locations in general will grow strongly due to increase of population. Thus, they have to face a challenge, whether to accept or resist these demands.
2. To face the traffic jam: Should more roads be available or could existing roads be better used? In all of this, what role can technological innovations or new forms of public and private transport play?
3. Environment-economy relationship: How can a sustainable economic and ecological balance be established? Is it possible to predict developments without damaging environmental quality and nature?
4. Lifestyles and housing needs: How can a division between the privileged and underprivileged be prevented? Should there be investments in human development or in the physical environment?
5. Relationship between government and individuals: How do governments and individuals engage in urban planning and land use? Does centralization or decentralization have to be emphasized?

When spatial planning is utilized, not only do these challenges arise, but there is also the issue of investors being discouraged from building due to construction delays caused by lengthy processes for obtaining planning approval associated with this tool (Krajewska, Zrobek, & Kovac, 2014). Nevertheless, considering the positive impacts on a country's economy and ecology, the benefits of using this planning tool far outweigh the challenges. Thus, successfully addressing these challenges and implementing spatial planning can lead to significant benefits in both ecological and economic aspects. In this context, to overcome the aforementioned challenges and achieve environmental and economic benefits, environmental impact assessment can be considered one of the most widely used tools across the world.

2.4 Environmental Assessment in Spatial Planning

The construction sector is regarded as one of the world's main sources of environmental pollution and has huge direct and indirect environmental effects (Enshassi, Kochendoerfer, & Rizq, 2014). Ecological systems provide human societies with various resources and contribute to life support systems (Cilliers & Timmermans, 2013). The Environmental Impact Assessment (EIA) has been established successfully throughout the world as an effective tool for dealing with the environmental impacts of project development proposals and depending on the system in place, the meaning of the word 'environment' in EIA may range from being exclusively bio/physical/natural environment, to cover also social and economic issues related to the impact of the proposal (Eggenberger & Partidario, 2014).

In some countries, environmental assessments link the economic and social implications of projects or plans with the physical effects that the development project is likely to have on the environment (Davoudi & Strange, 2009). This environmental impact assessment is also very helpful to enhance the biodiversity which is great significance to exist the living being including the human by preserving the green spaces (Cilliers & Timmermans, 2013). However, the project environmental assessment process needs considerable knowledge, resources, and data. This process would avoid or divert development that could have potentially significant adverse environmental impacts that cannot be compensated. Thus, developers may make a significant contribution to the environmental assessment process by offering knowledge, information, or evaluation resources (Lantitsou, 2017).

Latest statistics from a variety of international organizations indicate that natural disasters have caused a rise in deaths and economic losses and such factors are closely related to the increasing exposure of people and infrastructure to natural hazards due to population growth limited space and global climate change (Resurreccion, Sajor, & Fajber, 2008). Because population growth and urbanization say that more land is needed; some people have to live in disaster-prone areas (Cook, 2010). According to recent scientific literature, there is nothing wrong with landslide, because solid lands like clouds shape themselves move, and flow from another form (Landow, 2012). In this case, we can understand that undisturbed nature is orderly and harmonious and that ecological systems come back to their previous balance in terms of "natural balance" (Wu & Loucks, 1995). Therefore, because of wrong spatial settings of buildings or constructions definitely find a landslide. Indeed, an active volcanic island in the Pacific Ocean is not a disaster because it does not take place in human habitats (Jayalath & Jayalath, 2019).

Actually, if it hits remote areas uninhabited by people and there is no infrastructure in the area it will continue to be a common natural phenomenon and disaster impacts will depend on the type of natural disaster, geographical area, population density, and infrastructure conditions (Sutanta, Rajabifard, & Bishop, 2010). A hazard can be described as "a threat to something people value, such as life, safety, the environment, and lifestyle" (Mills, Yessis, Boyd, & Andrey, 2001). Thus, Disasters are defined in recent scientific literature as a result of an interaction of two variables which are hazards (e.g., triggering agents stemming from nature, as well as from human activity) and vulnerability (e.g., susceptibility to injury or loss influenced by physical, social, economic and cultural factors) (Mcentire, 2001). Reasons for why vulnerability is increasing in the future can be categorized under physical, social, cultural, political, economic, and technological headings. According to Mcentire (2001), a list of factors that augment vulnerability is indicated in Table 1.

Table 1. Reasons for increasing vulnerability

Factor	Reasons for increasing vulnerability
Physical	• improper construction of buildings • inadequate foresight relating to the infrastructure • degradation of the environment
Social	• limited education (including insufficient knowledge about disasters) • massive and unplanned migration to urban areas
Cultural	• defiance of safety precautions and regulations • dependency and an absence of personal responsibility
Political	• minimal support for disaster programs among elected officials • inability to enforce or encourage steps for mitigation • over-centralization of decision making • isolated or weak disaster-related institutions
Economic	• the pursuit of profit with little regard for consequences • scarce resources for disaster prevention, planning, and management • carelessness in industrial production

Thus, in environmental impact assessment can be identified the significant features related to the natural environment could impact the development projects, to make successful spatial planning. In fact, spatial planning should consider the entire ecosystem from the very beginning, prior to confining the spatial position of a building. For example, before constructing a residential building, it is essential to evaluate whether the location is prone to landslides and whether the structure will endure its expected lifetime. However, ancient civilizations implemented spatial planning in a thorough manner to prevent natural hazards.

Recent scientific literatures focus more on the question of how cities can recover after a disastrous event has happened (Vale, 2005). Spatial planning can play in mitigating multiple hazards by influencing urban structures and due to the character of spatial planning, it helps to strengthen urban resilience before an event happens (Fleischhauer, 2008). Spatial planning is extremely helpful in minimizing the exposure of people, socioeconomic activity, and infrastructure to natural disasters and to manage the long-term use of land (Sutanta, Rajabifard, & Bishop, 2013). Sri Lankan ancient spatial planning had been given a different meaning for the forms that last a longer period of time, compared to the landforms which have more tendency to change (Jayalath & Jayalath, 2019).

Figure 1. A schematic sketch of this ideology that had been adopted at Mirisawatiya.

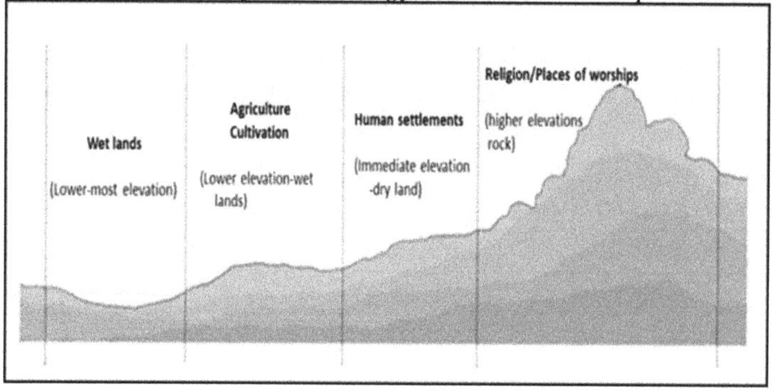

The hard rock for religious places was our ancient ideology for space planning. After establishing religious sites at peaks, no human settlements were found inside and around the landscape due to the utmost respect (Jayalath & Jayalath, 2019). Thus, they reduced the environmental impact on the construction industry by making significant identification of the landforms process and assessing their impacts on the construction industry. Consequently, these incidents advocate that proper spatial planning can minimize the exposure of people to natural disasters, by considering natural phenomena and landform characteristics. Accordingly, this approach will enhance the value of investments by addressing whole-life costs, including the maintenance and operational costs of the project.

Nevertheless, today, numerous drawbacks related to economy-focused construction and investment strategies have been encountered in society. Accordingly, in addition to environmental assessment, it is essential to focus on the connection between investments or projects and socio-economic aspects to achieve economic well-being alongside environmental sustainability.

2.5 Economic Assessment in Spatial Planning

2.5.1 Introduction

Spatial planning aims at achieving of socioeconomic outcomes – stronger local economies, more sustainable growth, and a better quality of life. Planning will assist in achieving these objectives by generating and focusing investment interest, enhancing and safeguarding land and property prices, and reducing investment risk by certainty, resolving current and new developments' external effects. Spatial planning may also impact land and property monetary values by encouraging investment (Bruin, Goosen, Ierland, & Groeneveld, 2013). With good preparation and planning of a project, one can avoid costly mistakes during the implementation phase by abandoning a project without incurring unnecessary further costs at the initial phase (e.g. a purchase of a land plot etc.) (Sobieraj, 2017).

2.5.2 Role of Spatial Planning in Economic Development

Expansion of the construction industry beyond the adaptive capacity of the economy will only waste national resources (Giang & Pheng, 2011). Thus, the planning stage of the construction industry plays a major role in the economic sector of the country. Therefore, the use of this planning tool which is called spatial planning, will coordinate not only the social factors but also environmental factors to develop the local economy (Biesbroek, Swart, & van der knaap, 2009). Environmental factors are mainly responsible for property valuation and those factors mainly influence real estate sector decision-making (Cellmer, Senetra, & Szczepanska, 2012).

The proximity to water bodies, forest covers, and land elevation can be identified as the environmental attributes that affect the property valuation (Cellmer, Senetra, & Szczepanska, 2012). Urban sustainability and regeneration strategies focus mainly on man-made components of the urban environment compared to the natural components. But the quality of the surrounding environment influences the value of property because the surrounding natural environmental factors of the property influence the property buyers' choice (Zrobek, Trojanek, Sokolnik, & Trojanek, 2015).

The real estate business which is one of the basic economic sectors in the world is affected and formed not only by economic and productive factors but also by various qualitative characteristics of the natural and human environment, in which each real estate activity is performed. Thus, it can be identified lakes, rivers and sea, topography as well as green areas, wetlands, and mountain areas, as they represent the most important influential factors. There are various benefits from these factors, some of which include global and local climate stabilization, a decrease in flood events, enhancement of habitats, etc. On the other hand, the risk of natural disasters can certainly influence negatively real estate market values (Nikoloas, Dimitra, & Agapi, 2014).

2.5.2.1 Real Estate and Topography

Although topography is the most difficult real estate factor to study and define, it can be identified the market value fluctuates respectively according to the land category within which the property is located (Nikoloas, Dimitra, & Agapi, 2014). It is proven that as the ground slope becomes less abrupt, real estate market value increases due to the willingness of the buyers to have more plain areas (Davies, et al., 2008). At the same time, the cost of natural disasters exceeds the amount of 60 billion dollars annually due to a lack of knowledge of topography and its improper usage (Nikoloas, Dimitra, & Agapi, 2014). It is widely known that human life and activities affect the course of the environmental life and circle and due to that effect, we have to face these kinds of natural disasters.

Thus, we must consider spatial planning which makes significant relationships with the social, economic, and environment of the region to make successful development projects that will ensure a better economic life. Spatial planning aims to allocate different land-use functions and activities as efficiently and effectively as possible for maximizing the 'benefits/profits' at a given location (Eggenberger & Partidario, 2014).Therefore, spatial planning can be used to identify the best use of the property for development projects. The highest and best-use concepts also can be used for this planning system.

2.5.2.2 Highest and Best-Use Concept

The value of a property could reasonably be determined after taking all factors into consideration and which is physically possible, financially feasible, legally permissible, and optimally productive. In the highest and best-use concept, the value should consider not only the land to be developed or left vacant but also if developed, what kind of improvement should be built (Boshoff, 2016). In many instances, a property's highest and best use may change in the foreseeable future (Dunn, 2010). Within the physical possibility test should address the size and accessibility of land and the risk of natural disasters such as floods or earthquakes (Dunn, 2010). Therefore, by using these concepts and knowledge in spatial planning our natural space can be best used and make good value for the property in the Sri Lankan context.

2.5.3 Cost Implication Over New Construction Projects, and Pros & Cons Over Externalities

There is always the possibility that a plan will not be executed, that parts of the plan will change and that the planning takes more time than expected. Therefore, as an example, if a road is expected to be constructed, the reduction in the value of the properties might already take place in an early stage of the planning process and the uncertainty about the implementation of the plan and when it will be executed can also influence the value of the relevant properties (Cotteleer & Peerlings, 2011). One of the reasons for that is according to the evidence; people who live in the proximity of a highway are at greater risk of serious health consequences (Bae, Sandlin, Bassok, & Kim, 2007). Moreover, it is found that a 1% increase in traffic noise was associated with a 1.3% decline in land price in Seoul, Korea after studying the effect of traffic noise on property values (Kim, Park, & Kweon, 2007). Another study in the city of Poznan showed, that plots located in the zone with noise exceedance at night were about 57% cheaper than those located outside this zone (Lowicki & Piotrowska, 2015). Therefore, it can be identified that properties with street views are priced lower (Jim & Chen, 2009). According to Lowicki & Piotrowska (2015), lack of spatial planning is the key explanation for road noise pollution.

In cost approach, there are different types of depreciations in terms of physical, functional and external (Ivanovskiy & Azmitov, 2015). Economic depreciation is an impairment and restriction of utilization of the property object resulting from the economic factors impact generated by the external market environment such as the location of the object, transport accessibility, environmental infrastructure, etc. For example, the occurrence of a permanent waste disposal site next to an apartment building will lead to the cost of an apartment reduction, whereas the physical and functional characteristics of the premises remain unchanged (Ivanovskiy & Azmitov, 2015).

Additionally, based on experience in society the proximity of schools, hospitals, and universities is positively related to property prices, and it is closer to the industrial use of land has a major adverse impact on property prices (Cotteleer & Peerlings, 2011). Thus, because of the negative externalities imposed on residential properties, many governments have to put regulations in place to compensate those who are negatively affected by construction works (Cotteleer & Peerlings, 2011). Therefore, these results can be helpful not only in especially for estimating the costs of road and other infra-structures investments but also in spatial planning

(Lowicki & Piotrowska, 2015). Therefore, if we adopt effective spatial planning, we can positively impact the cost of the construction project.

Consequently, these factors—such as natural environmental phenomena, built environmental features, and the value assigned by society to surrounding infrastructures and services—significantly impact not only the life cycle cost of the investment project but also, ultimately, the value of the investment itself. Therefore, proper spatial planning at the initial stage of investment and city development will help mitigate unexpected costs and ensure value for money in the investment.

However, In the evaluation of philosophies related to social, human, and environmental sciences, there has been a tendency to rethink the human role within modern social and ecological systems. In this context, the concept of the noosphere, which is described as the global sphere of collective wisdom, memory, culture, and human thought, has gained significant popularity. Noosphere represents one of the most important components of the spiritual response of humanity related to global problems in their society. Moreover, there is a profound interconnection origin between human beings and the geosphere through their experiences, memories, and folklore, and this sense of place gives respectively to the identity, values, and responsibilities. Therefore, consequently, this knowledge, way of thinking, and experiences aid in developing unique solutions at the right time and place. In this context, implementing successful spatial planning requires a thorough focus on social knowledge and its characteristics to develop a unique approach that addresses environmental, social, and economic challenges while ensuring value for money in investments. Consequently, by incorporating society's knowledge, this distinctive plan will safeguard investors' funds and promote sustainable development.

2.6 Benefits of Spatial Planning

According to Economic Commission for Europe (2008) using the knowledge of the social, natural environment flux and assess their impact on the development projects can fulfill not only economic benefits but also social and environmental benefits because of spatial planning.

Economic Benefits

- Provides more stability and confidence over investments
- Land identification in relevant areas to meet economic growth needs
- Promoting environmental quality in both urban and rural areas, and thus establishing better investment and development conditions
- Identifying development that meets the needs of local communities

Social Benefits

- Considering the needs of the local communities in policy development
- Promoting the re-use of vacant and derelict land
- To help create and maintain pleasant, safe and healthy environments.

Environmental Benefits

- Promoting regeneration and the appropriate use of land, buildings and infrastructure
- Conservation of important cultural, environmental and historical properties
- Addressing potential environmental risks (e.g. flooding, air quality)
- Protecting and enhancing areas for recreation and natural heritage
- Promoting energy efficiency in development layout and design

3. FACTORS INFLUENCING VALUE FOR MONEY FROM PROFESSIONALS' PERSPECTIVES

The initial study was carried out for the identification of the research problem, aim, objectives, scope, and limitations. The background study was done making reference to books, journal articles, and conference proceedings. A literature review is "an important chapter in the thesis, where its purpose is to provide the background to and justification for the research undertaken" (O'Gorman & Maclntosh, 2015). Therefore, information required for the literature review was collected from books, journals, research papers, conference papers, past dissertations, reports, and internet sources, and the initial steps of this study were covered through the literature survey.

The collection of data also plays a vital role in the research process. Unless the correct data is gathered there is no way of fully answering the research problem. Data can be either primary or secondary (Rabianski, 2003). Secondary data for the research was collected through the literature synthesis in the initial stage of this study. Primary data for the research was collected through a questionnaire survey research approach. There are many ways of gathering data each has its own merits and demerits. This study was focused on the specific purpose of assessing the environmental and economic impacts of using spatial planning in the construction industry in Sri Lanka. To achieve that aim, forty-two questionnaires were distributed among professionals who worked in the Sri Lankan construction industry, and collected thirty completed questionnaires. Considering the time constraints available for the completion of research, the Convenience sampling method was used for the questionnaire survey, and also generally professionals have a basic idea about the improper spatial planning and participants are often readily and easily available. SPSS was used for data analysis of this research as it covers a broad range of statistical procedures.

3.1 Level of Influence of Spatial Planning for Ensuring Value for Money

A Questionnaire survey has been conducted to find out the "level of influence" of "spatial planning" over the factors revealed from the literature findings and the impact of these improved factors over the cost of the project, ensuring value for money, as a result of carrying out proper spatial planning. In order to obtain a clear comparison between these factors, it is needed to consider the means of "their overall influence level". Thus, table no 2 shows the means of "the influence level of these factors" which were given by the respondents. In this section, it was attempted to identify how proper spatial planning affects the maintenance and construction cost of the built environment. Therefore, prior to this step, it was identified that some factors can be affected by proper spatial planning. Moreover, those factors are influenced not only by the maintenance and construction cost of the built environment but also by ensuring value for money of the project concerned. The cumulative impact of spatial planning on identified factors and the impact of those factors on the value for money of the project is given the exact impact of spatial planning on the value for money of the project. Thus, according to the above statement and the research findings the impact of proper spatial planning on value for money in the project can be represented by the formula illustrated in Figure 2.

Figure 2. Formula of the impact of proper spatial planning on value for money

Therefore, in order to obtain the impact of spatial planning over the value for money of the project in quantitative terms; the means of their "influence level" which is given by the respondents were calculated and represented in Figure 3.

Figure 3. Degree of influence expressed quantitatively using "mean"

	(1)	(2)	(3)	(4)	(5)	(6)	(7).	(8)	(9)	(10)
Mean (how spatial planning effect on identified factors)	4.10	4.27	4.13	4.33	4.17	3.80	3.50	4.27	4.03	3.97
Mean (how those factors effect on value for money)	4.43	3.83	3.73	4.43	4.37	4.07	4.33	4.17	3.70	3.90
Total mean out of 10	8.53	8.10	7.86	8.76	8.54	7.87	7.83	8.37	7.73	7.87

(1) Safeguard client rights
(2) Defending natural system
(3) Efficient usage of natural resources
(4) Safeguard on investments
(5) Enhance the economic life of the project
(6) Ability to withstand natural disasters
(7) Construction time saving
(8) Enhanced economic returns of the project
(9) Positive impact over the value of surrounding real estates
(10) Risks of disasters and its management

According to the "level of degree" it can be identified, factors having a mean more than 4 and a total mean more than 8 as high influence factors on value for money by adopting proper spatial planning. In here, one-sample t-test was used to compare the means and 8.00 is the hypothesized number which shows the "high level of influence". Eventually, how spatial planning affect over the value for money of the build environment is summarized below. These factors which were revealed from the literature findings are ordered by considering their level of influence according to respondents' point of view.

3.1.1 List of how spatial planning "highly affects" over value for money

1. By safeguarding the investments of the project
2. By enhancing the economic life of the project
3. By Safeguarding the client's rights in the project
4. By enhancing the economic returns of the project
5. By defending the natural system surrounding the built environment

Moreover, efficient usage of natural resources, ability to withstand natural disasters, construction time saving, positive impacts over the value of surrounding real estate, and risks of disasters and its management can be identified as means which were agreed by respondents that affect over value for money. Consequently, this primary data again confirmed and validated the literature findings by revealing how spatial planning affects the value for money of the investment. However, in order to gain manifold benefits from spatial planning, it warrants overcoming different challenges in Sri Lankan context.

4. IDENTIFY SUGGESTIONS TO OVERCOME THE CHALLENGES IN THE SPATIAL PLANNING PROCESS

Numerous suggestions to overcome the challenges in spatial planning were profoundly identified and validated through expert interviews. One of the respondents stated that there should be an economic value for the ecological balance in the society. Another view of a respondent was; in order to come across the above challenge a change in the thinking pattern of the society and every decision should be based on environmental concerns. Thus, by considering both these suggestions it could be identified that there should be a change in the thinking pattern in order to make an economical value for the ecological balance in the society. Another two respondents stated that there should be a complete site analysis by studying the existing area with special reference to landscape aspects. As a consequence of this statement and by considering previous suggestions, it could be concluded that every decision should be based on environmental concerns through a complete site analysis. The government has to develop its own national policy by considering its own social, economic, and environmental factors and their characteristics was another suggestion as the first step to establish a sustainable, economically and ecologically viable balance. Identifying the objectives of the project and its possibility of reaching project output, and engaging experienced and educated professionals are a few further suggestions provided by the professionals.

It means prior to deciding the meaning of development the particular context should be understood. It says that the meaning of development and sustainability should vary according to the context. The transdisciplinary approach is another view as the first step to establish a sustainable, economically and ecologically viable balance which was given by a respondent. A transdisciplinary approach means that every professional will contribute to making decisions in the construction industry. Actually, according to the literature finding not only the construction industry but also every process done by humans are linked to each other. But in order to become specialists in a particular area and to identify most bottom of the matters in that area, humans have been divided today into several areas. Even if it is so, the issues have been arising in other subject areas warranting decisions without considering the interdependency of those areas. Thus, according to the respondent's view, every professional who works in different subject areas such as social, economic, environmental, technological, etc should be given a chance to take part in making decisions in the construction industry.

Additionally, to address the challenges posed by continuing urbanization, establishing robust national policies or enhancing existing ones can be recommended as a critical initial step for the government. These policies should focus on sustainable development, efficient resource management, and balanced urban growth to meet the evolving needs of society. Furthermore, suggestions confirmed that creative and experienced professionals should be engaged in establishing the national policy and decision-making process. Moreover, this national policy should be based on the prepared physical plan by analyzing the population growth, and that policy should have a future plan in order to protect the environment. Making rural areas attractive by improving infrastructures can be identified as another suggestion that should be considered in making the national policy. Because it will help to decentralize urbanization. In addition, according to the suggestions when the government establishes the national policy it should be reconsidered policies in urban, suburban, and rural areas separately. Subsequently, it should make sure that the authorities adequately do their duties in order to achieve the ultimate goal of the established policies.

Additionally, among the suggestions, there was a different and considerable answer given by a respondent. According to that, they should re-invent a contextually defined meaning for the concept of 'urban'. As stated in the statement, urban is an attitude that is built up by the society. Thus, it could be a changeable concept. Humans believe that they will be able to get facilities when they live near to urban area. But if the facilities are decentralized by using a national policy according to the given recommendations, they will have all facilities wherever they live in the country. Then, it doesn't need to get confined to the concept of urban within the decision-making process.

Understand the Concept Called "Development"

The concept of development can be understood from multiple perspectives, with each viewpoint reflecting unique societal, environmental, and economic priorities. According to the comments made by one of the respondents, if most of the people can use public facilities comfortably it can be determined as the development. This perspective highlights the idea that development is fundamentally about meeting the needs and expectations of society. Thus, one of the respondents commented that development is the satisfaction of the society that they have. Another respondent mentioned that it should be the advancement of all, social, environmental, and economic aspects. Furthermore, the outcomes of the analysis of these comments given by the respondents makes some similarity to another idea given by a respondent. According to that, development begins with people's attitudes and it is an overall uplift of lifestyle, society, environment, and the economy. One respondent articulated that "development is maintaining the feeling of novelty," highlighting the need for continuous innovation and a fresh perspective to sustain societal interest and engagement. Interestingly, respondents also noted that development does not always represent a better position compared to the past but rather a shift from the existing context to a new state that aligns with societal needs. This subjective perspective of development implies that progress is relative and context-dependent. If governments can maintain a sense of novelty and fulfillment within society, it can be recognized as development. Additionally, fostering inclusivity, ensuring environmental preservation, and promoting economic growth are vital pillars that strengthen the foundation of sustainable development, ultimately ensuring long-term satisfaction and progress.

Understanding the Concept of "Value Addition" in Order to Ensure Value For Money Through Spatial Planning

Value is a multifaceted concept that has become a discourse across disciplines, from economics to philosophy. One of the respondents stated that value is an aspect of experience that should be understood as a human expression, which means value is a subjective concept and it depends on the point of view of the human. This view aligns with classical economic theories, such as those proposed by Adam Smith, who identified two key aspects of value: value in use and value in exchange. Value in use refers to the utility derived from an object or service, while value in exchange pertains to its market worth. Applying Smith's concepts to spatial planning highlights how value can be created both by the utility a project offers to society and its contribution to economic growth.

If value is created in the human mind by way of protecting the environment and meeting their needs by familiarizing the benefits of those processes, it could ensure value for money through spatial planning. For instance, preserving green spaces not only contributes to biodiversity but also enhances property values, provides recreational benefits, and reduces urban heat, thereby adding both use and exchange value. Respondents in the study further emphasized that value can be added for money by making well-planned and safe built environments that have economic viability through spatial planning. Proper spatial planning ensures that projects are not only viable but also in line with the demands of society and investors. The principle of Adam Smith regarding resource allocation applies here; if resources are used in the most economical way, then maximum benefits will be achieved from the projects and translate into value for money. For example, strategically planned urban infrastructure reduces travel time, lowers fuel consumption, and minimizes pollution, enhancing the quality of life for residents while reducing long-term operational costs. Furthermore, if the investment is a possible way to increase their wealth, it will ensure value for the money that was spent on that particular project. A well-planned project that aligns with market demands and minimizes risks provides investors with confidence in its profitability and sustainability.

An effective planning system is necessary to safeguard people's rights, use resources efficiently, establish a higher standard of government service, cooperate with various forms of actions and investments of public authority, and enhance co-operation between public and private acts. Such coordination ensures that the social, environmental, and economic factors are harmonized to maximize benefits. Spatial planning also addresses the broader concept of value addition by integrating environmental preservation into development processes. Protecting natural ecosystems within urban plans not only improves resilience to climate change but also attracts investments by enhancing the aesthetic and functional appeal of properties. Therefore, use of this planning tool which is called spatial planning, will coordinate not only the social factors but also environmental factors to develop the local economy (Biesbroek, Swart, & van der knaap, 2009).

In conclusion, value is a subjective concept that must be cultivated in the minds of stakeholders, including society, investors, and policymakers. Notably, spatial planning enhances the value of investments, safeguards natural ecosystems, and promotes economic growth, ultimately delivering value for money. This strategic approach highlights the importance of spatial planning as a cornerstone for sustainable development in an increasingly complex and resource-constrained world.

5. CONCLUSION

Spatial planning has become a critical tool for addressing the challenges of sustainable development in a rapidly urbanizing world. It ensures balancing the competing demands of economic growth and environmental conservation while optimizing resources, reducing risks, and enhancing the quality of life of the development projects. This chapter describes how spatial planning integrates environmental, social, and economic dimensions to achieve value for money and achieving long-term sustainability.

One of the key takeaways from this discussion is the role of spatial planning in addressing the subjective concept of value. It aligns development goals with societal needs and preferences, providing a framework for maximizing utility while protecting the natural environment. It ensures investments are planned to minimize costs and risks associated with improper land use and unplanned urbanization. Proper spatial planning not only secures the physical durability of projects but also makes them economically and socially more viable, while developing confidence among stakeholders.

Notably, Environmental Impact Assessments (EIAs) emerge as an important tool within the spatial planning framework. They provide a systematic way to evaluate the potential environmental consequences of development projects, enabling planners to incorporate ecological considerations from the outset. Moreover, EIAs contribute to the sustainability and resilience of built environments, by safeguarding natural ecosystems and mitigating risks such as landslides and resource depletion. Additionally, they help maintain the balance between development needs and environmental preservation toward long-term sustainable economic growth and social well-being. It further emphasizes the need for involvement in planning with societal knowledge. Development begins with people's attitudes and requires an inclusive approach that incorporates community perspectives and cultural values. Spatial planning ensures that projects are not only functional but also aligned with societal aspirations, by involving local communities and stakeholders in decision-making. This participatory approach develops ownership and acceptance by the communities themselves, which is vital for the success and sustainability of development initiatives.

The chapter also explores another key dimension of spatial planning called "economic viability". Proper planning optimizes the allocation of resources, reduces life cycle costs, and enhances the return on investments. Projects designed with a focus on sustainability and long-term efficiency yield significant economic benefits, including higher property values, reduced operational costs, and greater investor confidence. Therefore, Spatial planning becomes a tool for creating wealth and ensuring value for money within the built environment.

The challenges to spatial planning, such as lengthy approval processes, resistance to change, and insufficient regulatory frameworks, were also discussed in this chapter. These challenges can discourage investments and delay projects. This is actually indicative of the need for robust policies and streamlined procedures to support effective spatial planning. Governments and policymakers must consider in implementing comprehensive planning systems that integrate environmental assessments, stakeholder collaboration, and technological innovation. By addressing these challenges, spatial planning can be positioned as a cornerstone for achieving sustainable development.

However, the historical context of spatial planning provides valuable lessons for contemporary practice. Ancient civilizations demonstrated the significance of integrating environmental considerations into development, using spatial planning to mitigate natural hazards and optimize resource use. These practices highlight the timeless relevance of spatial planning in balancing human needs with environmental stewardship.

In conclusion, spatial planning is not just a technical process but a multidimensional approach where environmental, social, and economic objectives are integrated. It provides a pathway for achieving sustainable development by ensuring that investments are efficient, resilient, and aligned with societal goals. This chapter has identified that the holistic perspective in successful spatial planning combines environmental assessment, societal knowledge, and economic strategies t create value for money. By addressing the challenges and leveraging the opportunities presented by spatial planning, stakeholders can foster a sustainable future that balances development demands with environmental conservation and social well-being.

REFERENCES

Adams, V., Romero, J. A., & Capon, S. (2017). Making time for space: The critical role of spatial planning in adapting natural resource management to climate change. *Environmental Science & Policy, 74*, 57–67. DOI: 10.1016/j.envsci.2017.05.003

Allmendinger, P., & Haughton, G. (2010). Spatial planning, devolution, and new planning spaces. *Environment and Planning. C, Government & Policy, 28*(5), 803–818. DOI: 10.1068/c09163

Bae, C.-H. C., Sandlin, G., Bassok, A., & Kim, S. (2007). The exposure of disadvantaged populations in freeway air-pollution sheds: A case study of the Seattle and Portland. *Environment and Planning. B, Planning & Design, 34*(1), 154–170. DOI: 10.1068/b32124

Biesbroek, G. R., Swart, R. J., & van der Knaap, W. G. M. (2009). The mitigation-adapration dichotomy and the role of spatial panning. *Habitat International, 33*(3), 230–237. DOI: 10.1016/j.habitatint.2008.10.001

Boshoff, D. G. (2016). *Principles of Highest and Best Use Valuation;A South African Legal Perspective with Practical Implications*. Researchgate. DOI: 10.18638/quaesti.2016.4.1.283

Bruin, K., Goosen, H., Ierland, E. v., & Groeneveld, R. (2013). *Costs and benefits of adapting spatial planning to climate change: lessons learned from a large-scale urban development project in the Netherlands*. Springer.

Cellmer, R., Senetra, A., & Szczepanska, A. (2012). The Effect of Environmental Factors on Real Estate Value. *Knowing to manage the territory, protect the environment, evaluate the cultural heritage*, 1-13.

Cilliers, E. J., & Timmermans, W. (2013). Approaching value added planning in the green environment. *Journal of place Management and development*, 144-157.

Cook, B. (2010). *Knowledges, controversies and floods: national-scale flood management in Bangladesh*. Durham University.

Cotteleer, G., & Peerlings, J. H. (2011). Spatial planning procedures and property prices: The role of expectations. *Landscape and Urban Planning, 100*(1-2), 77–86. DOI: 10.1016/j.landurbplan.2010.11.010

Davies, R., Barbosa, O., Fuller, R., Tratalos, J., Burke, N., Lewis, D., & Gaston, K. (2008). *City-wide relationships between green spaces, urban land*. Springer Science and Business Media publications.

Davoudi, S., & Strange, I. (2009). *conceptions of space and place in strategic spatial planning*. Abingdon: Routledge.

Department of economic and social affairs. (2014). *World urbanization prospects: The 2014 revision*. Population Devision, United Nation.

Dunn, C. (2010). The Appraisal of real estate 3rd canadian edition. *SAUDER; school of Business* .

Economic Commission for Europe. (2008). *Spatial planning; key instrument for development and effective governance with spacial reference to countries in Transition*. United Nations.

Eggenberger, M., & Partidario, M. (2014). Development of a framework to assist the integration of environmental, social and economic issues in spatial planning. *Impact Assessment and Project Appraisal*. Advance online publication. DOI: 10.3152/147154600781767448

Enshassi, A., Kochendoerfer, B., & Rizq, E. (2014). An evaluation of environmental impacts of construction projects. *Revista ingeieria de construccion*, 234-254.

Fleischhauer, M. (2008). *The Role of Spatial Planning in Strengthening Urban Resilience*. Springer. DOI: 10.1007/978-1-4020-8489-8_14

Giang, D. T., & Pheng, L. S. (2011). Role of construction in economic development; review of key concepts in the past 40 years. *Habitat International, 35*(1), 118–125. DOI: 10.1016/j.habitatint.2010.06.003

Hersperger, A., Oliveira, E., Pagliarin, S., Palka, G., Verburg, P., Bolliger, J., & Gradinaru, S. (2018). Urban land-use change: The role of strategic spatial planning. *Global Environmental Change, 51*, 32–42. DOI: 10.1016/j.gloenvcha.2018.05.001

Holling, C. (1973). Resilience and Stability of ecological systems. *Annual Review of Ecology and Systematics, 4*(1), 1–23. DOI: 10.1146/annurev.es.04.110173.000245

Ivanovskiy, I., & Azmitov, R. (2015). Economic Depreciation in the Property Objects Valuation. *Mediterranean Journal of Social Sciences*, •••, 512–518. DOI: 10.5901/mjss.2015.v6n1s3p512

Janssen, M., & Scheffer, M. (2004). Overexploitation of Renewable Resources by Ancient Societies and the Role of Sunk-Cost Effects. *Ecology and Society, 9*(1), art6. DOI: 10.5751/ES-00620-090106

Jayalath, G., & Jayalath, S. (2019). Landform process centered approach for planning the build environment with particular reference to managing the impacts of floods and landslides.

Jim, C. Y., & Chen, W. Y. (2009). Value of scenic views: Hedonic assessment of private housing in Hong Kong. *Landscape and Urban Planning, 91*(4), 226–234. DOI: 10.1016/j.landurbplan.2009.01.009

Kim, K., Park, S. J., & Kweon, Y.-J. (2007). Highway traffic noise effects on land price in an urban area. *Transportation Research Part D, Transport and Environment, 12*(4), 275–280. DOI: 10.1016/j.trd.2007.03.002

Kobani, D., & Douglas, P. (2021). Indigeneous people, Multiculturalism and inclusiveness in portharcourt city developmnt planning. 2-5.

Koresawa, A., & Konvitz, J. (2001). *Towards a New Role for Spatial Planning.* Organisation for Economic Co-operation and Development.

Krajewska, M., Zrobek, S., & Kovac, M. S. (2014). *The role of spatial planning in the investment process in poland and slovenia.* De Gruyter. DOI: 10.2478/remav-2014-0017

Landow, G. P. (2012, 03 25). Retrieved from The Victorian Web: http://www.victorianweb.org/science/darwin/geochange.html

Lantitsou, K. (2017). Eco development and environmental spatial planning. *Fresenius Environmental Bulletin*, 1291–1300.

Lowicki, D., & Piotrowska, S. (2015). Monetary valuation of road noise. Residential property prices as an indicator of the acoustic climate quality. *Ecological Indicators, 52*, 472–479. DOI: 10.1016/j.ecolind.2015.01.002

Mcentire, D. A. (2001). Triggering agents, vulnerabilities and disaster reduction: Towards a holistic paradigm. *Disaster Prevention and Management, 10*(3), 189–196. DOI: 10.1108/09653560110395359

Mills, B., Yessis, J., Boyd, D., & Andrey, J. (2001). The urban environment a hazard source and sink. *Environments*, 17–38.

Ministry of the Environment. D. (2001). *Spatial planning as an instrument for promoting sustainable development in the Nordic countries; Action programme for 2001-2004.* Denmark: Nordisk Miljomaerkning.

Nadin, V. (2007). The emergence of the spatial planning approach in England. *Planning Practice and Research, 22*(1), 43–62. DOI: 10.1080/02697450701455934

Nadin, V., & Stead, D. (2008). European Spatial Planning Systems, Social Models and Learning. *disP - The Planning Review*, 35-47.

Natarajan, L. (2015). Socio-spatial learing: A case study of community knowledge in participatory spatial planning. *progress planning*.

Nikoloas, K., Dimitra, V., & Agapi, X. (2014). Real estate values and environment: A case study on the effect of the environment on residential real estate values. *International Journal of Academic Research*, 861–866.

O'Gorman, K., & Maclntosh, R. (2015). *Research Methods for Business & Management* (2nd ed.). Goodfellow Publishers Limited. DOI: 10.23912/978-1-910158-51-7-2736

Olatunji, S. O., Olawumi, T. O., & Awodele, O. A. (2017). Achieving value for money in construction projects. *Civil and environmental Research*, 54-61.

Rabianski, J. (2003). *Primary and secondary data: Concepts, concerns, errors, and issues*. The Appraisal Journal.

Redman, C. (1999). *human impact on ancient environments.* university of Arizona press.

Reid, A., Teamey, K., & Dillon, J. (2007). Valuing and utilizing traditional ecological knowledge: Tensions in the context of education and the environment. *Environmental Education Research*, 1–19.

Resurreccion, B. P., Sajor, E. E., & Fajber, E. (2008). *Climate Adaptation in Asia: Knowledge Gaps and Research Issues in South East Asia*. ISET-International and ISET.

Senaratne, A. (2019). Ensuring Sustainable Development in post-conflict Sri Lanka. In *Managing Domestic and International Challenges and opportunities in post conflict development*. DOI: 10.1007/978-981-13-1864-1_5

Silva, P. d. (2008, 02 07). *On Relativity of knowledge*. Retrieved from Sri lanka Guardian: http://www.srilankaguardian.org/2008/02/on-relativity-of-knowledge .html

Singh, P. K. (2019). Spatial planning for socio-economic developoment . *International Journal of Geography and Regional Planning*, 102-110.

Sobieraj, J. (2017). Impact of spatial planning on the pre-investment phase of the development process in the residential construction field. *Archives of civil engineering*, 113-131.

Sutanta, H., Rajabifard, A., & Bishop, I. (2013). Disaster risk reduction using acceptable risk measures for spatial planning. *Journal of Environmental Planning and Management*, *56*(6), 761–785. DOI: 10.1080/09640568.2012.702314

Sutanta, H., Rajabifard, A., & Bishop, I. D. (2010). Integrating Spatial Planning and Disaster Risk Reduction at the Local Level in the Context of Spatially Enabled Government. *Spatially Enabling Society Research, Emerging Trends and Critical Assessment*, 55-68.

The world Bank. (2022, October 6). *The world Bank in Sri Lanka*. Retrieved from worldbank.org: https://www.worldbank.org/en/country/srilanka/overview

Tiwari, I., & Gupta, B. (2023). Environment and Sustainable Development. 213-215.

Vale, L. (2005). *The resilient city: How modern cities recover from disaster*. Oxford university press. DOI: 10.1093/oso/9780195175844.001.0001

Vileniske, i. g., & Arturas, N. (2012). environmental ethics and sustainability in housing design. 1-13.

Vileniske, i. g., & Arturas, N. (2012). environmental ethics and sustainability in housing design. 1-13.

Wijesundara, J. (2015). Re-invigorating the spirit of place of kandy; urban Design and city planning strategies. *Built- environment: Sri lanka*, 1-11.

Wu, J., & Loucks, O. (1995). from balace of nature to hierarchical patch dynamics; A paradigm shift in ecology. 1-2.

Zhou, X., Lu, X., Lian, H., Chen, Y., & Wu, Y. (2017). Construction of a spatial planning system at city level: Case study of "integration of multi-planning" in Yuanqing Wu. *Habitat International*, *65*, 32–48. DOI: 10.1016/j.habitatint.2017.04.015

Zrobek, S., Trojanek, M., Sokolnik, A. Z., & Trojanek, R. (2015). The influence of environmental factors on property buyers' choice of residential location in Poland. *Journal of International Students*, 164–174.

Zsolnai, L. (2011). Ethical Principles and Economic Transformation – A Buddhist Approach, Issues in Business Ethics. *Springer Science+Business Media* .

Chapter 7

Is Transport Infrastructure Attractive to Labor Settlement: Evidence From China's High-Speed Rail Cities

Lishuang Huang
https://orcid.org/0009-0000-4297-2914
Jinan University, Guangzhou, China

Shengming Li
https://orcid.org/0009-0006-6338-7398
Guizhou University, Guiyang, China

Shixi Kang
https://orcid.org/0000-0003-4847-5171
University of Barcelona, Spain

Meng Wei
https://orcid.org/0009-0007-7696-8463
Huaqiao University, Quanzhou City, China

Jingwen Tan
https://orcid.org/0000-0002-3452-959X
University of Barcelona, Spain

ABSTRACT

This paper empirically analyzes the impact of high-speed rail (HSR) on migrants' settlement intentions, using data from the 2016 and 2017 China Migrants Dynamic

DOI: 10.4018/979-8-3373-1554-6.ch007

Survey (N=294,425). The results of the linear probability model (LPM) suggest that cities with HSR increase migrants' settlement intentions by 5.58%. Mechanism analysis shows that the more relatives accompanying the migrants, the smaller the effect of HSR on the settlement intention. If migrants move across province or with their own land assets in hometown, the settlement intention is easily affected by HSR. In addition, HSR cities can provide higher income and business environment for migrants. Regional heterogeneity analysis shows that the effect of HSR cities in Northeast China on the settlement intention of migrants is not significant, while it is significant in other regions of China. This study linking transport infrastructure development with micro-level migration decisions, offering insights for urban planning and labor mobility policies.

I. INTRODUCTION

Urbanization is a necessary path for developing countries to transform into developed countries (Chen et al., 2013). The agglomeration of industries and economic activities with easy access to transportation has led to a concentration of rural migrants in large cities (Gakenheimer, 1999). In 2022, China's urban migrants reached 283 million, this group making a huge contribution to urban development as the main workforce in cities (CNBS, 2023). The settlement of migrants can transform the labor force that stays in the city for short period into permanent, providing more stable human resources for economic growth.

Global development experience shows that transport is a prerequisite for economic development and population mobility (Clark et al., 2004; Donaldson, 2018; Faber, 2014; Bernard et al., 2019). Transport infrastructure has catalytic effect on the economic growth of cities (Gospodini, 2005). A common view is that a well-developed transport infrastructure can indirectly contribute to local economic growth through mechanisms such as improved investment and savings (Aschauer, 1989). As a city's transport infrastructure builds economic growth as well as environmental changes, there will be an influx of more labor into the city, and building up human capital(Desmet and Rossi-Hansberg, 2013). Based on this, this paper explores the factors influencing the settlement intention of migrants from the perspective of urban transportation.

Over the past decade China government has invested heavily in the HSR (High-Speed Railway) network. By 2022, the operating mileage of China's railway reached 155,000 kilometers, of which the operating mileage of HSR will reach 42,000 kilometers, accounting for 27.10% of the total railway traffic (CNRB, 2023). By 2035, China plans to achieve a national railway network of 200,000 km, including about 70,000 km of HSR, and cities with populations more than 500,000 will be covered by HSR (CNRG, 2020). The maximum design speed of China's HSR has reached 350 kilometers per hour, with an average operating speed of around 250 kilometers per hour (Yuan et al., 2023). The speed, convenience and reliability of HSR provide more choices and opportunities for population mobility, and reduce the economic distance between migrants' work cities and hometowns (Hanley et al., 2022).

Based on the above discussion, this paper empirically analyses the impact of cities with HSR on the migrants' settlement intention. The results show that HSR cities are more attractive to migrants for settlement. This paper also finds that a larger number of accompanying relatives mitigates the influence of HSR on migrants' settlement choices; while the greater the distance migrants move or migrants own land assets in their hometown, the greater their settlement intention is affected by HSR; HSR attracts migrants by enhancing the cities' economic environment.

Many studies have analyzed the impact of transportation infrastructure on the urban economic environment, but few studies have analyzed the impact of transportation infrastructure development on people's micro-behavior. This paper focuses on China's migrants and uses survey data to enrich the research on the micro-behavior of the population. We also investigate the mechanism of the impact of HSR construction and find the regional heterogeneity impact of HSR on the settlement intention of the migrants.

Overview of this paper, Section I lays out the relevant background; Section II summarizes the main literature; Section III describes the source of data and introduces the empirical design in detail; Section IV summarizes the empirical results, including baseline results, robustness check and mechanism analysis; Section V concludes the paper.

II. REVIEW OF THE LITERATURE

The construction of HSR brings a lot of economic and social benefits (Huang et al., 2019; Campos and De Rus, 2009; Zhang et al., 2020b). HSR reduces the cost of inter-city commuting, which facilitates knowledge spillovers with more face-to-face communication, and reduces the social costs of innovation as well as inter-city trade (Donaldson and Hornbeck, 2016; Bernard et al., 2019). The construction of HSR can break down the barrier of geographical distance, form closer regional integration, and promote the optimal allocation of resources and balanced economic development (Sun et al., 2023; Li et al., 2023). The capital and technology required for the construction of HSR promote the development of related industrial chains and create employment opportunities, attracting more investment (Lin et al., 2023; Chang et al., 2021).

HSR makes traveling more convenient for residents, improves the quality of life and increases the attractiveness of cities (Carten`ı et al., 2017). In addition, HSR not only strengthens the links between cities, but also provides convenient transport for urban and rural population mobility and integration (He et al., 2023). From the perspective of labor supply, HSR affects population migration and promotes rational allocation of human resources (Kim, 2000). Thus, HSR is important in promoting regional economic growth (Ahlfeldt and Feddersen, 2018; Chen and Silva, 2013).

There are multiple factors affecting the settlement intentions of migrants. The economic factor is the key element (Au and Henderson, 2006). Under China's urban-rural binary economics system, there is a large income gap between rural and urban areas, and migrants can have better employment opportunities and higher wages in big cities (Zhang et al., 2016). The 'sharing, matching and learning' effect of big cities results in higher productivity, skills matching, accumulation of human capital and wage premiums, attracting migrants to settle in cities (Glaeser and Mare, 2001; Duranton and Puga, 2019). Wang and Shen (2022) suggests that job opportunity also affects settlement intentions, with developed cities usually providing more jobs. In addition, public services and comfortable environment are factors that influence migration (Banzhaf and Walsh, 2008). Education services are also closely related to labor migration decisions (Dustmann and Glitz, 2011; Xiong, 2015). The price of housing affects the settlement choices of migrants, which tend to leave rather than settle when there is a large gap between house prices and wages in a city (Huang and Chen, 2022). The abundance of cultural and recreational facilities usually makes cities more attractive, especially for young people and culture lovers (Antonova et al., 2020).

Under China's household registration system, a lot of research has explained the motivations for the settlement intention of migrants in terms of the value of urban household registration (Song, 2014; Chan, 2015; Zhang et al., 2020a; Vortherms and Liu, 2022). China household registration system takes many barriers to migration (Chan and Buckingham, 2008; Bosker et al., 2012). The existence of household registration discrimination in cities affects family reunification, and the problem of children left behind is highlighted (Fan et al., 2011; Fang and Shi, 2018). Family reunification is closely related to commuting costs and spatial distances between urban and rural areas, with ease of access to transport between cities facilitating population mobility, reducing commuting costs, and leading to talent agglomeration effects and high-skill preferences (Hodgson, 2018; Fretz et al., 2017). Additionally, the land rights and benefits associated with rural household registration are seen as increasingly valuable. As a result, many migrant workers choose to move back and forth between urban and rural areas rather than give up their rural household registration (Chen and Fan, 2016).

Transport facilities can reduce the cost of commuting to and from migrants residence and workplace when the migrants is separated (Monte et al., 2018). Guirao et al. (2020) finds that HSR commuting significantly improves the distribution of employment opportunities between regions by analyzing regional wage differentials and housing rental prices. Green et al. (1999) explores long-distance commuting as an alternative to relocation in the UK and argues that it can mitigate regional labor market imbalances. Garmendia et al. (2008) examines the impact of HSR on the integration of isolated small cities in large metropolitan areas, where the introduction of HSR has facilitated residential development. Guirao (2013) based on Spain's HSR operations, shows that HSR not only improves regional transport, but also influences the urban fabric and regional development, although HSR faces high costs and sustainability. Haas and Osland (2014) examined the complex interactions between commuting, migration, housing and the labor market, which were altered by the introduction of HSR, enhancing inter-city accessibility, and affecting housing and labor market dynamics. The fixed costs of population migration differ between regions and the basic transport facilities are an important factor affecting the fixed costs of migration (Ma et al., 2021). As an important part of public services, transport facilities affect migrants' settlement intention.

While most studies focus on the impact of HSR on macroeconomic indicators, this paper focuses on China's migrants and uses survey data to enrich the research on the micro-behavior of the population. We conclude that the development of transport infrastructure is conducive to enhancing the settlement intention of the migrants, which can transform the migrants into a long-term urban labor force, and facilitate the city urbanization process. This paper also investigates the mechanism of the impact of HSR construction and finds the regional heterogeneity impact of HSR on the settlement intention of the migrants.

III. DATA AND METHODOLOGY

A. Data Sources

This paper uses data from the 2016 and 2017 China Migrants Dynamic Survey (CMDS), organized by the China National Health Planning Commission. The survey covered 31 provinces (municipalities and autonomous regions) in mainland China, and the sample was selected from the migrants who stayed in the city for more than one month. The average age of the sample ranged from 15-60 years. This survey uses a stratified, multi-stage, large-scale probability proportionate to size sampling method to investigate the developmental status, individual characteristics, social integration and employment of the migrants in China. Because the survey on settlement intention was only available in the 2016 and 2017 CMDS data, these two years were selected in this paper. The data relating to the urban control variables were obtained from the China Urban Statistical Yearbook. The variables in this paper are shown in Table 1.

In this paper, the data were cleaned before empirical analysis. This paper matches the survey data with city economics data from the China Urban Statistical Yearbook and retained the successfully matched sample. In addition, due to the small sample size of the Xinjiang Production and Construction Corps, it was excluded from the empirical analysis. The processed sample size was distributed among 292 municipalities. There are 338,989 samples before data cleaning and 294,425 samples after data cleaning.

B. Baseline Model

This paper focuses on the impact of the cities' access HSR on the migrants' settlement intention. With the dependent variable as dummy, models that can be used include the linear probability model (LPM) based on ordinary least squares (OLS) estimation and Logit model based on likelihood estimation. This paper chooses the linear probability model in the baseline regression to estimate the effect of the city's access to HSR on the intention to settle among the migrants, while the estimation result of the Logit is reported in the robustness check. The interpretation of the coefficients of the linear probability model is more intuitive, and with reasonable model setup, the direction and significance of the coefficients obtained using least squares estimation are not significantly different from the likelihood estimation (Ferrer- i-Carbonell and Frijters, 2004; Angrist and Pischke, 2009). The linear probability model in this paper can be expressed as

$$settle_{ict} = \beta_0 + \beta_1 hsr_{ict} + \beta_2 X_{ict} + \mu_c + \lambda_t + (\mu_c \times \lambda_t) + u_{ict} \quad (1)$$

where $settle_{ict}$ is the settlement intention of migrant i of year t in city c, hsr_{ict} is the HSR access dummy variable, which equals 1 if the inflow city c of migrant i has HSR access in year t, and 0 otherwise. X_{ict} represents other control variables. μ_c denotes the provincial fixed effects, as λ_t is the time fixed effects, and $\mu_c \times \lambda_t$ the interaction fixed effects, μ_{ict} is the random error.

Table 1. Variable Description.

Variables	Variables Description
settle	Migrants with settlement intention =1, without = 0.
gender	Male = 1, female = 0.
age	Age.
edu	Years of education.
hukou	Rural household registration=1, urban household registration=0.
marriage	Married = 1, unmarried = 0.
job	With work = 1, without work = 0.
lnincome	Logarised monthly household income.
hometown	Rural area = 1, township = 2, county town = 3, prefecture city = 4, provincial capital = 5, municipality = 6.
lnteacher	Logarised number of primary school teachers of the city.
lnbed	Logarised number of hospital beds of the city.
lnpgdp	Logarised GDP per capita.
structure	Proportion of urban tertiary sector employees to total number of employees.

Notes: This table provides a detailed description of the variables used in the analysis. All variables are sourced from the 2016 and 2017 China Migrants Dynamic Survey (CMDS) and the China Urban Statistical Yearbook. Among these variables, "lnbed" and "lnteacher" are both measured per 10,000 population at the province level.

C. Matching Model

Linear probability model may not address the problem of estimation bias due to self-selection. In addition, the dependent and independent variables in this paper are dummies, this data structure is suitable to use matching model to obtain a more accurate treatment effect. In this paper, the entropy-balanced matching (EBM) is chosen to establish a counterfactual framework.

In addition to EBM, the more common model is Propensity Score Matching (PSM). Compared to PSM, there are three main advantages of EBM (Hainmueller, 2012). First, EBM ensures that the experimental and control groups are similarly distributed over the higher order moments of each covariate, whereas PSM focuses on balancing the propensity scores of the experimental and matched control groups do not ensure this effect. Secondly, in the implementation of EBM, the only parameters that need to be set subjectively by the researcher are almost the tolerance of the results of the iterative algorithm, which circumvents the disadvantages of having more subjective settings in the PSM implementations, including variable selection and model setting for propensity score calculation, and whether observations are put back to matching. In addition, EBM can retain all samples for estimation, and larger numbers of samples are lost in the PSM.

The average treatment effect when using EBM to construct a counterfactual framework equation can be represented by

$$\tau = E[settle(1)|hsr = 1] - E[settle(0)|hsr = 1] \tag{2}$$

$E[settle(1) \mid hsr = 1]$ indicates the effect of cities accessing HSR in the treatment group on migrants' settlement intention. $E[settle(0) \mid hsr = 1]$ denotes the size of the effect of the cities absence of HSR of the treatment group on migrants' settlement intention, which is a counterfactual indicator. the expression for EBM is:

$$E[settle(0)|hsr = 1] = \frac{\sum i|hsr = 0^{settle_i \omega_i}}{\sum i|hsr = 0^{\omega_i}} \tag{3}$$

$$minH(\omega_i) = \sum_{i|hsr=0} \omega_i log\ (\omega_i/q_i) \tag{4}$$

where $q_i = 1/n$ and n denotes the number of samples in the control group. The core idea of ω_i with certain constraints is that the control covariates are given an additional set of moment constraints to balance with the covariates in the treatment group. To ensure the rigor of the conclusions, the PSM estimator is also reported in this paper for robustness check.

D. Instrumental Variable

The HSR construction is likely an endogenous variable. When the settlement intention of migrants in a region is high, the potential movement of people may increase local passenger traffic and prompt the local government to open HSR, thus creating possible reverse causation. The instrumental variable is widely used in the treatment of reverse causality problems.

The use of geographical instrumental variables is well established in empirical studies related to transport infrastructure. This paper chooses the city's average slope and elevation as the instrumental variables for city access to HSR. The average slope and elevation of a city affect the construction of HSR, which is consistent with the correlation hypothesis of the instrumental variable. Additionally, the average slope and elevation of a city are not considered in the integration process of the migrants in the inflow area, which is consistent with the exogeneity hypothesis of the instrumental variable. Based on these instrumental variables, this paper performs regression analysis in Two-Stage least squares (2SLS). The first stage expression is

$$hsr_{ct} = \varphi_0 + \varphi_1 elevation_c + \varphi_2 slope_c + \mu_c + \lambda_t + (\mu \times \lambda_t) + u_{ct} \quad (5)$$

where hsr_{ct} is the city's HSR access, $elevation_c$ is the average elevation of the city c, $slope_c$ is the average slope of the city. μ is the provincial fixed effects, λ_t is the time fixed effects, and $\mu \times \lambda_t$ is the interaction fixed effect.

E. Descriptive Statistics

The dependent variable in this paper is migrants' settlement intention (settle), and the independent variable is whether the city access HSR (hsr). The control variables are selected by two dimensions: individual and city. The demographic characteristics are often used as control variables. In human capital theory, factors such as education and work experience can help individuals obtain better jobs and income, thus enabling them to integrate into the cities. In this paper, gender (gender), years of education (edu), age (age), household register (hukou), marital status (marriage), whether have a job (job) and logarithmic monthly household income (lnincome) are chosen as individual control variables.

Table 2. Descriptive Statistics.

Variable	N	Mean	Std. dev.	Min	Max
settle	294,425	0.369	0.482	0	1
hsr	294,425	0.837	0.369	0	1
gender	294,425	0.515	0.500	0	1
age	294,425	35.340	9.447	18	60
edu	294,425	10.310	3.229	0	19
hukou	294,425	0.165	0.371	0	1
marriage	294,425	0.815	0.388	0	1
job	294,425	0.846	0.361	0	1
lnincome	294,425	8.697	0.668	0	15.623
hometown	294,425	1.419	0.889	1	6
lnteacher	294,425	4.396	1.571	1.977	16.071
lnbed	294,425	6.350	2.175	1.425	13.657
lnpgdp	294,425	11.011	1.009	3.969	13.156
structure	294,425	54.228	14.012	17.880	91.300

Notes: This table presents the descriptive statistics for the main variables used in the analysis. The sample includes 294,425 observations after data cleaning. The mean settlement intention (settle) is 0.369, indicating that about 36.9% of the migrants have the intention to settle in the city. The mean HSR access (hsr) is 0.837, indicating that 83.7% of the cities in the sample have HSR access.

City characteristics may affect the intention of the migrants to stay in the city. The urban control variables selected in this paper mainly reflect the level of economic development and other infrastructure development. These include the number of tertiary sector employees (structure), to measure the city's industrial structure; logarithmic gross domestic product per capita (lnpgdp), to measure the city's economic situation; the number of hospital beds in the city (lnbed), to measure the city's health care resources; the number of primary school teachers in the city (lnteacher), to measure the city's educational resources. The economic situation of migrants' hometowns is also likely to affect their settlement intention in the inflow area, and this paper uses the administrative level of migrants' hometowns (hometown) as control variable.

Table 2 reports the descriptive statistics and Table 11 reports the results of t-tests for differences in means between HSR cities and non HSR cities. All these control variables are significant in the t-test of differences between groups, demonstrating that the control variables can better reflect the factors other than the access of the HSR on migrants' settlement intention in the city.

IV. EMPIRICAL RESULTS

A. Baseline Regression Results

In this section, this paper reports the LPM regression results of estimating the impact of HSR cities on migrants' settlement intention. According to Table 3, there is a significant positive effect of cities with HSR on the settlement intention of the migrants in all models. With the city and individual control variables, the treatment effect of HSR cities on migrants' settlement intention is 5.71% (column 4). Following the results of column 5 with the inclusion of control variables and interaction fixed effects, the impact of HSR cities on the intention of migrants to settle is 5.58%. All results reached the confidence level at 99%.

A city's well-developed transport infrastructure can improve travel efficiency. Migrants usually consider employment, education and healthcare when choosing where to live, and cities with convenient and efficient transport infrastructure can offer more options, such as relying on easy access to the resources of neighboring cities (Wang and Chen, 2019). Additionally, the construction of transport infrastructure can drive economic development, attract more talent and capital to the city, and promote the upgrading and transformation of the city's industrial structure (Zhang and Qi, 2021). As a result, the migrants are more willing to settle in cities with developed transport infrastructure.

B. Robustness

In this paper, robustness checks were conducted in the following methods: (1) Cluster robust standard errors. In the baseline model, we use ordinary standard errors, the robustness check uses robust standard error clustering by city. Cluster robust standard errors can accurately reflect the correlation between variables within cities and avoid excessive significance level. (2) Logit model. (3) Replacement of dependent variable. We replaced the dependent variable for robustness check, the dependent variable "intention to settlement" is replaced with "intention to long stay". (4) The availability of HSR in a city does not equate to the accessibility of HSR, and the location of HSR stations in a city may affect the accessibility. We include an interaction term between the opening of HSR and road mileage in the city as a control variable. We assume that migrants reach the HSR station by road, and the road mileage can reflect the convenience of road transport in a city, and the effect of accessibility can be controlled by the interaction term between road mileage and HSR opening.

As can be seen from the regressions in Table 4, the results of column (1) using the cluster-robust standard errors by cities do not change significantly compared with the baseline regression. In column (2) Logit model as well as column (3) marginal effect results, the significance of the HSR cities remains unchanged. The coefficients decrease after replacing the dependent variables in column (4), but the sign and confidence level do not change. The result shows a subtle change after column (5) controls the interaction term between road mileage and HSR opening in model. The above results suggest that the findings of this paper are robust.

Table 3. Linear Probability Model Regression Results.

	(1)	(2)	(3)	(4)	(5)
	OLS	OLS	OLS	OLS	OLS
hsr	0.121***	0.0879***	0.0801***	0.0571***	0.0558***
	(0.00225)	(0.00229)	(0.00243)	(0.00244)	(0.00293)
gender		-0.0234***		-0.0215***	-0.0191***
		(0.00182)		(0.00179)	(0.00174)
age		0.000957***		0.000647***	0.000801***
		(0.000109)		(0.000107)	(0.000105)
edu		0.0197***		0.0172***	0.0174***
		(0.000322)		(0.000317)	(0.000313)
hukou		0.153***		0.144***	0.140***

continued on following page

Table 3. Continued

	(1)	(2)	(3)	(4)	(5)
	OLS	**OLS**	**OLS**	**OLS**	**OLS**
		(0.00320)		(0.00314)	(0.00308)
marriage		0.0203***		0.0260***	0.0198***
		(0.00258)		(0.00254)	(0.00251)
job		-0.0347***		-0.0230***	-0.0230***
		(0.00254)		(0.00250)	(0.00246)
lnincome		0.0521***		0.0425***	0.0197***
		(0.00153)		(0.00148)	(0.00140)
hometown		0.0427***		0.0351***	0.0302***
		(0.00109)		(0.00106)	(0.00104)
lnteacher			0.00836***	0.00984***	0.0107***
			(0.000623)	(0.000617)	(0.000949)
lnbed			0.0154***	0.0135***	0.0101***
			(0.000534)	(0.000523)	(0.000732)
lnpgdp			0.0448***	0.0390***	0.0126***
			(0.00130)	(0.00125)	(0.00145)
structure			0.00520***	0.00449***	0.00223***
			(0.0000650)	(0.0000651)	(0.0000993)
constant	0.268***	-0.430***	-0.608***	-1.091***	0.247***
	(0.00202)	(0.0132)	(0.0136)	(0.0170)	(33.10)
Province FE	No	No	No	No	Yes
Time FE	No	No	No	No	Yes
Interactive FE	No	No	No	No	Yes
N	294425	294425	294425	294425	294425
R^2	0.009	0.048	0.053	0.081	0.133

Notes: This table reports the results of the linear probability model (LPM) regression, estimating the impact of HSR access on migrants' settlement intention using the data of 2016 and 2017 China Migrants Dynamic Survey (CMDS) . Robust standard errors are reported in parentheses. *p < 0.1, **p < 0.05, ***p < 0.01.

Table 4. Robustness Checks

	(1)	(2)	(3)	(4)	(5)
	Clustercity	**Logit**	**Marginal**	**Longstay**	**Road**
hsr	0.056***	0.309***	0.071***	0.016***	0.456***
	(0.0133)	(0.0162)	(0.0037)	(0.0031)	(0.0239)
Control	Yes	Yes	Yes	Yes	Yes
City FE	Yes	Yes	Yes	Yes	Yes
Time FE	Yes	Yes	Yes	Yes	Yes
Interactive FE	Yes	Yes	Yes	Yes	Yes
N	294425	294425	294425	294425	294425
R^2	0.1328	0.1026	0.1026	0.1891	0.1411

Notes: This table presents robustness checks for the main regression results. Column (1) uses cluster-robust standard errors by city to account for within-city correlation. Column (2) reports the Logit model results, and column (3) shows the marginal effects. Column (4) replaces the dependent variable "settle" with "long stay" to test the robustness of the results. Column (5) includes an interaction term between road mileage and HSR access to control for accessibility. The results remain consistent across all robustness checks, confirming the robustness of the main findings. *p < 0.1, **p < 0.05, ***p < 0.01.

C. Instrumental Variables and Matching Model

In this paper, city slope and elevation are chosen as instrumental variables for HSR cities. The TSLS estimator in column (1) shows the 9.75% effect of HSR cities, and the limited information maximum likelihood (LIML) estimator in column (2) is similar to TSLS model, demonstrating that there is no weak instrumental variables problem. The exogeneity of instrumental variables needs to be justified before conducting instrumental variable regression. No research demonstrates that urban slope and elevation affect migrants' settlement intentions, and this paper conducts the over-identification test for two instrumental variables. The results of the over-identification test rejected the original hypothesis that the instrumental variables are endogenous, indicating that these two instrumental variables are exogenous (Appendix Table 12).

Column (3) in Table 5 shows the EBM estimator, which calculates the treatment effect of HSR cities on the migrants' settlement intention is 9.5%. Column (4) shows the PSM estimator, which is similar to the EBM at 8.1%. The tests of the matched model are presented in the Appendix, and Tables 13 and 14 show that there is a large difference between the treatment and control group's means before EBM, while there is almost no difference after EBM, which proves that the matching is successful. Figures 1 and 2 show a reduction in the difference between the treatment group control group after PSM.

Table 5. Instrumental variables and matching model results

	(1)	(2)	(3)	(4)
	TSLS	**LIML**	**EBM**	**PSM**
hsr	0.0975***	0.104***	0.095***	0.081***
	(0.01900)	(0.01500)	(0.00400)	(0.00600)
Control	Yes	Yes	Yes	Yes
Province FE	Yes	Yes	Yes	Yes
Time FE	Yes	Yes	Yes	Yes
Interactive FE	Yes	Yes	Yes	Yes
N	294425	294425	294425	294425
R^2	0.131	0.131	0.0672	-

Notes: This table presents the results of the instrumental variables (IV) and matching models. Columns (1) and (2) show the Two-Stage Least Squares (2SLS) and Limited Information Maximum Likelihood (LIML) estimators, respectively, using city slope and elevation as instruments for HSR access. Columns (3) and (4) report the Entropy Balance Matching (EBM) and Propensity Score Matching (PSM) estimators. The results from both IV and matching models confirm the positive impact of HSR access on migrants' settlement intention. Standard errors are reported in parentheses. *p < 0.1, **p < 0.05, ***p < 0.01.

D. Excluding Competing Hypotheses

To accurately identify the impact of HSR on the settlement intention of migrants, a competing hypothesis needs to be falsified: only HSR and no other transportation affect the settlement intention of migrants. Cities with HSR tend to be economically developed and with flat land, so HSR access not only reflects the construction of HSR, but may also reflect the construction of other transport facilities. In addition, transport infrastructure in China's provincial capitals is generally more developed, and the distance between cities and provincial capitals may also affect the commuting experience of migrants (Liu et al., 2022).

Table 6 reports the results of controlling other transport infrastructures, with variables measuring water transportation (column 1), air transportation (column 2), general railway transportation (column 3), and road transportation (column 4), and controlling for the distance from each city to the provincial capital (column 5). It can be seen that the coefficients on the HSR variable do not change significantly, this proves that the impact of the HSR on the settlement intention of the migrants is not disturbed by other transport facilities.

Table 6. Excluding Competing Hypotheses

	(1)	(2)	(3)	(4)	(5)	(6)
	OLS	**OLS**	**OLS**	**OLS**	**OLS**	**OLS**
hsr	0.0549***	0.0509***	0.0560***	0.0556***	0.0543***	0.0501***
	(0.00293)	(0.00292)	(0.00293)	(0.00293)	(0.00305)	(0.00308)
water	0.0117***					0.0106***
	(0.000535)					(0.000564)
air		0.0103***				0.00786***
		(0.000423)				(0.000482)
rail			0.000599			-0.00150***
			(0.000428)			(0.000450)
road				-0.0310***		-0.0237***
				(0.00213)		(0.00232)
distance					0.000778	0.000814
					(0.000800)	(0.000884)
Control	Yes	Yes	Yes	Yes	Yes	Yes
Province FE	Yes	Yes	Yes	Yes	Yes	Yes
Time FE	Yes	Yes	Yes	Yes	Yes	Yes
Interactive FE	Yes	Yes	Yes	Yes	Yes	Yes
N	294425	294425	294425	294425	294425	294425
R^2	0.134	0.135	0.133	0.134	0.136	0.139

Note: As water transportation, air transportation and general railway transportation is public transport, this paper uses logarithmic passenger traffic to measure the level of infrastructure for these three modes of transport. For road transport, total road mileage is used to measure the level of infrastructure, *p < 0.1, **p < 0.05, ***p < 0.01.

E. Mechanism Analysis

Impact of Homesickness

This paper considers the moderating effect of homesickness. The homesickness of the migrants can be influenced by two factors: the number of relatives accompanying and the distance between work city and hometown. When migrants have more relatives accompanying, they will miss their family members less, and homesickness will diminish (Scharp et al., 2016). Thus, the need for transportation to return to their hometown is decreased. The distance of mobility also impacts the demand for transport infrastructure. When migrants are far away from their hometowns, there is likely to be a greater emphasis on rapid long-distance transportation.

Table 7 columns (1) report the result for the interaction term of HSR city and migration scope. The coefficients on the interaction terms are positive and significant, indicating that large scope migration increases the effect of HSR city on the settlement intention of migrants, with positive moderating effect. Table 7 columns (2) report the result of the interaction term between HSR city and the number of accompanying relatives. The coefficients on the interaction terms are negative, indicating that the number of accompanying relatives weakens the effect of HSR city on migrants' settlement intention.

Table 7. Interaction Term Results

	(1)	(2)
	Migrant Scope	**Accompanying Relatives**
hsr	0.0490***	0.0751***
	(0.00334)	(0.00714)
hsr*scope	0.0274***	
	(0.00472)	
scope	-0.0888***	
	(0.00422)	
hsr*families		-0.00600***
		(0.00198)
families		0.0140***
		(0.00186)
Control	Yes	Yes
Province FE	Yes	Yes
Time FE	Yes	Yes
Interactive FE	Yes	Yes
N	294425	294425
R^2	0.136	0.140

Note: Migration scope is a dummy variable (1 = cross-provincial migration, 0 = intra-provincial migration). Accompanying relatives is an ordered variable that is the number of relatives accompanying the migrants, *p < 0.1, **p < 0.05, ***p < 0.01.

Impact of Hometown Land Assets

If migrants own housing (homestead) or cultivated land (contracted land) in their hometown, they will have more demand (e.g. for agricultural production) to commute to and from hometown (Zou et al., 2022). Therefore, migrants with land assets in their hometown may attach more importance to HSR. Table 8 reports the results for the migrants grouped by whether the migrants owned homestead or contracted land. Column (1) (2) shows that the coefficient of HSR city on settlement intention is 7.18% for the migrants with homesteads, and the coefficient of migrants without homesteads is lower. Column (3)(4) shows that the coefficient is 7.13% for migrants with contracted land, and the coefficient for migrants without contracted land is not significant. The above results indicate that migrants with land assets in their hometowns are more likely to be affected by the HSR cities on their settlement intention.

Table 8. Homestead, Contracted Land Ownership Subgroup Results

	(1) Without Land	(2) Homestead Land	(3) Without Land	(4) Contracted Land
hsr	0.0137**	0.0718***	0.00452	0.0713***
Control	(0.00553) Yes	(0.00276) Yes	(0.00507) Yes	(0.00283) Yes
Province FE	Yes	Yes	Yes	Yes
Time FE	Yes	Yes	Yes	Yes
Interactive FE	Yes	Yes	Yes	Yes
N	60857	85097	79575	66379
R^2	0.090	0.076	0.092	0.077

Notes: This table presents subgroup analysis results for migrants with and without homestead or contracted land in their hometowns. *p < 0.1, **p < 0.05, ***p < 0.01.

HSR and Economic Environment

The construction of HSR can strengthen the economic connectivity of cities, facilitate the flow and allocation of resources, and improve the overall competitiveness of cities, thereby positively affecting the economic environment, wage income and business environment of cities (Wang et al., 2022; Guo et al., 2022). Table 9 analyses the impact of HSR cities on the economic environment using the migrants' income satisfaction as well as business environment satisfaction as dependent variable and HSR cities as independent variable. The coefficients of column (1) (2) are both positive and significant, it proves that HSR cities can provide higher wages and good business environment for migrants. This paper also uses KHB (Karlson–Holm–Breen) method to analyze the mediating effect of income satisfaction and business environment satisfaction. Appendix Tables 15 and 16 show that the coefficients of wage satisfaction and business environment satisfaction are both significant, proving the existence of the mediating effect. The above results show that HSR cities can attract migrants to settle by providing higher income and decent business environment.

Table 9. HSR and City Economic Environment

	(1) **Income Satisfaction**	**(2)** **Business Environment Satisfaction**
hsr Control	0.0229*** (0.00439) Yes	0.00774* (0.00465) Yes
Province FE	Yes	Yes
Time FE	Yes	Yes
Interactive FE	Yes	Yes
N	145954	145954
R^2	0.083	0.088

Note: The economic environment variables are only available for the 2017 CMDS data, so we use data from this year.

F. Regional Heterogeneity Analysis

In this section, all samples were divided into four economic regions according to the China government for heterogeneity analysis (Qi et al., 2013). The results (Table 10) show an insignificant effect of HSR cities on migrants' settlement intention in the northeast region. The effect of HSR cities in the eastern region is higher, with treatment effect about 4.95%, significant at the 99% confidence level. The effect of HSR cities in the central region is low, with treatment effect about 2.57%. The treatment effect in the western region is about 6.28%, the highest among all regions.

The main reasons for the different effects are the geographical locations of cities, climate, and patterns of economic development, which result in regional heterogeneity in the impact of HSR cities on migrants' lower settlement intention. Additionally, the northeast region of China has shown a net outflow of population in recent years (You et al., 2021). The construction of HSR has facilitated the movement of labor across the region, which may lead to a more pronounced outflow of population.

Table 10. Regional Heterogeneity Analysis Results

	(1) Northeast	(2) East Region	(3) Central Region	(4) West Region
hsr	-0.00953	0.0495***	0.0257***	0.0628***
Control	(0.0159) Yes	(0.00701) Yes	(0.00549) Yes	(0.00532) Yes
Province FE	Yes	Yes	Yes	Yes
Time FE	Yes	Yes	Yes	Yes
Interactive FE	Yes	Yes	Yes	Yes
N	17828	130310	55481	87938
R^2	0.093	0.184	0.069	0.050

Notes: This table presents the regional heterogeneity analysis results, dividing the sample into four economic regions according to the official classification of Chinese cities into eastern, central, western, and northeastern regions. *p < 0.1, **p < 0.05, ***p < 0.01.

V. CONCLUSION

Based on the 2016 and 2017 China Migrants Dynamic Survey (CMDS), we empirically analyze the impact of HSR cities on the settlement intention of migrants. The result of linear probability model (LPM) shows that cities with HSR can increase migrants' settlement intention by 5.58%. The findings can provide empirical evidence for policy evaluation that the development of transport infrastructure can attract more human resources to stay in cities and provide human capital for economic development.

The development of transport infrastructure can attract labor to settle in cities, and cities still have both short- and long-term benefits from building HSR. Local governments can continue to invest and develop transport infrastructure, especially HSR networks. This will help to attract more labor and increase urbanization. Additionally, local governments need to balance the short-term costs and long-term benefits of investment in the HSR network. The short-term benefits of HSR construction include stimulating economic growth and promoting tourism, while the long-term benefits can improve the quality of life of the local population and improve the balance of regional development.

China's land ownership system is still an important force hindering urbanization, and the construction of HSR network can mitigate the negative impact of land ownership on the migrants working in the city and settlement. The results of this paper demonstrate that the migrants' settlement intention with land assets is more influenced by the construction of HSR, but land assets can negatively affect migrants' settlement intention. China's land system restricts land to be transferred, and China's peasants only have the right to use land, but not the right to transfer it (Hong and Sun, 2020). This means that peasants are unable to sell or lease their land to others in the market, limiting the efficient allocation and mobility of land. If peasants choose to settle in cities and obtain an urban household registration, their land assets in their hometowns are reclaimed collectively, resulting in a loss of assets, which affects their settlement intention in places of inflow. When cities have a more widespread and convenient HSR network, migrants can quickly travel between their hometowns and the city where they work and better maintain their land assets. On the one hand, it can facilitate the settlement of migrants in cities and promote urban mobility. On the other hand, it can improve the convenience of travelling between home and work cities for the migrants, considering hometown land assets and reducing the negative economic costs of settling in cities due to hometown land assets.

The construction of HSR does not have the effect of migrants settling in Northeast China. The economic development of Northeast China is relatively lagging behind, especially when compared to the eastern region (Ren et al., 2020). Even though the construction of HSR has provided faster transportation, the lack of adequate employment opportunities and development prospects in parts of the region has limited interest in settling in the region (Cho, 2013). In areas of population outflow, the development of transport infrastructure cannot attract migrants to settle. Attracting labor requires a combination of measures to promote economic development, adjust the industrial structure and improve living conditions, to enhance the attractiveness and competitiveness of the region and thus attract more labor to settle in the region.

REFERENCES

Ahlfeldt, G. M., & Feddersen, A. (2018). From periphery to core: Measuring agglomeration effects using high-speed rail. *Journal of Economic Geography*, *18*(2), 355–390. DOI: 10.1093/jeg/lbx005

Angrist, J. D., & Pischke, J.-S. (2009). *Mostly harmless econometrics: An empiricist's companion*. Princeton university press. DOI: 10.1515/9781400829828

Antonova, N., Abramova, S., & Polyakova, V. (2020). "cities are good for us": Assessment of urban attractiveness by students. [IOP Publishing.]. *IOP Conference Series. Earth and Environmental Science*, *459*(3), 032090. DOI: 10.1088/1755-1315/459/3/032090

Aschauer, D. A. (1989). Is public expenditure productive? *Journal of Monetary Economics*, *23*(2), 177–200. DOI: 10.1016/0304-3932(89)90047-0

Au, C.-C., & Henderson, J. V. (2006). Are chinese cities too small? *The Review of Economic Studies*, *73*(3), 549–576. DOI: 10.1111/j.1467-937X.2006.00387.x

Banzhaf, H. S., & Walsh, R. P. (2008). Do people vote with their feet? an empirical test of tiebout's mechanism. *The American Economic Review*, *98*(3), 843–863. DOI: 10.1257/aer.98.3.843

Bernard, A. B., Moxnes, A., & Saito, Y. U. (2019). Production networks, geography, and firm performance. *Journal of Political Economy*, *127*(2), 639–688. DOI: 10.1086/700764

Bosker, M., Brakman, S., Garretsen, H., & Schramm, M. (2012). Relaxing hukou: Increased labor mobility and china's economic geography. *Journal of Urban Economics*, *72*(2-3), 252–266. DOI: 10.1016/j.jue.2012.06.002

Campos, J., & De Rus, G. (2009). Some stylized facts about high-speed rail: A review of hsr experiences around the world. *Transport Policy*, *16*(1), 19–28. DOI: 10.1016/j.tranpol.2009.02.008

Carten'ı, A., Pariota, L., & Henke, I. (2017). Hedonic value of high-speed rail services: Quantitative analysis of the students' domestic tourist attractiveness of the main italian cities. *Transportation Research Part A, Policy and Practice*, *100*, 348–365. DOI: 10.1016/j.tra.2017.04.018

Chan, K. W. (2015). Five decades of the chinese hukou system. In *Handbook of Chinese migration* (pp. 23–47). Edward Elgar Publishing. DOI: 10.4337/9781783476640.00009

Chan, K. W., & Buckingham, W. (2008). Is china abolishing the hukou system? *The China Quarterly*, *195*, 582–606. DOI: 10.1017/S0305741008000787

Chang, Z., Diao, M., Jing, K., & Li, W. (2021). High-speed rail and industrial movement: Evidence from china's greater bay area. *Transport Policy*, *112*, 22–31. DOI: 10.1016/j.tranpol.2021.08.013

Chen, C., & Fan, C. C. (2016). China's hukou puzzle: Why don't rural migrants want urban hukou? *China Review*, *16*(3), 9–39.

Chen, G., & Silva, J. A. e. (2013). Regional impacts of high-speed rail: A review of methods and models. *Transportation Letters*, *5*(3), 131–143. DOI: 10.1179/1942786713Z.00000000018

Chen, M., Liu, W., & Tao, X. (2013). Evolution and assessment on china's urbanization 1960–2010: Under-urbanization or over-urbanization? *Habitat International*, *38*, 25–33. DOI: 10.1016/j.habitatint.2012.09.007

Cho, M. Y. (2013). *The specter of" the people": urban poverty in northeast China*. Cornell University Press. DOI: 10.7591/cornell/9780801451652.001.0001

Clark, X., Dollar, D., & Micco, A. (2004). Port efficiency, maritime transport costs, and bilateral trade. *Journal of Development Economics*, *75*(2), 417–450. DOI: 10.1016/j.jdeveco.2004.06.005

CNBS. (2023). *China Statistical Yearbook 2023, by China National Bureau of Statistics*. CNBS.

CNRB. (2023). *Statistical Bulletin of National Railways of China 2022, by China National Railway Bureau*. CNRB.

CNRG (2020). In *Outline of Planning for Railway Pioneering in the New Era of a Strong Transportation State, by China National Railway Group (CNRG).*

Desmet, K., & Rossi-Hansberg, E. (2013). Urban accounting and welfare. *The American Economic Review*, *103*(6), 2296–2327. DOI: 10.1257/aer.103.6.2296

Donaldson, D. (2018). Railroads of the raj: Estimating the impact of transportation infrastructure. *The American Economic Review*, *108*(4-5), 899–934. DOI: 10.1257/aer.20101199

Donaldson, D., & Hornbeck, R. (2016). Railroads and american economic growth: A "market access" approach. *The Quarterly Journal of Economics*, *131*(2), 799–858. DOI: 10.1093/qje/qjw002

Duranton, G., & Puga, D. (2019). *Urban growth and its aggregate implications. Technical report*. National Bureau of Economic Research. DOI: 10.3386/w26591

Dustmann, C., & Glitz, A. (2011). Migration and education. In *Handbook of the Economics of Education* (Vol. 4, pp. 327–439). Elsevier. DOI: 10.1016/B978-0-444-53444-6.00004-3

Faber, B. (2014). Trade integration, market size, and industrialization: Evidence from china's national trunk highway system. *The Review of Economic Studies, 81*(3), 1046–1070. DOI: 10.1093/restud/rdu010

Fan, C. C., Sun, M., & Zheng, S. (2011). Migration and split households: A comparison of sole, couple, and family migrants in beijing, china. *Environment & Planning A, 43*(9), 2164–2185. DOI: 10.1068/a44128

Fang, Y. and Shi, Z. (2018). Children of migrant parents: Migrating together or left behind.

Fang, Y., & Shi, Z. (2018, June). Children of migrant parents: Migrating together or left behind. *Habitat International, 76*, 62–68. DOI: 10.1016/j.habitatint.2018.05.001

Ferrer-i-Carbonell, A., & Frijters, P. (2004). How Important is Methodology for the Estimates of the Determinants of Happiness? *Economic Journal (London), 114*(497), 641–659. DOI: 10.1111/j.1468-0297.2004.00235.x

Fretz, S., Parchet, R., and Robert Nicoud, F. (2017). Highways, market access, and spatial sorting.

Gakenheimer, R. (1999). Urban mobility in the developing world. *Transportation Research Part A, Policy and Practice, 33*(7-8), 671–689. DOI: 10.1016/S0965-8564(99)00005-1

Garmendia, M., de Uren˜a, J. M., Ribalaygua, C., Leal, J., & Coronado, J. M. (2008). Urban residential development in isolated small cities that are partially integrated in metropolitan areas by high speed train. *European Urban and Regional Studies, 15*(3), 249–264. DOI: 10.1177/0969776408090415

Glaeser, E. L., & Mare, D. C. (2001). Cities and skills. *Journal of Labor Economics, 19*(2), 316–342. DOI: 10.1086/319563

Gospodini, A. (2005). Urban development, redevelopment and regeneration encouraged by transport infrastructure projects: The case study of 12 european cities. *European Planning Studies, 13*(7), 1083–1111. DOI: 10.1080/09654310500242121

Green, A. E., Hogarth, T., & Shackleton, R. E. (1999). Longer distance commuting as a substitute for migration in britain: A review of trends, issues and implications. *International Journal of Population Geography*, *5*(1), 49–67. DOI: 10.1002/(SICI)1099-1220(199901/02)5:1<49::AID-IJPG124>3.0.CO;2-O

Guirao, B. (2013). Spain: Highs and lows of 20 years of hsr operation. *Journal of Transport Geography*, *31*, 201–206. DOI: 10.1016/j.jtrangeo.2013.05.010

Guirao, B., Casado-Sanz, N., & Campa, J. L. (2020). Labour opportunities provided by spanish high-speed rail (hsr) commuting services in a period of financial crisis: An approach based on regional wage disparities and housing rental prices. *Regional Studies*, *54*(4), 539–549. DOI: 10.1080/00343404.2018.1530751

Guo, Y., Cao, L., Song, Y., Wang, Y., & Li, Y. (2022). Understanding the formation of city-hsr network: A case study of yangtze river delta, china. *Transport Policy*, *116*, 315–326. DOI: 10.1016/j.tranpol.2021.12.014

Haas, A. and Osland, L. (2014). Commuting, migration, housing and labour markets: Complex interactions.

Hainmueller, J. (2012). Entropy balancing for causal effects: A multivariate reweighting method to produce balanced samples in observational studies. *Political Analysis*, *20*(1), 25–46. DOI: 10.1093/pan/mpr025

Hanley, D., Li, J., and Wu, M. (2022). High-speed railways and collaborative innovation.

Hanley, D., Li, J., & Wu, M. (2022, March). High-speed railways and collaborative innovation. *Regional Science and Urban Economics*, *93*, 103717. DOI: 10.1016/j.regsciurbeco.2021.103717

He, D., Chen, Z., Pei, T., & Zhou, J. (2023). Analysis of structural evolution and its influencing factors of the high-speed railway network in china's three urban agglomerations. *Cities (London, England)*, *132*, 104063. DOI: 10.1016/j.cities.2022.104063

Hodgson, C. (2018). The effect of transport infrastructure on the location of economic activity: Railroads and post offices in the american west. *Journal of Urban Economics*, *104*, 59–76. DOI: 10.1016/j.jue.2018.01.005

Hong, Z. and Sun, Y. (2020). Power, capital, and the poverty of farmers' land rights in china.

Hong, Z., & Sun, Y. (2020, March). Power, capital, and the poverty of farmers' land rights in China. *Land Use Policy*, *92*, 104471. DOI: 10.1016/j.landusepol.2020.104471

Huang, X., & Chen, M. (2022). Understanding the role of housing in rural migrants' intention to settle in cities: Evidence from china. *Habitat International, 128*, 102650. DOI: 10.1016/j.habitatint.2022.102650

Huang, Y.. (2019). The correlation between hsr construction and economic development– empirical study of chinese cities. *Transportation Research Part A, Policy and Practice, 126*, 24–36. DOI: 10.1016/j.tra.2019.05.017

Kim, K. S. (2000). High-speed rail developments and spatial restructuring: A case study of the capital region in south korea. *Cities (London, England), 17*(4), 251–262. DOI: 10.1016/S0264-2751(00)00021-4

Li, R., Xu, M., & Zhou, H. (2023). Impact of high-speed rail operation on urban economic resilience: Evidence from local and spillover perspectives in china. *Cities (London, England), 141*, 104498. DOI: 10.1016/j.cities.2023.104498

Lin, C., Liu, J., & Li, W. (2023). Influence of the high-speed railway (hsr) construction on industrial structure transformation. *Enterprise Information Systems, 17*(2), 1942998. DOI: 10.1080/17517575.2021.1942998

Liu, Z., Zeng, S., Jin, Z., & Shi, J. J. (2022). Transport infrastructure and industrial agglomeration: Evidence from manufacturing industries in china. *Transport Policy, 121*, 100–112. DOI: 10.1016/j.tranpol.2022.04.001

Ma, L., Niu, D., & Sun, W. (2021). Transportation infrastructure and entrepreneurship: Evidence from high-speed railway in china. *China Economic Review, 65*, 101577. DOI: 10.1016/j.chieco.2020.101577

Monte, F., Redding, S. J., & Rossi-Hansberg, E. (2018). Commuting, migration, and local employment elasticities. *The American Economic Review, 108*(12), 3855–3890. DOI: 10.1257/aer.20151507

Qi, Y., Yang, Y., & Jin, F. (2013). China's economic development stage and its spatiotemporal evolution: A prefectural-level analysis. *Journal of Geographical Sciences, 23*(2), 297–314. DOI: 10.1007/s11442-013-1011-0

Ren, W., Xue, B., Yang, J., & Lu, C. (2020). Effects of the northeast china revitalization strategy on regional economic growth and social development. *Chinese Geographical Science, 30*(5), 791–809. DOI: 10.1007/s11769-020-1149-5

Scharp, K. M., Paxman, C. G., & Thomas, L. J. (2016). "i want to go home" homesickness experiences and social-support-seeking practices. *Environment and Behavior, 48*(9), 1175–1197. DOI: 10.1177/0013916515590475

Song, Y. (2014). What should economists know about the current chinese hukou system?

Song, Y. (2014, June). What should economists know about the current Chinese hukou system? *China Economic Review*, *29*, 200–212. DOI: 10.1016/j.chieco.2014.04.012

Sun, X., Yan, S., Liu, T., & Wang, J. (2023). The impact of high-speed rail on urban economy: Synergy with urban agglomeration policy. *Transport Policy*, *130*, 141–154. DOI: 10.1016/j.tranpol.2022.11.004

Vortherms, S. A., & Liu, G. G. (2022). Hukou as benefits: Demand for hukou and wages in china. *Urban Studies (Edinburgh, Scotland)*, *59*(15), 3167–3183. DOI: 10.1177/00420980221074911

Wang, C., & Shen, J. (2022). Settlement intention of migrants in urban china: The effects of labor-market performance, employment status, and social integration. *Applied Geography (Sevenoaks, England)*, *147*, 102773. DOI: 10.1016/j.apgeog.2022.102773

Wang, F., Liu, Z., Xue, P., & Dang, A. (2022). High-speed railway development and its impact on urban economy and population: A case study of nine provinces along the yellow river, china. *Sustainable Cities and Society*, *87*, 104172. DOI: 10.1016/j.scs.2022.104172

Wang, Z., & Chen, L. (2019). Destination choices of chinese rural–urban migrant workers: Jobs, amenities, and local spillovers. *Journal of Regional Science*, *59*(3), 586–609. DOI: 10.1111/jors.12444

Xiong, Y. (2015). The broken ladder: Why education provides no upward mobility for migrant children in china. *The China Quarterly*, *221*, 161–184. DOI: 10.1017/S0305741015000016

You, H., Yang, J., Xue, B., Xiao, X., Xia, J., Jin, C., & Li, X. (2021). Spatial evolution of population change in northeast china during 1992–2018. *The Science of the Total Environment*, *776*, 146023. DOI: 10.1016/j.scitotenv.2021.146023

Yuan, H., Guan, Z., Jiang, C., & Yang, H. (2023). Peer effect in the construction of china's high-speed rail stations: Empirical evidence from spatial econometric analysis. *Transportation Research Part A, Policy and Practice*, *173*, 103720. DOI: 10.1016/j.tra.2023.103720

Zhang, B., & Qi, R. (2021). Transportation infrastructure, innovation capability, and urban economic development. *Transformations in Business & Economics*, •••, 20.

Zhang, J., Huang, J., Wang, J., & Guo, L. (2020a). Return migration and hukou registration constraints in chinese cities. *China Economic Review*, *63*, 101498. DOI: 10.1016/j.chieco.2020.101498

Zhang, L., Sharpe, R. V., Li, S., & Darity, W. A.Jr. (2016). Wage differentials between urban and rural-urban migrant workers in china. *China Economic Review*, *41*, 222–233. DOI: 10.1016/j.chieco.2016.10.004

Zhang, P., Zhao, Y., Zhu, X., Cai, Z., Xu, J., & Shi, S. (2020b). Spatial structure of urban agglomeration under the impact of high-speed railway construction: Based on the social network analysis. *Sustainable Cities and Society*, *62*, 102404. DOI: 10.1016/j.scs.2020.102404

Zou, J., Chen, J., & Chen, Y. (2022). Hometown landholdings and rural migrants' integration intention: The case of urban china. *Land Use Policy*, *121*, 106307. DOI: 10.1016/j.landusepol.2022.106307

Appendix

Table 11. t-test for Mean Between Groups.

Variables	Non HSR City		HSR City		Difference	p - Value
	N	Mean	N	Mean		
settle	48015	0.27	246412	0.39	-0.12	0.00***
gender	48015	0.52	246412	0.51	0.01	0.00***
age	48015	36.07	246412	35.2	0.87	0.00***
edu	48015	9.51	246412	10.47	-0.96	0.00***
hukou	48015	0.13	246412	0.17	-0.04	0.00***
marriage	48015	0.83	246412	0.81	0.01	0.00***
job	48014	0.8	246411	0.85	-0.05	0.00***
lnincome	48015	8.5	246412	8.74	-0.23	0.00***
hometown	48015	1.33	246412	1.44	-0.11	0.00***
lnteacher	48015	4.49	246412	4.38	0.11	0.00***
lnbed	48015	5.08	246412	6.6	-1.52	0.00***
lnpgdp	48015	10.37	246412	11.14	-0.77	0.00***
structure	48015	56.84	246412	53.72	3.12	0.00***

Notes: This table reports the results of the t-tests for differences in means between HSR and non-HSR cities. All control variables show significant differences between the two groups, confirming that the control variables can effectively capture the factors influencing migrants' settlement intention. *$p < 0.1$, **$p < 0.05$, ***$p < 0.01$.

Table 12. Instrumental Variables Tests.

Underidentification test (Anderson canon. corr. LM statistic): Chi-sq(2) P=	1.00E+04 0
Weak identification test (Cragg-Donald Wald F statistic):	5285.529
Stock-Yogo weak ID test critical values: 10% maximal IV size	19.93
15% maximal IV size	11.59
20% maximal IV size	8.75
25% maximal IV size	7.25
Source: Stock-Yogo (2005). Reproduced by permission.	
Sargan statistic (overidentification test of all instruments):	0.257
Chi-sq(1) P=	0.6121

Notes: This table presents the results of the instrumental variables tests. The underidentification test (Anderson canon. corr. LM statistic) and weak identification test (Cragg-Donald Wald F statistic) confirm that the instruments (city slope and elevation) are strong and valid. The Sargan statistic (overidentification test) indicates that the instruments are exogenous. *p < 0.1, **p < 0.05, ***p < 0.01.

Table 13. Entropy Balance Matching Equilibrium Test (Before).

	mean	Treat variance	skewness	mean	Control variance	skewness
gender	0.514	0.250	-0.057	0.522	0.250	-0.087
age	35.200	88.360	0.422	36.070	93.190	0.317
edu	10.470	10.190	-0.007	9.507	10.860	-0.275
hukou	0.171	0.142	1.744	0.133	0.115	2.164
marriage	0.813	0.152	-1.606	0.825	0.144	-1.714
job	0.854	0.125	-2.008	0.801	0.160	-1.505
lnincome	8.735	0.424	-2.402	8.503	0.512	-4.431
hometown	1.437	0.833	2.409	1.328	0.567	2.636
lnteacher	4.378	2.722	4.218	4.490	1.152	0.860
lnbed	6.597	4.748	0.129	5.079	2.699	0.466
lnpgdp	11.140	0.968	-3.419	10.370	0.786	-0.376
structure	53.720	201.900	-0.018	56.840	159.900	-0.278
year	0.495	0.250	0.021	0.501	0.250	-0.004

Notes: This table presents the results of the Entropy Balance Matching (EBM) equilibrium test. The table shows the mean, variance, and skewness of the covariates before matching. The results indicate that the matching is successful, as the differences in means and higher-order moments are minimized after matching.

Table 14. Entropy Balance Matching Equilibrium Test (After).

	mean	Treat variance	skewness	mean	Control variance	skewness
gender	0.514	0.250	-0.057	0.514	0.250	-0.057
age	35.200	88.360	0.422	35.200	88.490	0.453
edu	10.470	10.190	-0.007	10.470	10.760	-0.063
hukou	0.171	0.142	1.744	0.171	0.142	1.744
marriage	0.813	0.152	-1.606	0.813	0.152	-1.606
job	0.854	0.125	-2.008	0.854	0.125	-2.008
lnincome	8.735	0.424	-2.402	8.735	0.326	0.570
hometown	1.437	0.833	2.409	1.437	0.780	2.225
lnteacher	4.378	2.722	4.218	4.378	0.781	0.673

continued on following page

Table 14. Continued

	Treat			Control		
	mean	variance	skewness	mean	variance	skewness
lnbed	6.597	4.748	0.129	6.597	2.373	-0.475
lnpgdp	11.140	0.968	-3.419	11.140	0.427	-0.458
structure	53.720	201.900	-0.018	53.720	105.900	-0.806
year	0.495	0.250	0.021	0.495	0.250	0.021

Notes: This table presents the results of the EBM equilibrium test. The table shows the mean, variance, and skewness of the covariates after matching. The results indicate that the matching is successful, as the differences in means and higher-order moments are minimized after matching.

Table 15. KHB mediating effect analysis: income satisfaction.

Model	Coefficient	Std. err.	z	P>z	[95% conf. interval]	
Reduced	0.0432124	0.0044867	9.63	0.00	0.0344187	0.0520061
Full	0.0433217	0.0044871	9.65	0.00	0.0345272	0.0521163
Diff	-0.0001094	0.0000659	-1.66	0.097	-0.0002385	0.0000197

Notes: This table presents the results of the KHB mediating effect analysis for income satisfaction. The results show that income satisfaction partially mediates the effect of HSR access on migrants' settlement intention. The difference in coefficients between the reduced and full models is significant, indicating the presence of a mediating effect.

Table 16. KHB intermediary effect analysis: business environment satisfaction.

Model	Coefficient	Std. err.	z	P>z	[95% conf. interval]	
Reduced	0.0432124	0.0044844	9.64	0.00	0.0344232	0.0520016
Full	0.0435354	0.0044844	9.71	0.00	0.0347461	0.0523248
Diff	-0.000323	0.0001489	-2.17	0.03	-0.0006149	-0.0000312

Notes: This table presents the results of the KHB mediating effect analysis for business environment satisfaction. The results show that business environment satisfaction partially mediates the effect of HSR access on migrants' settlement intention. The difference in coefficients between the reduced and full models is significant, indicating the presence of a mediating effect.

Figure 1. Pre-matching kernel density

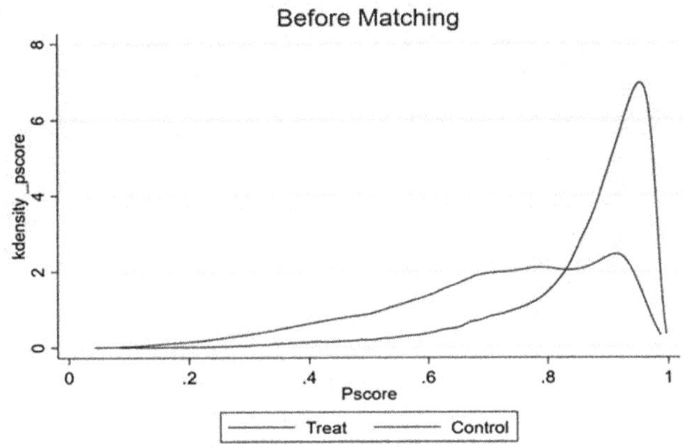

Figure 2. Post-matching kernel density

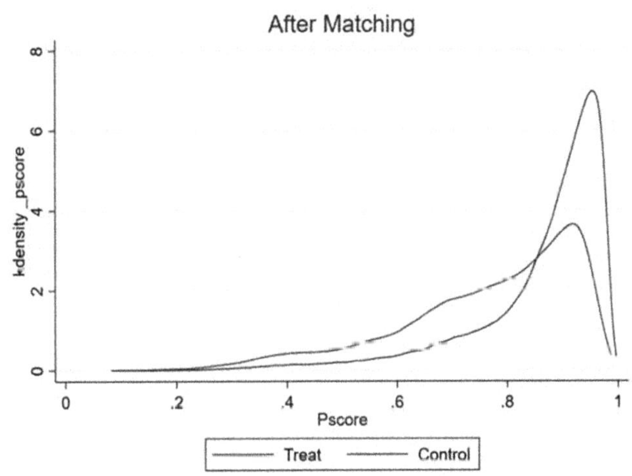

Chapter 8
Impact of Built-Up Area Dispersion on Natural Heritage Sites in the South Aegean, Greece

Efstratia Chatzi
https://orcid.org/0000-0001-7586-2607
University of the Aegean, Mytilene, Greece

Georgios Tsilimigkas
University of the Aegean, Greece

ABSTRACT

Natural heritage sites and areas are vital for preserving biodiversity and ecological integrity, especially on islands with limited size and isolation. This study investigates the impact of uncontrolled built-up area sprawl on protected areas and surrounding forest ecosystems in the Southern Aegean Region, where development pressures like tourism and infrastructure expansion threaten fragile ecosystems. The analysis highlights the importance of forest and protected areas in maintaining biodiversity. Using geospatial datasets from the Copernicus Land Monitoring Service and the Ministry for the Environment, the study examines spatial relationships between urban expansion and Natura 2000 sites and forested regions. The findings reveal significant conflicts from built-up dispersion, highlighting the inadequacy of current management frameworks in protecting these natural resources. This underscores the urgent need for integrated spatial planning that balances ecological preservation with development, ensuring the sustainability of biodiversity on islands amidst ongoing urban pressures.

DOI: 10.4018/979-8-3373-1554-6.ch008

1. INTRODUCTION

The preservation of biodiversity within natural heritage areas, including Natura 2000 sites and forest areas, is essential for maintaining ecological integrity, especially in isolated ecosystems such as those in the South Aegean Islands. These areas conserve unique plant and animal species adapted to the Mediterranean's distinctive climate and geography (McDonald et al., 2008). However, the Southern Aegean region of Greece, which is home to many protected zones, faces growing pressures from the dispersion of built-up areas, primarily driven by tourism. This rapid development disrupts habitats and threatens the ecological services critical for both local sustainability and global environmental stability (Bolund & Hunhammar 1999).

Protected areas stabilize ecosystems, support hydrological cycles, provide habitats for native species and mitigate climate change impacts (Andam et al., 2008). Natura 2000 sites, under EU Law, include Sites of Community Importance (SCIs) and Special Areas of Conservation (SACs) under the Habitats Directive (Directive 92/43/EEC), as well as Special Protection Areas (SPAs) under the Birds Directive (Directive 2009/147/EC) (Directive 92/43/EEC; Directive 2009/147/EC; Lai & Zoppi 2017). These designations aim to protect habitats and species of high conservation value, requiring EU Member States to implement conservation measures that prevent habitat degradation and fragmentation (Evans 2006; Mammides et al.,2024). Projects impacting these sites undergo an "appropriate assessment" to ensure ecological integrity, aligning with EU's goals like the "no net land take" target by 2050 and a 30% increase in protected areas by 2030 (EU biodiversity strategy).

The South Aegean, one of the Mediterranean's biodiversity hotspots, hosts a wide variety of flora and fauna uniquely adapted to its topography and climate (Luck et al., 2004). However, expanding urbanization and tourism growth fragment habitats and degrade ecosystems (Tsilimigkas et al.,2022; Gonzalez-Garcia et al., 2022; Beriatos 2005). The region's fragmented geography, steep terrain, limited arable land and fragile island ecosystems heighten its vulnerability to land-use changes (Mücher et al., 2006). Protected forests not only support biodiversity but also play vital roles in soil stabilization, flood control and water recharge, which are essential in resource environments such as the South Aegean (Bonilla-Bedoya et al., 2020; Canetti et al.,2017).

Tourism infrastructure and second-home demand have intensified pressures on these ecosystems, leading to habitat degradation, higher fire risk and boundary effects that disrupt ecological stability (Tsilimigkas & Derdemezi 2020; Kovács et al., 2015; Kizos et al., 2010; Geldmann et al.,2013). Unregulated development often affects protected areas and forests, undermining ecological resilience in the islands (Tsilimigkas & Gourgiotis 2023, Romero et al., 2018). Despite Natura 2000 protections, Greece's implementation faces challenges, including administrative inefficiencies and regulatory constraints (Kallimanis et al., 2015; Terkenli 2005). Unchecked tourism-driven growth exacerbates "coastalization" and diminishes ecological connectivity, reducing ecosystem resilience and vital services like carbon storage and water regulation (Derdemezi et al.,2021; Chatzi et al.,2025; Chape et al.,2005).

Balancing built-up dispersion with nature conservation is a complex challenge in fragile island regions where land is limited (Kallimanis et al., 2015). Dispersed urban growth in the Southern Aegean raises concerns about the adequacy of current conservation policies to protect biodiversity effectively (Elmqvist et al., 2015; Seto et al., 2017; Dimopoulos et al.,2017). In high-tourism areas, such as the Southern Aegean, these impacts of urban sprawl are exacerbated by dispersed development and increased demand from visitors. This highlights the importance of obtaining detailed, site-specific data to evaluate the impact of built-up areas dispersion on conservation (Gonzalez-Garcia et al., 2022; Kati et al.,2015; Mammides et al.,2024; Stratigea & Katsoni 2015). Effective spatial planning regulations are critical for mitigating these impacts, directing growth away from sensitive habitats and promoting sustainable land use (Theodora 2019; Kizos et al., 2017). Properly enforced spatial planning frameworks can balance development with conservation, safeguarding ecological integrity (Tsilimigkas et al., 2021; Chorianopoulos et al., 2014). The South Aegean Islands, characterized by fragile coastal and forested ecosystems, illustrate the need for robust spatial planning to protect these ecologically valuable areas from the pressures of tourism and urban growth (Papageorgiou & Vogiatzakis 2006; Matos et al.,2002; Chatzi et al.,2025).

This study examines the spatial impact of built-up area dispersion on protected areas in Greece's Southern Aegean, where pressures from rapid development, especially from tourism and infrastructure expansion, threaten sensitive ecosystems. The protected areas analyzed include Natura 2000 sites, areas designated for the conservation of important bird populations by the Hellenic Ornithological Society as part of an international network, wildlife areas regulated under Greece's forest legislation based on data from forestry services and forest maps from the study region.

Using geospatial data provided by the Copernicus Land Monitoring Service and Greece's Ministry for the Environment, the study maps patterns of built-up areas dispersion within protected areas. By analyzing spatial interactions, the study underscores the importance of integrated spatial planning approaches that balance development with conservation, while preserving biodiversity and strengthening the resilience of local communities.

The research hypothesis suggests that unregulated dispersion of built-up areas disrupts biodiversity and critical ecosystem services, challenging ecological resilience. To explore this, the study focuses on two key questions:

a) How does built-up area dispersion impact protected areas such as Natura 2000 sites and forest areas?
b) How does the extent of built-up areas, both inside and outside established settlement boundaries, affect the integrity of surrounding protected areas?

These questions are approached through quantitative analysis and geospatial mapping, identifying priority areas where urban pressures most threaten ecological balance. The findings aim to inform conservation policies that support sustainable development in these ecologically fragile regions. A central concern is the urban expansion beyond settlement boundaries, which exacerbates edge effects, fragments habitats and disrupts ecological connectivity. This weakens protected areas' resilience, highlighting the need for effective spatial planning to manage the dispersion of built-up areas and safeguard ecosystems and their services.

The article is structured as follows: Section 2 introduces the study's background and objectives, focusing on spatial planning, protected areas, and the policy considerations at various spatial levels. Section 3 reviews the data sources and methodology used in the study. Section 4 outlines the study region, providing context on the geographic and ecological characteristics of the area. Section 5 presents the results, highlighting key findings related to built-up areas dispersion and its impact on protected areas. Finally, Section 6 concludes with key insights and suggestions for future research on sustainable planning for island ecosystems.

2. SPATIAL PLANNING AND PROTECTED AREAS – POLICY CONSIDERATIONS AT VARIOUS SPATIAL LEVELS

At the international and European levels, spatial planning for protected areas is guided by several key policies, conventions and directives that aim to conserve biodiversity and promote sustainable land-use practices. The Convention on Biological Diversity (CBD), adopted in 1992 at the Earth Summit in Rio de Janeiro, emphasizes the importance of establishing and managing protected areas as a critical tool for biodiversity conservation (McDonald et al.,2009; Directive 92/43/EEC). This international agreement urges countries to integrate biodiversity conservation into their spatial planning and development strategies, ensuring long-term ecosystem health and species protection globally (Directive 2009/147/EC).

A milestone at the European level was the establishment of the Natura 2000 network under the EU Habitats Directive (92/43/EEC), adopted in 1992 and the Birds Directive (2009/147/EC), which came into force in 2009 (Directive 2009/147/EE; Natura 2000). The Habitats Directive aims to protect the most endangered species and habitats in Europe by designating Special Areas of Conservation (SACs), while the Birds Directive focuses on the protection of wild birds through the designation of Special Protection Areas (SPAs). Together, these form the Natura 2000 network which now covers approximately 18% of the EU's land area and 6% of its marine waters, forming the backbone of EU biodiversity protection efforts (European Commission 2011).

In 2011, the EU Biodiversity Strategy to 2020 was adopted as part of the EU's commitment to halt biodiversity loss by 2020. One of the strategy's key targets was the further expansion of protected areas, with a focus on strengthening the management of Natura 2000 sites and ensuring their integration into land-use and spatial planning (Princé et al., 2021; European Commission 2020). These objectives were further strengthened with the EU Green Deal (2019), which set an ambitious target to protect 30% of Europe's land and marine areas by 2030, with at least a third under strict protection. This initiative underscores the urgent need to address the pressures of urbanization on sensitive ecosystems (Prioritized Action Framework 2024 Greece).

Greece has integrated these international and European directives into its national frameworks, becoming a party to CBC in 1994 and incorporating its principles into national laws. Greece's commitment to biodiversity conservation is reflected in its adherence to the Habitats and Birds Directives, leading to the designation of 446 Natura 2000 sites covering 27,59% of its land area and 19,6% of its marine territory. This makes Greece an important center of biodiversity in the EU (OGG 2383/2014; McKenna et al., 2014; OGG 1650/1986)

Law 1650/1986 on environmental protection provided the foundation for the creation and management of protected areas, including national parks and Natura 2000 sites. This law introduced guidelines for protecting nature reserves, wildlife habitats and forests, requiring protection from disruptive human activities. It established management frameworks and zoning to limit development, but enforcement has faced bureaucratic hurdles and ineffective policy execution, especially in high tourism areas (OGG4759/2020).

The Greek Spatial Planning Law (Law 4759/2020) aimed to update spatial planning processes to foster sustainable development, addressing the challenges of urbanization and environmental protection. It emphasizes the integration of environmental considerations, especially with regard to biodiversity conservation and the safeguarding of protected areas, within spatial and urban planning frameworks, while also ensuring better coordination between local and national authorities (EEA 2018).

Despite legislative progress, continue to exert pressure on ecologically sensitive regions, especially in the South Aegean. The National Strategy for Biodiversity 2021-2027, with input from the LIFE-IP 4 Natura project, focuses on integrating Natura 2000 sites into spatial planning, addressing knowledge gaps, improving governance and enhancing habitat and species management (OGG 2383/2014)

However, as the demand for tourism-related infrastructure and second homes continues to rise, the challenge of balancing built-up dispersion with conservation remains a pressing issue. The expansion of tourism infrastructure, especially in island/coastal areas, continues to threaten protected habitats, highlighting the need for stronger spatial planning tools, more effective land-use controls and better co-ordination in order to protect Greece's natural heritage.

3. DATA AND METHODOLOGY

This study examines the spatial interactions between urban development and protected ecosystems in the Southern Aegean region. Using a combination of quantitative analysis and geospatial mapping, the study assesses the dispersion of built-up areas.

Built-up area data were obtained from the Copernicus Land Monitoring Service (EEA 2018), which provides high-resolution satellite imagery for land-use classification. The Natura 2000 sites, including Special Areas of Conservation (SACs), Special Protection Areas (SPAs) and forested areas were sourced from the Greek Ministry of Environment and Energy. These datasets comply with EU regulations, specifically Directive 92/43/EEC for habitats, Directive 2009/147/EC for SPAs and the updated Natura 2000 sites from 2021. These areas have been established in accordance with the provisions of the forestry legislation and demarcated using the elements provided by the Forestry Services. Additional data from 2022 forest maps and wildlife park boundaries were sourced through the Forest Maps platform, were also used to boundaries via the e-poleodomia GIS platform. Furthermore, important bird areas were sourced from the Hellenic Ornithological Society.

For settlement data, the study used information from e-poleodomia online platform and the Hellenic Cadastre, which provided official settlement boundaries, including digitized traditional settlements based on official government gazettes (NCMA 2023). These boundaries were digitized from topographic charts and orthophoto maps using the Greek Geodetic Reference System 1987 (GGRS87) at a scale of 1:5,000, following the standards of the National Cadastre and Mapping Agency.

To analyze built-up areas dispersion, the built-up areas within and outside settlement boundaries were mapped. The base maps and geospatial datasets were processed at a scale of 1:10,000 to ensure consistency with spatial planning requirements and align with the Impervious Built-up (IBU) layer, a key dataset for evaluating urban development impacts. Raster analysis was applied to convert protected areas into raster format, allowing the overlay of these areas with built-up zones. This process enabled the identification and quantification of urban encroachment into Natura 2000 and forest conservation zones. The resulting maps and visualizations provide a clear understanding of how built-up areas dispersion intersects with and impacts the ecological integrity of protected areas across the Southern Aegean islands.

4. THE STUDY REGION

The Southern Aegean islands, especially the Cyclades and Dodecanese, host significant natural heritage sites including Natura 2000 areas and Important Bird Areas (IBAs), which safeguard biodiversity and ecosystem services. However, tourism and infrastructure development threaten these zones, leading to habitat loss and ecosystem disruption (Watson et al., 2014). Covering approximately 5,286 square kilometers, the region has a population of around 400,000 as of 2020 (EL-STAT 2020). The region includes 74 wildlife areas, 26 important bird areas (IBAs) and 53 Natura 2000 sites. These provide shelters for local biodiversity. However, unregulated urban growth, particularly in coastal and tourist areas, fragments habitats and disrupts ecological processes, threatening the long-term resilience of ecosystems (Watson et al., 2014; Chatzi et al., 2025).

The region's ecosystems, including wetlands, forests, marine areas and islands, are vital for Greece's biodiversity but face pressures from encroaching tourism and infrastructure projects (Terkenli 2005). Built-up areas dispersion near protected zones transforms landscapes and threatens ecological functions. The absence of an integrated spatial planning framework exacerbates these impacts potentially undermining conservation efforts (Tsilimigkas et al., 2021). Implementing effective strategies is crucial to mitigate these threats. This includes mapping urban expansion, assessing its ecological impacts, and guiding sustainable development policies.

5. RESULTS

This section presents and analyzes the results of the study, focusing on the spatial relationship between built-up dispersion and protected areas across the Southern Aegean region. Table 1 details the extent of built-up areas (in hectares) and their overlap with Important Bird Areas (IBAs), Natura 2000 sites, and Wildlife Areas. It reveals that while most urban development is concentrated within settlement boundaries, a substantial portion occurs outside settlements, often encroaching on protected areas. For example, Amorgos has 59,367.96 hectares of IBA-designated land, but the built-up footprint within these protected zones remains relatively limited.

Figure 1 visualizes the distribution of built-up areas relative to protected regions across the islands, while Map 1 focuses on four specific islands—Kos, Paros, Naxos, and Rhodes—highlighting the geographic context of urban expansion into sensitive ecological zones.

Table 1. Built-up Areas dispersion with Protected Areas (IBA, Wildlife, Natura 2000) Across Southern Aegean Islands

Island	Built-up Outside %	Built-up Inside %	IBA area Hectares	Built-up Outside IBA %	Built-up Inside IBA %	WILDLIFE Hectares	Built-up Outside WILDLIFE %	Built-up Inside WILDLIFE %	NATURA 2000 Hectares	Built-up Outside NATURA 2000%	Built-up Inside NATURA 2000%
AMORGOS	46,43	53,57	59367,96	12,26	2,82				62047,26	7,53	2,75
ANAFI	32,39	67,61	19669,53	32,39	67,61				12522,85		
ANDROS	53,53	46,47	29817,94	3,87	3,05	4341,64	5,44	0,28	59456,2	4,04	3,00
ANTIPAROS	70,61	29,39									
ARKOI	21,21	78,79				745,39	21,21	78,79	12460,99	100	100
ASTIPALAIA	26,43	73,57	37117,66	27,55	80,99	5896,3	5,50	13,69	35347,82	2,46	7,21
CHALKI	1,74	98,26							35320,29	12,83	100
DONOUSA	45,68	54,32									
FOLEGANDROS	24,34	75,66				405,68			7047	9,90	23,81
IOS	51,26	48,74	32363,17	68,78	100	3658,92	15,83	1,08			0,00
IRAKLEIA	68,11	31,89	6225,488	18,52	100	917,52			12588,69	100	100
KALIMNOS	6,83	93,17	23485,77	1,76	1,71	726,78	1,07	2,47	18847,29	1,54	0,80
KARPATHOS	53,17	46,83				5799,16	8,91	3,20	20614,72	5,12	5,68
KASOS	10,02	89,98							37405,43	11,45	89,98
KEA	79,97	20,03				2384,65	11,23	4,12	7151,73	5,90	
KIMOLOS	13,20	86,80				831,99	2,00		13871,30	3,23	0,23
KOS	52,12	47,88	9108,258	7,66	0,83	6778,56	12,85	0,63	10556,98	6,74	0,83
KOUFONISI	43,43	56,57	5778,755	71,13	80,94						
KYTHNOS	56,45	43,55							2904,43		

continued on following page

Table 1. Continued

Island	Built-up Outside %	Built-up Inside %	IBA area Hectares	Built-up Outside IBA %	Built-up Inside IBA %	WILDLIFE Hectares	Built-up Outside WILDLIFE %	Built-up Inside WILDLIFE %	NATURA 2000 Hectares	Built-up Outside NATURA 2000%	Built-up Inside NATURA 2000%
LEIPSOI	25,78	74,22							47123,18	98,17	20,82
LEROS	33,22	66,78	48880,21	7,19	5,95	726,26	1,58	0,24			
MILOS	33,03	66,97	10456,29	2,04		750,39	1,81	0,72	19778,97	1,97	
MYKONOS	89,16	10,84	18564,42	0,50	0,00	826,38	2,18		18508,58	0,47	
NAXOS	48,29	51,71	25597,95	7,94	13,75	2843,45	2,92	0,38	20996,96	5,78	2,98
NISYROS	21,69	78,31	4730,825	21,69	78,31	1633,30			22570,99	17,81	78,31
PAROS	60,40	39,60	31917,73	17,10	5,14	1210,66	2,46	0,23	25800,54	6,74	0,74
PATMOS	86,58	13,42	2793,9	23,74	0,08				4447,16		
RHODES	46,67	53,33	64761,15	2,55	1,15	10233,36	1,32	0,20	68074,20	3,18	0,87
SANTORINI	42,87	57,13				864,653	2,92	4,29	1219,44	1,11	0,59
SCHOINOUSA	41,57	58,43	15982,85	10,27							
SERIFOS	35,93	64,07	15333,73	11,00	15,73	1347,152	7,96	30,80	10144,93	2,64	3,20
SIFNOS	46,21	53,79	3212,911	3,04	0,27	691,185	9,15	11,15	2091,93	0,17	0,00
SIKINOS	24,36	75,64							9874,97		
SIMI	3,32	96,68	14106,14	3,61	96,68					0,46	0,82
SYROS	40,10	59,90	3319,567	4,73	4,24	251,095	0,04		3707,33	2,69	2,14
THIRA	20,59	79,41									
TILOS	12,37	87,63	17776,04	20,63	89,79	3485,757	5,90	24,73	17773,42	20,63	89,79
TINOS	42,48	57,52	5055,94	1,99		1110,777	0,36		7052,61	2,10	

Figure 1. Built-up Areas dispersion in Relation to Protected Areas (IBA, Wildlife, Natura 2000) Across Southern Aegean Islands

Figure 2. Map of Built-up Area Dispersion on Protected Areas in Kos, Naxos, Paros, and Rhodes

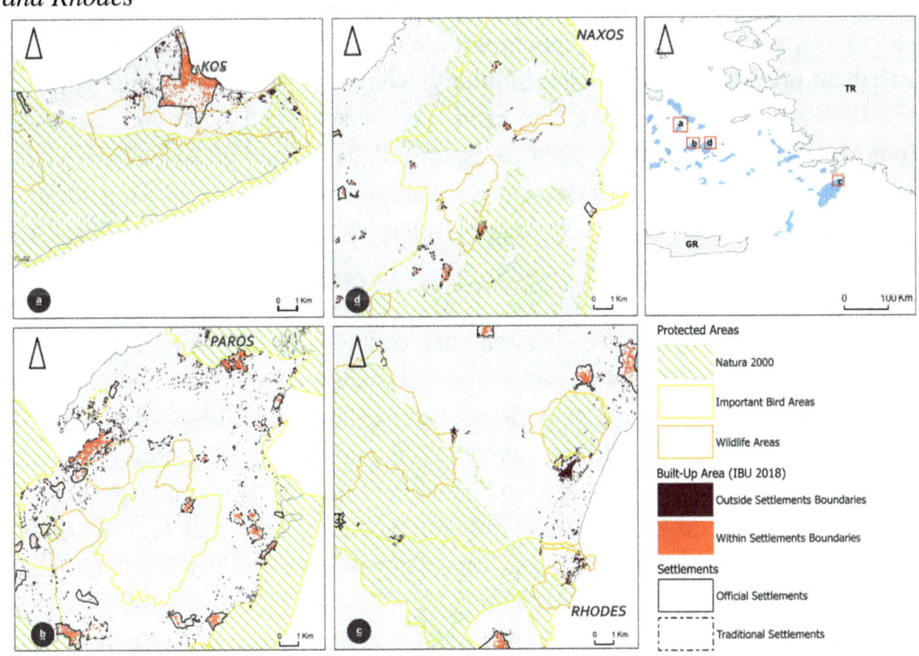

Built-up outside settlements stands out on islands like Mykonos, where 89.16% of built-up areas are located outside settlements. Despite this extensive sprawl, Mykonos has minimal overlap with IBAs (0.50%) and Natura 2000 zones (0.47%). Conversely, Patmos, where 86.58% of built-up areas are outside settlements, experiences substantial ecological pressure, with 23.74% of these areas overlapping with IBAs. On the other hand, small islands like Chalki and Simi, with only 1.74% and 3.32% of built-up areas outside settlements, respectively, show minimal urban sprawl. However, even these small percentages still intersect with protected areas, such as Chalki's 12.83% overlap with Natura 2000 zones.

The ecological vulnerability of certain islands is underscored by high overlap percentages despite limited total development. Leipsoi is a prime example: while only 25.78% of its built-up area lies outside settlements, 98.17% of that development overlaps with Natura 2000 zones. Similarly, in Tilos, 12.37% of development outside settlements intersects with both IBAs and Natura 2000 zones (20.63% overlap), indicating the excessive ecological impact of even low-density sprawl. These patterns highlight the urgent need for targeted spatial planning strategies tailored to island-specific development pressures and ecological sensitivities.

Astypalaia and Serifos also show significant overlap between built-up areas dispersion and ecological zones. On Astypalaia, 26.43% of built-up areas are outside settlements, with 5.50% overlapping with Wildlife Areas and 2.46% with Natura 2000 zones. Serifos, meanwhile, exhibits 35.93% of built-up areas outside settlements, with 30.80% overlapping with Wildlife Areas, indicating substantial ecological pressure on this island. Similarly, Koufonisi and Nisyros highlight the overlap between built-up areas, both inside and outside settlements and IBAs. On Koufonisi, 71.13% of built-up areas outside settlements and 80.94% inside intersect with IBAs, while Nisyros shows similar proportions, with 21.69% outside and 78.31% inside overlapping. These islands emphasize that development of built-up areas, whether inside or outside settlements, can create significant ecological pressure in sensitive habitats.

Patterns of built-up areas dispersion vary widely across other islands. For instance, Andros demonstrates a more balanced distribution, with 53.53% of built-up areas outside settlements and 46.47% inside. Overlaps with IBAs (3.87% outside and 3.05% inside) and Wildlife Areas (5.44% outside and 0.28% inside) are relatively low, reflecting a more contained ecological impact. Kos similarly shows a relatively balanced urban sprawl pattern, with 52.12% outside and 47.88% inside, though overlaps are more significant, especially with IBAs (7.66% outside) and Natura 2000 zones (6.74% outside).

At the regional level, there are significant distinctions between the Cyclades and Dodecanese islands in terms of built-up areas dispersion patterns and their ecological impacts. The Cyclades tend to exhibit higher levels of built-up areas dispersion outside settlements. Mykonos and Kea, for instance, have 89.16% and 79.97% of built-up areas outside settlements, respectively, though their overlaps with protected zones vary widely. Mykonos has minimal overlap with protected areas, whereas Kea has 11.23% of built-up areas outside intersecting with Wildlife Areas. In contrast, the Dodecanese islands generally display a more balanced distribution of built-up areas inside and outside settlements. Islands like Kos and Rhodes show almost equal proportions of built-up areas inside and outside settlements, with overlaps in protected zones being more moderate compared to the Cyclades.

Overall, the Southern Aegean islands exhibit complex patterns of urbanization and ecological impact. While the Cyclades show higher dispersion of built-up areas outside settlements, the Dodecanese maintain a more balanced development of built-up areas. These distinctions highlight the need for tailored land-use planning approaches to balance development and conservation across the region, recognizing the unique ecological and urban dynamics of each island.

Table 2, provides an overview of how built-up areas are distributed across various Greek islands, categorized by their location relative to settlements and land type, such as forest, non-forest and grassland areas. The results highlight significant patterns, as well as outliers, offering insights into land use and development practices on these islands. Figure 2 visualizes the results of Table 2, while Map 2, illustrates the distribution of built-up areas specifically on the islands of Kos, Paros, Naxos and Rhodes.

Table 2. Dispersion of Built-up Areas in Forest, Non-Forest, and Grassland Zones across the Southern Aegean Islands, Inside and Outside Settlements

Island	Built-up Outside Settlements in Forest Areas %	Built-up Inside Settlements in Forest Areas %	Built-up Outside Settlements in No Forest Areas %	Built-up Inside Settlements in No Forest Areas %	Built-up Outside Settlements in Grasslands Areas %	Built-up Inside Settlements in Grasslands Areas %
AMORGOS	1,29	0	90,54	6,99	1,18	0
ANAFI	0	0	31,25	65,38	3,08	0,29
ANDROS	5,73	0,55	82,34	5,79	5,55	0,03
ANTIPAROS	7,98	0,02	85,41	4,53	2,06	0
ARKOI	0	0	100	0	0	0
ASTIPALAIA	14,76	0	52,20	3,96	28,85	0,22
CHALKI	0	0	1,79	98,21	0	0
DONOUSA	23,81	0,26	72,75	2,65	0,53	0
FOLEGANDROS	0	0	25,52	73,01	1,38	0,09

continued on following page

Table 2. Continued

Island	Built-up Outside Settlements in Forest Areas %	Built-up Inside Settlements in Forest Areas %	Built-up Outside Settlements in No Forest Areas %	Built-up Inside Settlements in No Forest Areas %	Built-up Outside Settlements in Grasslands Areas %	Built-up Inside Settlements in Grasslands Areas %
IOS	2,62	0	47,25	48,81	1,29	0,03
IRAKLEIA	38,19	2,08	58,33	1,39	0	0
KALIMNOS	9,77	0,25	49,08	14,94	22,20	3,76
KARPATHOS	32,90	0,23	66,10	0,63	0,03	0,10
KASOS	5,88	0	91,18	0,42	2,52	0
KEA	0,08	0	92,87	1,19	5,86	0
KIMOLOS	0	0	84,40	14,00	1,60	0
KOS	0,69	0,02	89,69	9,32	0,28	0
KOUFONISI	12,51	0	85,51	1,98	0	0
KYTHNOS	0,00	0	95,13	2,24	2,64	0
LEIPSOI	27,82	0	65,41	6,02	0,75	0
LEROS	1,06	0	89,36	3,45	5,88	0,25
MILOS	1,87	0	89,33	2,11	6,70	0
MYKONOS	0	0	93,48	4,12	2,40	0
NAXOS	10,92	0,02	84,15	1,26	3,64	0
NISYROS	1,39	0	90,28	8,33	0	0
PAROS	1,98	0,13	86,14	10,56	1,15	0,04
PATMOS	42,26	0,31	45,17	5,82	6,44	0,00
RHODES	7,02	0,15	82,64	8,39	1,67	0,12
SANTORINI	0	0	94,20	4,00	1,55	0,25
SCHOINOUSA	17,70	0	81,07	0,00	1,23	0
SERIFOS	0	0	36,81	62,73	0,46	0
SIFNOS	3,13	0,04	86,32	9,20	0,88	0,43
SIKINOS	5,91	0,77	25,96	67,36	0	0
SIMI	9,17	0,92	66,97	22,94	0	0
SYROS	0,81	0,01	86,97	8,69	3,51	0,01
THIRA	0	0	65,02	31,84	2,69	0,45
TINOS	1	0,00	90,28	7,80	1,33	0,08

Figure 3. Dispersion of Built-up Areas in Forest, Non-Forest, and Grassland Zones of the Southern Aegean Islands: Inside vs. Outside Settlements

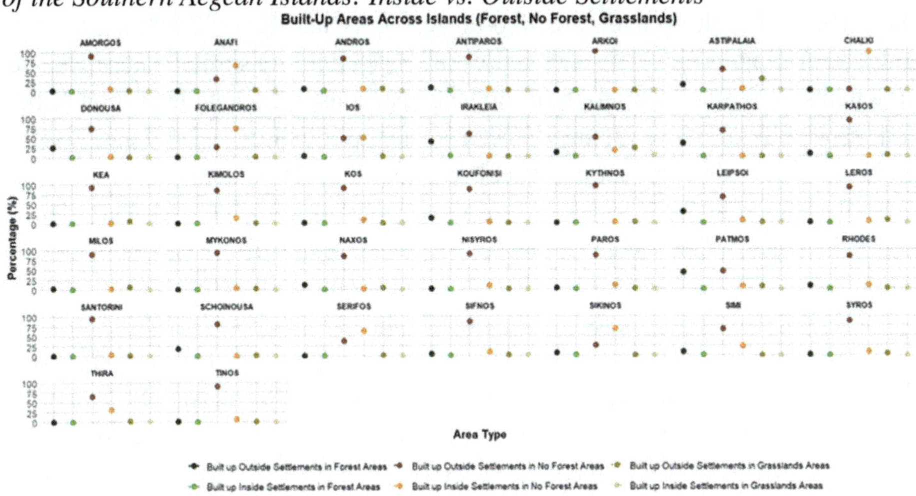

Figure 4. Map of Built-up Area Dispersion in Forest Areas of Kos, Naxos, Paros and Rhodes

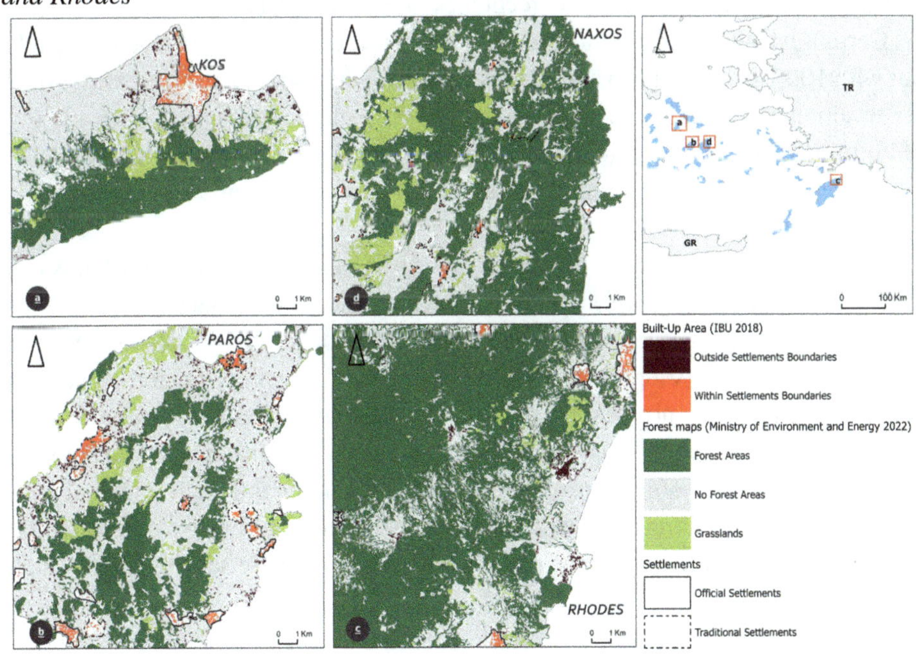

The majority of built-up areas on most islands are located outside settlements in non-forest areas. This trend is particularly noticeable in islands such as Arkoi, where 100% of built-up land falls into this category and Kea (92.87%). Other islands like Antiparos (85.41%) and Andros (82.34%) follow a similar pattern. Inside settlements in non-forest areas, built-up land also constitutes a large portion of the total, especially on more urbanized islands such as Chalki (98.21%) and Serifos (62.73%) where development is concentrated inside settlements.

Built-up areas in forest zones are uncommon, but not completely absent. Islands such as Donousa, Irakleia and Patmos stand out in this regard. Donousa has 23.81% of its built-up area outside settlements in forest zones, a proportion that is significantly higher than most other islands. Irakleia (38.19%) and Patmos (42.26%) also have notable percentages of built-up area in forest zones outside settlements. Development inside settlements in forest zones is almost nonexistent across all islands, with most percentages near zero, indicating that forest areas are largely protected from concentrated urban growth, which is crucial approach for preserving these ecosystems.

Grassland areas host even fewer built-up zones than forests, with some exceptions. Astipalaia has the highest percentage of built-up areas outside settlements in grassland zones, at 28.85%, followed by Kalimnos with 22.20%. These outliers highlight that some islands use grasslands for development more extensively than others. However, for most islands, built-up areas in grasslands are minimal, especially inside settlements, where percentages rarely exceed 1%.

Certain islands display unique or extreme patterns. For example, Arkoi stands out with 100% of its built-up areas outside settlements and in non-forest areas, with no development in forests or grasslands. Patmos is another exception, with 42.26% of built-up areas outside settlements in forest zones. Urbanization patterns are also evident in islands with high proportions of built-up areas inside settlements, particularly in non-forest zones.

In conclusion, the data reveal significant differences in how Greek islands use their land for development. Most islands have built-up areas concentrated in non-forest zones, which is typical of urban expansion. Forest and grassland areas see relatively limited development, though some islands, such as Donousa, Patmos and Astipalaia, show notable exceptions with substantial built-up areas in these sensitive zones. These findings highlight the diversity of land-use practices across the islands, offering valuable insights for spatial planning and environmental conservation. Understanding these patterns is crucial for managing development while safeguarding natural ecosystems, particularly in ecologically sensitive areas such as forests and grasslands.

6. CONCLUSIONS

Landscapes are constantly evolving due to natural processes and human activities, such as urban expansion, tourism infrastructure and transport development. When these changes occur rapidly and without strategic and physical planning, they can significantly impact the environment, especially in protected areas. In island regions, these pressures are heightened by fragile socio-spatial systems and the intense demands of mass tourism. Our findings indicate that uncontrolled growth of built-up areas poses significant threats to critical ecological zones, including Natura 2000 sites, Wildlife Areas, Important Bird Areas (IBAs) and forest areas. The overlap between built-up areas and these protected regions highlights the urgent need for strategic and physical land-use planning to safeguard the integrity of these vital habitats.

The study reveals a complex relationship between built-up areas growth and ecological preservation across the Southern Aegean islands. Urbanization patterns vary significantly between islands. Some experience concentrated development inside settlements, while others show extensive growth outside settlements, encroaching upon protected areas. This highlights the importance of understanding specific urbanization patterns and their unique ecological consequences.

Smaller islands, despite exhibiting minimal built-up areas growth, still intersect with protected zones, demonstrating that even limited development can disrupt ecological integrity. Islands with higher overlaps between built-up areas and Natura 2000 zones illustrate the vulnerability of areas with small-scale development but high ecological value.

The study also highlights key regional distinctions between the Cyclades and Dodecanese islands. The Cyclades exhibit higher levels of built-up areas dispersion outside settlements, while the Dodecanese islands show more balanced development of cities and settlements. These differences emphasize the need for an integrated spatial planning framework that considers both ecological preservation and urban growth across regions.

In conclusion, the findings reinforce the need for targeted land-use planning strategies that integrate urban development with conservation efforts. While many islands show trends of built-up areas dispersion, the extent of ecological impact varies based on the island's unique characteristics and development patterns. The Southern Aegean islands, with their diverse landscapes and ecological significance, require carefully crafted policies that ensure development does not undermine the protection of critical natural habitats. Effective spatial planning should balance urban growth with the preservation of protected areas, ensuring the long-term sustainability of both human and ecological systems.

REFERENCES

Andam, K. S., Ferraro, P. J., Pfaff, A., Sanchez-Azofeifa, G. A., & Robalino, J. A. (2008). Measuring the effectiveness of protected area networks in reducing deforestation. *Proceedings of the National Academy of Sciences of the United States of America, 105*(42), 16089–16094. DOI: 10.1073/pnas.0800437105

Beriatos, E. (2005). Planning and management of protected areas in Greece. Sustainable Development and Planning II, Vol. 1, WIT Transactions on Ecology and the Environment, 84, 389. WIT Press. ISSN 1743-3541

Bolund, P., & Hunhammar, S. (1999). Ecosystem services in urban areas. *Ecological Economics, 29*(2), 293–301. DOI: 10.1016/S0921-8009(99)00013-0

Bonilla-Bedoya, S., Mora, A., Vaca, A., Estrella, A., & Herrera, M. Á. (2020). Modelling the relationship between urban expansion processes and urban forest characteristics: An application to the Metropolitan District of Quito. *Computers, Environment and Urban Systems, 79*, 101420. DOI: 10.1016/j.compenvurbsys.2019.101420

Canetti, A., Cordeiro, M., Póvoa, P., Mattos, D., & Muñoz, E. (2017). Understanding multi-temporal urban forest cover using high-resolution images. *Urban Forestry & Urban Greening, 29*, 106–112. DOI: 10.1016/j.ufug.2017.10.020

Chape, S., Harrison, J., Spalding, M., & Lysenk, I. (2005). Measuring extent and effectiveness of protected areas as an indicator for meeting global biodiversity targets. *Philosophical Transactions of the Royal Society of London. Series B, Biological Sciences, 360*(1454), 433–455. DOI: 10.1098/rstb.2004.1592

Chatzi, E., Derdemezi, E.-T., & Tsilimigkas, G. (2025). The impact of built-up area dispersion on the cultural heritage of the region of the South Aegean, Greece. *ISPRS International Journal of Geo-Information, 14*(3), 97. DOI: 10.3390/ijgi14030097

Chorianopoulos, I., Tsilimigkas, G., Koukoulas, S., & Balatsos, T. (2014). The shift to competitiveness and a new phase of sprawl in the Mediterranean city: Enterprises guiding growth in Messoghia–Athens. *Cities (London, England), 39*, 133–143. DOI: 10.1016/j.cities.2014.03.005

Council of the European Union. (1992). Directive 92/43/EEC of 21 May 1992 on the conservation of natural habitats and of wild fauna and flora. *Official Journal of the European Union, L, 206*(7).

Derdemezi, E.-T., Tsilimigkas, G., & Kizos, T. (2021). Mining activity and island landscape issues: Evidence from Cyclades islands, Greece. *European Planning Studies, 30*(2), 384–404. DOI: 10.1080/09654313.2021.1958172

Dimopoulos, P., Drakou, E., Kokkoris, I. P., Katsanevakis, S., Kallimanis, A., Tsiafouli, M., Bormpoudakis, D., Kormas, K., & Arends, J. (2017). The need for the implementation of an ecosystem services assessment in Greece: Drafting the national agenda. *One Ecosystem*, *2*, e13714. DOI: 10.3897/oneeco.2.e13714

e-poleodomia. GIS platform for urban and settlement boundaries. Retrieved from http://gis.epoleodomia.gov.gr/v11#/

EEA (European Environment Agency). (2018). *Impervious Built-up 2018*. Copernicus Land Monitoring Service.

EL-STAT (Hellenic Statistical Authority). (2020). Retrieved from https://www.statistics.gr/en/home

Elmqvist, T., Zipperer, W. C., & Güneralp, B. (2015). Urbanization, habitat loss, and biodiversity decline: Solution pathways to break the cycle. In Mounsey, H., & Tomlinson, R. (Eds.), *Building Databases for Global Science* (1st ed.). Routledge., DOI: 10.4324/9781315849256

European Commission. (2011). *Our life insurance, our natural capital: An EU biodiversity strategy to 2020*. European Commission.

European Commission. (2020). The Natura 2000 network. Retrieved from https://ec.europa.eu/environment/nature/natura2000/index_en.htm

European Commission. (2020). *EU biodiversity strategy for 2030: Bringing nature back into our lives. COM, 2020*. European Commission.

European Parliament & Council of the European Union. (2009). Directive 2009/147/EC of the European Parliament and of the Council of 30 November 2009 on the conservation of wild birds (codified version). *Official Journal of the European Union, L, 20*, 7–25. https://data.europa.eu/eli/dir/2009/147/oj

European Parliament & Council of the European Union. (2009). Directive 2009/147/EC of 30 November 2009 on the conservation of wild birds (replacing Council Directive 79/409/EEC). *Official Journal of the European Union, L, 20*(7).

European Union. (1992). Council Directive 92/43/EEC of 21 May 1992 on the conservation of natural habitats and of wild fauna and flora. Official Journal of the European Communities, L 206/7.

Evans, D. (2006). The habitats of the European Union Habitats Directive. *Biology and Environment*, *106B*(3), 167–173. DOI: 10.1353/bae.2006.0032

Forest maps and wildlife park data. Retrieved from http://geodata.gov.gr

Geldmann, J., Barnes, M., Coad, L., Craigie, I. D., Hockings, M., & Burgess, N. D. (2013). Effectiveness of terrestrial protected areas in reducing habitat loss and population declines. *Biological Conservation, 161*, 230–238. DOI: 10.1016/j.biocon.2013.02.018

González-García, A., Palomo, I., Arboledas, M., González, J. A., Múgica, M., Mata, R., & Montes, C. (2022). Protected areas as a double-edged sword: An analysis of factors driving urbanization in their surroundings. *Global Environmental Change, 74*, 102522. DOI: 10.1016/j.gloenvcha.2022.102522

Greek Ministry of Environment and Energy. Natura 2000 areas. Retrieved from http://geoportal.ypen.gr/

Hellenic Ornithological Society. (2022). Important Bird Areas data. Retrieved from https://ornithologiki.gr/el/

Kallimanis, A. S., Touloumis, K., Tzanopoulos, J., Mazaris, A. D., Apostolopoulou, E., Stefanidou, S., Scott, A. V., Potts, S. G., & Pantis, J. D. (2015). Vegetation coverage change in the EU: Patterns inside and outside Natura 2000 protected areas. *Biodiversity and Conservation, 24*(3), 579–591. DOI: 10.1007/s10531-014-0837-9

Kati, V., Hovardas, T., Dieterich, M., Ibisch, P. L., Mihok, B., & Selva, N. (2015). The challenge of implementing the European network of protected areas Natura 2000. *Conservation Biology, 29*(1), 260–270. DOI: 10.1111/cobi.12366

Kizos, T., Koulouri, M., Vakoufaris, H., & Psarrou, M. (2010). Preserving characteristics of the agricultural landscape through agri-environmental policies: The case of cultivation terraces in Greece. *Landscape Research, 35*(6), 577–593. DOI: 10.1080/01426397.2010.519434

Kizos, T., Tsilimigkas, G., & Karampela, S. (2017). What drives built-up area expansion on islands? Using soil sealing indicators to estimate built-up area patterns on Aegean islands, Greece. *Tijdschrift voor Economische en Sociale Geografie, 108*(1), 35–52. DOI: 10.1111/tesg.12244

Kovács, E., Kelemen, K., Kalóczkai, A., Margóczi, K., Pataki, G., Gébert, J., Málovics, G., Balázs, B., Roboz, A., Krasznai Kovács, E., & Mihók, B. (2015). Understanding the links between ecosystem service trade-offs and conflicts in protected areas. *Ecosystem Services, 12*, 117–127. DOI: 10.1016/j.ecoser.2014.09.012

Lai, S., & Zoppi, C. (2017). The influence of Natura 2000 sites on land-taking processes at the regional level: An empirical analysis concerning Sardinia (Italy). *Sustainability (Basel), 9*(2), 259. DOI: 10.3390/su9020259

Luck, G. W., Ricketts, T. H., Daily, G. C., & Imhoff, M. (2004). Alleviating spatial conflict between people and biodiversity. *Proceedings of the National Academy of Sciences of the United States of America, 101*(1), 182–186. DOI: 10.1073/pnas.2237148100

Mammides, C., Zotos, S., & Martini, F. (2024). Quantifying the amount of land lost to artificial surfaces in European habitats: A comparison inside and outside Natura 2000 sites using a quasi-experimental design. *Biological Conservation, 293*, 110556. DOI: 10.1016/j.biocon.2024.110556

Mammides, C., Zotos, S., & Martini, F. (2024). Quantifying the amount of land lost to artificial surfaces in European habitats: A comparison inside and outside Natura 2000 sites using a quasi-experimental design. *Biological Conservation, 293*, 110556. DOI: 10.1016/j.biocon.2024.110556

Matos, A. T., Santos, C. J. F., & Chevalier, D. R. (2002). Fire and restoration of the largest urban forest of the world in Rio de Janeiro City, Brazil. *Urban Ecosystems, 6*(3), 151–161. DOI: 10.1023/A:1026164427792

McDonald, R. I., Forman, R. T. T., Kareiva, P., Neugarten, R., Salzer, D., & Fisher, J. (2009). Urban effects, distance, and protected areas in an urbanizing world. *Landscape and Urban Planning, 93*(1), 63–75. DOI: 10.1016/j.landurbplan.2009.06.002

McDonald, R. I., Kareiva, P., & Forman, R. T. T. (2008). The implications of current and future urbanization for global protected areas and biodiversity conservation. *Biological Conservation, 141*(6), 1695–1703. DOI: 10.1016/j.biocon.2008.04.025

McKenna, D., Naumann, S., McFarland, K., Graf, A., & Evans, D. (2014). Literature review, the ecological effectiveness of the Natura 2000 Network. ETC/BD report to the EEA, 30 pp. DOI: 10.13140/RG.2.1.4358.9288

Ministry of Environment and Energy, Directorate of Natural Environment Management and Biodiversity. (2024). Prioritized Action Framework (PAF) for Natura 2000 in Greece for the Multiannual Financial Framework period 2021–2027. Greece.

Mücher, S., Gerard, F., Olschofsky, K., Hazeu, G., Luque, S., Pino, J., Gregor, M., Wachowicz, M., Halada, L., Tompo, E., Kohler, R., Petit, S., Smith, G., & Kolar, J. (2006). Spatial impact of conservation sites (Natura 2000) on land cover changes. *Proceedings of the 2nd Workshop of the EARSeL SIG on Land Use and Land Cover.*

. National Strategy on the Protection of Biodiversity, April 2014. Government Gazette B 2383/08.09.2014.

NCMA (National Cadastre and Mapping Agency). (2023). Viewing orthophotos. Greek Cadastre. Retrieved from https://www.ktimatologio.gr

. (1986). OGG (Official Government Gazette) A′ 160/16.10.1986, Law No. 1650 on the Protection of the Environment. *Greek Official Gazette, 160*(Part I), 3257–3272.

OGG (Official Government Gazette) A' 245/2020, Law No. 4759/2020 on Spatial Planning, Urban Planning, and Other Provisions, Greek Official Gazette, 245, Part I, 2020. Available at: https://www.et.gr/

Papageorgiou, K., & Vogiatzakis, I. N. (2006). Nature protection in Greece: An appraisal of the factors shaping integrative conservation and policy effectiveness. *Environmental Science & Policy, 9*(5), 476–486. DOI: 10.1016/j.envsci.2006.03.003

Princé, K., Rouveyrol, P., Pellissier, V., Touroult, J., & Jiguet, F. (2021). Long-term effectiveness of Natura 2000 network to protect biodiversity: A hint of optimism for common birds. *Biological Conservation, 253*, 108871. DOI: 10.1016/j.biocon.2020.108871

Romero, A. C., Issii, T. M., Pereira-Silva, E. F. L., & Hardt, E. (2018). Effects of urban sprawl on forest conservation in a metropolitan water source area. *Revista Árvore, 42*(1), e420114. DOI: 10.1590/1806-90882018000100014

Seto, K. C., Golden, J. S., Alberti, M., & Ii, B. L. T. (2017). Sustainability in an urbanizing planet. *Proceedings of the National Academy of Sciences of the United States of America, 114*(34), 8935–8938. DOI: 10.1073/pnas.1606037114

Stratigea, A., & Katsoni, V. (2015). A strategic policy scenario analysis framework for the sustainable tourist development of peripheral small island areas: The case of Lefkada-Greece Island. *European Journal of Futures Research, 3*(1), 5. DOI: 10.1007/s40309-015-0063-z

Terkenli, T. (2005). Human activity in landscape seasonality: The case of tourism in Crete. *Landscape Research, 30*(2), 221–239. DOI: 10.1080/01426390500044408

Theodora, Y. (2019). Aegean Sea: Challenges and dilemmas in management and planning for local development in fragmented insular regions. *Heritage, 2*(3), 108. Advance online publication. DOI: 10.3390/heritage2030108

Tsilimigkas, G., & Derdemezi, E.-T. (2020). Unregulated built-up area expansion on Santorini Island, Greece. *European Planning Studies, 28*(9), 1790–1811. DOI: 10.1080/09654313.2019.1687656

Tsilimigkas, G., & Gourgiotis, A. (2023). Unregulated urban sprawl and spatial distribution of fire events: Evidence from Greece. *Environmental Hazards, 17*(5), 436–455. DOI: 10.1080/17477891.2018.1430554

Tsilimigkas, G., Gourgiotis, A., & Derdemezi, E.-T. (2022). Spatial planning incompetence to discourage urban sprawl on Greek islands: Evidence from Paros, Greece. *Journal of Coastal Conservation*, *26*(2), 11. Advance online publication. DOI: 10.1007/s11852-022-00859-2

Tsilimigkas, G., Kizos, T., & Derdemezi, E.-T. (2021). Urbanization and land cover change in Aegean islands: Assessing the socio-economic impact of tourism development. *Land (Basel)*, *10*(5), 545. DOI: 10.3390/land10050545

Watson, J. E. M., Dudley, N., Segan, D. B., & Hockings, M. (2014). The performance and potential of protected areas. *Nature*, *515*(7525), 67–73. DOI: 10.1038/nature13947

Chapter 9
The Impact of Tourism on the Traditional Settlement of Spetses

Despina Dimelli
http://orcid.org/0000-0002-0234-1293
Technical University of Crete, Chania, Greece

Alexandra Ntousi
http://orcid.org/0009-0002-6353-0204
Technical University of Crete, Greece

Eleni Prifti
http://orcid.org/0009-0007-8458-665X
Technical University of Crete, Kalivia, Greece

ABSTRACT

The traditional settlement of Spetses is a rich palimpsest of historical and architectural value, on which the social, political and economic details of the society are reflected throughout the last centuries. The seafaring character of the Spetsian society is clearly visible in the 18th century, when the traditional character of the settlement was formed, its transformation into a high society tourist resort in the 19th century, as well as its subsequent inseparable connection with tourism until this day. The research examines how the urban structure of the settlement, the typologies and the morphological characteristics of the traditional architecture are affected by the phenomenon of tourism, as it changes its character and form. Also, the research focuses on the strict policies that targets the conservation of the traditional settlement, arising from the great influx of the mass tourism of the 20th century and the consequent uncontrolled mass building construction.

DOI: 10.4018/979-8-3373-1554-6.ch009

INTRODUCTION

Spetses is a small island in the Argosaronic Gulf of great historical significance due to its contribution to the Greek Revolution of 1821. The island's primary residential complex is the traditional settlement of Spetses *(Figure 1)*, which is the subject of this study. Today, Spetses's economy is primarily based on tourism, with the tertiary and secondary economic sectors being particularly well-developed as a result (services, craft industries, and construction activities to meet the needs of tourism).

This chapter investigates how tourism has influenced the configuration of the settlement, focusing on the architectural and urban planning changes that it has brought to the built environment and the traditional forms of the settlement. The sections to be developed include the historical context and the formation of the traditional settlement, as well as the history of tourism on the island. Furthermore, the impact of tourism over time on the settlement's architecture and urban planning is analyzed, along with the current relationship between tourism and Spetses. Finally, the conclusions of the research are presented.

Figure 1. Aerial view of Spetses

https://www.flickr.com/photos/160866001@N07/48451224667/in/photostream/ (retrieved on 19/6/2023)

Methodology

The method of research material collection for this study was mainly based on primary and secondary literature review, through books, scientific articles, and academic chapters, and secondarily on online resources. Particularly useful was the historical material collected in the National Library of Greece, which mainly concerned old or rare books. Part of the research also involved the collection of architectural and urban plans, diagrams, maps, and photographs of the settlement in various historical phases up to the present, as well as their comparison. Subsequently, the research team conducted research where the key existing urban characteristics of the settlement under study were mapped and diagrammed. This research proved crucial for the delineation of the current condition of the town, its public spaces and building stock, aiming to draw conclusions related to the preservation or alteration of its traditional character. All of these methods provided a comprehensive understanding of the relationship between tourism and the architectural transformation of the settlement.

Historical Context and the Formation of the Traditional Settlement

The traditional character of Spetses was shaped during the 18th century, with its significant development largely attributed to the close relationship of its inhabitants with the sea. Until the early decades of the 19th century, the majority of the population was directly or indirectly involved in shipping and maritime trade (Charitatos, 2004). Spetses's experienced merchant fleet enabled the island to be the first to participate in the Greek Revolution of 1821, and along with ships from Hydra and Psara to form Greece's naval warfare force. During the early years of the independent Greek state, the island's economic prosperity led to further social and intellectual development. The population of the settlement grew, and Spetses flourished both economically and culturally. However, after 1870, with the prevalence of steam-powered ships, the decline of sail-based shipping began. As a result, Spetses entered a period of decline, with a significant portion of the population migrating, due to the island's diminished economic opportunities. Despite this downturn, the legacy of Spetses' maritime past continued to shape the settlement's architectural and urban character, with many of the island's traditional buildings reflecting the prosperity of its earlier naval era.

During the period of flourishing maritime activity, the center of the settlement's activities developed near the sea, primarily around the port. The maritime community that formed on the island had a clear social structure based on the privileges of each group (Galettas & Mpouzoumpardi, 2004). This distinction determined the arrangement of the settlement into districts with different characteristics, leading to the grouping of residents according to their professional occupations and social activities.

Typically, around the large houses (mansions), smaller residences are clustered with clearly defined property boundaries, ensuring proximity among family members or groups with common interests, as well as to their workplaces (Galettas & Mpouzoumpardi, 2004) *(Figure 2)*. In the district of the seafarers, the houses are small; however, their need for space leads to the appropriation, but not privatization, of the public space between them by different families, transforming parts of it into semi-public areas (Mpia, 2018) *(Figure 3)*.

Figure 2. Typical layout of a Captain's home

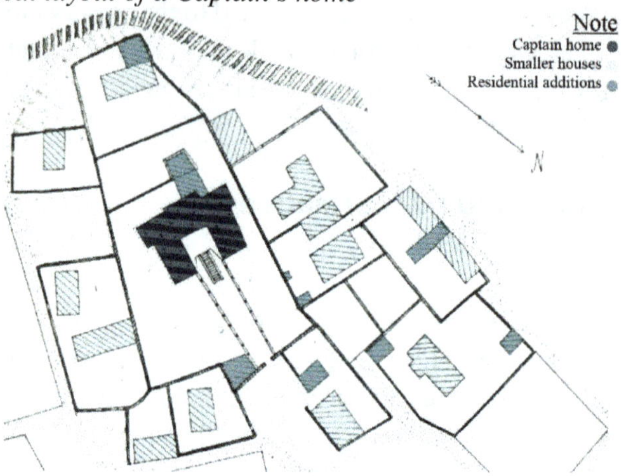

Figure 3. Shared public spaces with the surrounding smaller houses in mariner's quarters

The traditional architecture of Spetses has shaped and evolved its form over the years according to the prevailing political, economic, and social conditions. The residence includes all the buildings that serve the daily needs of the family, characterized by simple and strict geometry (Galettas & Mpouzoumpardi, 2004) *(Figure 4)*. The house is inextricably linked to its surrounding environment, which is typically protected by high stone walls, preventing direct contact with the public spaces. The materials used in the constructions are sourced from various locations within Spetses, neighboring areas of the Peloponnese and the surrounding islands (Filindra, 1985).

Figure 4. Traditional house

https://www.facebook.com/photo/?fbid=10212736437899603&set=g.656940781088377 (retrieved on 28/9/2024)

Traditional residences are categorized into three types based on the shape of their floor plans: iota-shaped (I or "makrinari"), gamma-shaped (Γ), and pi-shaped (Π). After 1830, classicism led to a square-shaped floor plan. Classicism was gradually and smoothly integrated into the settlement's environment, ultimately being assimilated into it, marking the end of the traditional period in the early 20th century.

THE HISTORY OF TOURISM ON THE ISLAND

The relationship between the island of Spetses and tourism can be historically situated in the early twentieth century, specifically with the construction and subsequent operation of the emblematic "Poseidónion" hotel in 1914. This landmark investment marked a turning point in the island's socio-economic trajectory, serving not merely as a luxurious accommodation facility but also as a symbol of Spetses' potential integration into the wider networks of elite European tourism. In the decades that followed, the island gradually emerged and consolidated its status as a privileged leisure destination, attracting members of the upper classes from both Greece and various European countries (Charitatos, 2004). The island's social and cultural evolution during this formative period underscored the increasing significance of tourism as a principal economic activity, capable of shaping both the urban fabric and the socio-economic dynamics of the local community.

Nevertheless, the profound economic crisis of the 1930s, exacerbated by the catastrophic impacts of the Second World War, severely disrupted tourism-related activities on Spetses. The broader context of economic hardship, material destruction, and political instability inevitably hindered the development of tourism infrastructure and services during this era. However, in the post-war period, the rapid acceleration of Greece's national economy created new conditions conducive to the revitalization of tourism activity on the island. This phase was marked by the expansion of hospitality infrastructure, the diversification of tourism services, and the progressive adaptation of the island's tourism offer to the evolving preferences and expectations of contemporary visitors. This resurgence represented a critical juncture in the evolution of Spetses' tourism sector, aligning the island with broader processes of modernization and internationalization within the global tourism industry.

A significant institutional development occurred during the 1950s, with the establishment of the Spetses Tourist Association, a formal body dedicated to the systematic promotion and organization of tourism on the island. Concurrently, the construction of large-scale hotel facilities, including the "Xenía" and "Nisiá" hotel units, signaled the island's entry into the era of mass tourism, characterized by increasing visitor flows and the intensification of tourism development. This expansion culminated in the period between the 1970s and the 1990s, when both domestic and international tourist arrivals reached unprecedented levels. During this time, the total capacity of the island's hospitality sector proved barely sufficient to accommodate the approximately 10,000 visitors who arrived during peak tourism periods, particularly in the summer months (Delenda, 2013). Nevertheless, from 1995 onwards, Spetses experienced a gradual decline in mass tourism, initiating a period of transition towards a different tourism model that sought to reposition the island within a changing tourism landscape.

Since the early 2000s, the island's tourism dynamics have been predominantly shaped by the domestic market, largely due to its geographical proximity to Athens. This spatial advantage has facilitated an increase in short-term visits, particularly on weekends, public holidays, and during the summer season. In recent years, new forms of tourism have progressively emerged on Spetses, reflecting broader trends in the pursuit of sustainable and differentiated tourism experiences. Specifically, the development of cultural tourism, sports tourism, and event based tourism has aimed to enrich the overall visitor experience, providing alternative modes of engagement with the island's natural environment, historical heritage, and local culture. These strategic initiatives aspire not only to diversify the island's tourism offer but also to address issues of seasonality by extending the tourist season beyond the traditional summer months. In doing so, they contribute to the promotion of a more sustainable, balanced, and resilient tourism development model, aligned with contemporary principles of place-based tourism and cultural valorization.

THE IMPACT OF MASS TOURISM OVER TIME ON THE ARCHITECTURE AND URBAN PLANNING OF THE SETTLEMENT

Throughout the period of intense tourism development, the settlement was shaped to meet new needs, with the built environment densifying and expanding to the west and southeast *(Figure 5)*. New buildings occupied available vacant land and public spaces, while construction continued on fragmented existing properties too *(Figure 6)*. As a result, a significantly denser urban fabric was created in many areas, leading even to the complete disappearance of public spaces in some cases.

Figure 5. The development of the settlement enduring the 2nd half of the 20th century

Figure 6. Densification through the construction of new residences in the empty private and public spaces

The majority of new buildings accommodated hospitality businesses, including large hotel units and smaller accommodations such as apartments, as well as low-capacity dining and entertainment establishments. The large hotel units concentrated along the coastal front of the settlement, where their significant mass and unique morphological characteristics distinguished them from the prevailing architectural forms *(Figure 7)*.

Figure 7. Hotel "Spetses", a typical example of tourist development of the 20th century

https://cf.bstatic.com/xdata/images/hotel/max1024x768/44238242.jpg?k=12fbfdefe1acb92936dccb88a50cc7b9d24d1ff20740d6d8f9f498dee2a98502&o=&hp=1 (retrieved on 23/6/2023)

For the housing of smaller businesses, existing structures were restored and adapted according to the contemporary practices of the time, while many were often extended vertically with the addition of extra floors *(Figure 8)*. Notably, the modern construction standards and rapid pace of building did not allow for the adherence to traditional practices in terms of typologies, morphological characteristics, and materials.

Figure 8. Densification through the construction of additional floors in an existing traditional building

https://www.google.com/maps/@37.2659348,23.1556223,3a,90y,133.47h,104.77t/data=!3m8!1e1!3m6!1sAF1QipOJL9o1lbBQhOY90L5jgo1FLvr6AWzOx9fTY!2e10!3e11!6shttps:%2F%2Flh5.googleusercotent.com%2Fp%2FAF1QipOJL9o1lbBQhO-Y90L5j-go1FLvr6AWzOx9fTY%3Dw203-h100-k-no-pi-1.4554754-ya283.95984-ro-5.636077-fo100!7i5120!8i2560?entry=ttu (retrieved on 28/9/2024)

The rapid and often unregulated development of mass tourism on the island of Spetses, particularly during the latter half of the twentieth century, was accompanied by an increasingly pronounced trend towards sprawling and often incompatible construction activity. This phenomenon, driven by the intensification of tourism flows and the corresponding demand for infrastructure and accommodation, posed significant threats to the architectural integrity, spatial coherence, and traditional character of the island's built environment. In response to these emerging challenges, it became evident that the implementation of regulatory frameworks and protective measures was essential in order to mitigate the risks of irreversible alteration and degradation of the island's historical and architectural heritage.

Consequently, beginning in 1969, a systematic effort was undertaken by the competent authorities to introduce a series of planning regulations, construction controls, and mobility restrictions, specifically designed to safeguard the distinct urban morphology and cultural landscape of Spetses. These regulatory interventions sought to manage development pressures while ensuring the preservation of the island's traditional character and aesthetic identity. In particular, given the organically evolved and compact urban fabric of the settlement, characterized by narrow streets and small-scale structures, vehicular traffic was significantly restricted. Private car circulation was prohibited within the core urban area, limiting motorized transport primarily to two-wheeled vehicles, such as motorcycles and mopeds, which were considered more compatible with the spatial constraints of the traditional settlement.

Further regulatory developments took place in the early 1980s, when a comprehensive urban planning framework was formally established, aiming to provide a coherent basis for the future development and management of the island's urban space (Figure 9). This planning framework introduced specific and detailed restrictions regarding the preservation and enhancement of the architectural and cultural character of Spetses. A central feature of this regulatory approach was the division of the settlement into distinct sectors, each characterized by particular morphological, functional, and historical features. The delineation of these sectors was based on existing conditions and patterns of land use, urban form, and architectural typology.

Corresponding to this sectoral division, differentiated building regulations and land-use restrictions were defined and implemented. These planning measures aimed to preserve not only the existing residential structure and architectural typology but also the distinctive features, spatial arrangements, and environmental qualities of each sector. Specifically, the regulatory framework distinguished between the commercial core of the settlement, the primary residential areas, and the peripheral zones surrounding the central nucleus of the town. In the residential sectors, particular emphasis was placed on restoring the historical relationship between buildings and their outdoor spaces — a relationship that had been considerably disrupted during the preceding decades of unregulated tourism-driven development. Efforts were made to re-establish the traditional spatial organization of courtyards, gardens, and semi-open spaces, which had historically constituted integral elements of the island's architectural identity. In this manner, the urban planning interventions of the 1980s represented a critical turning point in the attempt to reconcile tourism development with the preservation of Spetses' cultural heritage, architectural authenticity, and overall spatial coherence.

Figure 9. Urban plan – 1981

At the same time, restrictions were established regarding the typological (composition, scale, proportion) and morphological characteristics of new constructions and existing buildings undergoing renovation, which should adhere to traditional standards.

THE IMPACT OF TOURISM ON THE TRADITIONAL SETTLEMENT OF SPETSES TODAY

Since the 2000s, there have been no large-scale expansions. Residential development has focused on new constructions, scattered throughout the urban fabric, as well as individual buildings in the surroundings, which are intended for seasonal use. The commercial uses of the town form a cluster in the harbor area and small linear zones between the residences. The placement of these uses follows the traditional development pattern of commercial activity within the settlement, and in many cases, it is observed that the buildings for such purposes maintain their commercial use over time *(Figure 10)*. The dining and recreational services, primarily serving tourism, also develop linearly along the coastal axis.

Figure 10. Traditional cafe in the port of Ntapia throughout time

https://www.facebook.com/917531414996998/photos/a.921058201310986.1073741849.917531414996998/1012347008848771/?type=3&theater (retrieved on 28/9/2024) https://www.facebook.com/photo/?fbid=386176848390699&set=g.656940781088377 (retrieved on 28/9/2024) https://www.google.com/maps/@37.2673509,23.1558202,3a,90y,112.05h,98.81t/data=!3m7!1e1!3m5!1sAF1QipMortbM--lE8bisypZDEgJM6Rpye18WFYYnaeE!2e10!3e11!7i5120!8i2560?entry=ttu (retrieved on 18/6/2023)

To meet the ongoing accommodation needs of the visitors, alongside the old large hotels and smaller room complexes, a multitude of individual accommodations have emerged *(Figure 11)*. The large hotel units of the previous century remain active, but no new ones are being created. Additionally, the smaller-scale complexes, mainly remnants of the past tourist era, are scattered throughout the interior of the town *(Figure 12)*.

Figure 11. Tourist accommodation in 2023, recorded by authors

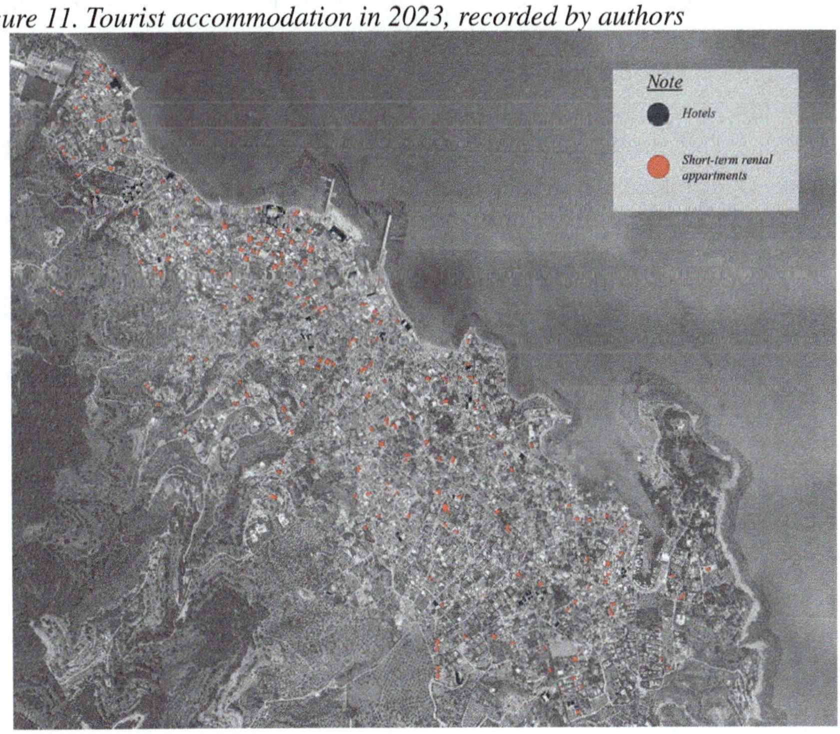

Figure 12. Small establishment of the 20th century attempting to imitate the traditional architectural forms

https://www.facebook.com/photo/?fbid=387886896833697&set=a.387886863500367 (retrieved on 23/6/2023)

In recent years, the urban and socio-economic landscape of Spetses has undergone substantial transformations, primarily driven by the intensified expansion of short-term rental accommodations within the boundaries of the settlement. A growing number of such properties have emerged throughout the urban area, responding to the increasing demand generated by both domestic and international tourism. Notably, a significant proportion of these rental units are housed within pre-existing vacant buildings, many of which possess considerable historical, architectural, or cultural value. The adaptation of these structures for tourism purposes often involves restoration or maintenance interventions, contributing — albeit selectively — to the preservation of the built heritage, albeit framed within a market-oriented logic centered on visitor accommodation and economic exploitation.

This trend towards the proliferation of short-term rental units is further accentuated by the widespread presence of luxury vacation residences — primarily private villas — which are dispersed both within the urban fabric of the main settlement and across peripheral or semi-rural areas of the island. These high-end properties aim to provide exclusive and customized hospitality services, targeting affluent segments of the tourist population. Their increasing number and prominence within the spatial and economic structure of Spetses has rendered them a critical factor in shaping the island's tourism dynamics, particularly in relation to the intensification of seasonality. Indeed, the clustering of such luxury accommodations amplifies the cyclical nature of tourism flows, concentrating economic activity and population presence almost exclusively within the summer period.

The pronounced dependence of the local economy on tourism-related activities exerts a profound influence not only on the island's spatial organization but also on the patterns of social life, demographic composition, and the availability of services throughout the year. According to the 2021 national census (ELSTAT), the permanent population of Spetses amounts to 3,748 residents. However, during the summer months, this figure more than doubles, with the total number of visitors frequently surpassing 11,000 individuals. This dramatic seasonal fluctuation significantly alters the rhythms of everyday life and reshapes the socio-economic profile of the settlement. The vast majority of commercial establishments — including retail businesses, hospitality venues, and dining facilities — operate on a seasonal basis, directly responding to the fluctuating demand generated by tourism. As a result, the permanent residents of Spetses face considerable limitations regarding access to essential services, commercial functions, and recreational opportunities during the off-season months. Notably, key areas such as the waterfront, which serve as focal points of tourist activity during the summer, become largely deserted in the winter period, contributing to a sense of spatial and social fragmentation.

Moreover, the concentration of population and tourist activity within a relatively short time frame places substantial pressure on the island's infrastructure and natural resources. During peak tourist periods, issues related to the adequacy of basic provisions — most notably water supply — become increasingly critical. The island's limited natural resources, coupled with the surge in temporary population, lead to recurring problems of water scarcity, necessitating reliance on external resources, particularly from neighboring regions of the Peloponnese.

In terms of urban structure and spatial organization, the current layout of Spetses is characterized by a notable scarcity of systematically designed and managed public spaces. The few instances of organized public areas are predominantly situated along the coastal front, functioning as key reference points and possessing a supra-local significance (Figure 13). However, these spaces face significant challenges during the summer season, as they are often appropriated by commercial dining establishments, which install temporary shading structures and expand their seating arrangements into the public realm (Figure 14). This privatization of public space reduces the availability and functionality of these areas for residents and non-consumptive uses.

Beyond the coastal front, the remaining public spaces within the urban area appear largely degraded, under-maintained, and insufficiently integrated into the everyday life and functioning of the settlement. Their fragmented distribution and low-quality design limit their capacity to support social interaction, recreation, or public life beyond tourism-driven consumption.

Simultaneously, mobility within the urban area is shaped by the regulatory restrictions imposed on vehicle circulation. While the prohibition of private cars within the dense urban fabric was intended to preserve the traditional character of the settlement, it has led to the widespread and often excessive use of motorized two-wheelers — particularly motorcycles — by both residents and tourists. During peak hours in the summer season, the coastal road is temporarily converted into a pedestrian zone, accommodating the high density of tourist activities and enhancing walkability (Vassi & Christodoulopoulou, 2014). Nevertheless, the overall functionality of the public space network remains compromised by the unregulated parking of vehicles and motorcycles in every available vacant spot, both in formal on-street locations and in informal, often inappropriate spaces. This situation exacerbates the pressures on the urban environment, contributing to visual clutter, spatial fragmentation, and a deterioration of the island's public realm.

Figure 13. Supra-local public square in front of "Poseidónion" hotel

https://www.skantzostravel.com/wp-content/uploads/2023/06/%CE%A7%CF%89%CF%81%CE%AF%CF%82%CF%84%CE%AF%CF%84%CE%BB%CE%BF.jpg (retrieved on 24/6/2023)

Figure 14. Occupation of public space by restaurants and cafes in the port of Ntapia

https://hotel-roumani-spetses-island.ibooked.gr/data/Photos/OriginalPhoto/12553/1255363/1255363237/Hotel-Roumani-Spetses-Island-Exterior.JPEG (retrieved on 24/6/2023)

CONCLUSION

Since the early 20th century, tourism has constituted a fundamental driving force in the spatial, social, and economic development of Spetses, profoundly shaping both its urban morphology and its socio-cultural dynamics. The evolution of tourism on the island can be broadly distinguished into two dominant periods, each exerting distinct influences on the political, economic, and social landscape of the settlement, as well as on its architectural and urban configuration.

The first significant period of tourism-related transformation emerged during the 1950s, coinciding with the rise of mass tourism across many Mediterranean destinations. The increasing influx of visitors to Spetses during this time rendered the adaptation of the existing urban fabric to the new demands of tourism not only necessary but inevitable. However, this period was characterized by a lack of comprehensive spatial planning and limited awareness of the potential long-term impacts of unregulated development. The response to the growing tourism flows was largely shaped by ad-hoc interventions and fragmented decisions, often driven by the immediate economic interests of landowners and entrepreneurs rather than by coherent planning strategies or preservation concerns.

As a consequence, this early phase of tourism expansion resulted in considerable transformations of the traditional built environment. Alterations affected multiple dimensions of the settlement, including its architectural typologies, morphological features, and overall urban structure (Figure 15). In particular, large-scale constructions were introduced, frequently diverging from the prevailing vernacular architectural forms and the spatial logic of the traditional urban fabric. These developments often disrupted the aesthetic coherence and typological continuity of the settlement, reflecting the dominance of tourism-driven development over heritage conservation priorities during this era.

Nevertheless, despite their initial incongruity with the traditional architectural language of Spetses, many of the structures erected during this early phase of mass tourism have, over time, acquired a particular symbolic and functional significance. They have become integrated into the spatial identity of the island, serving as enduring markers of its post-war tourism development trajectory and representing the complex interplay between modernization pressures and local heritage.

From the 1980s onwards, the tourism development model of Spetses underwent a gradual but notable shift towards more regulated and preservation-oriented approaches. This transition was facilitated by the introduction of specific urban planning tools, regulatory frameworks, and building restrictions aimed at safeguarding the architectural and cultural heritage of the settlement. New construction activities were required to comply with these protective regulations, ensuring a greater degree of compatibility between contemporary architectural interventions and the existing traditional built environment (Figure 16).

This period marked the institutionalization of heritage conservation principles within the planning and development processes of the island. New buildings were designed to respect the traditional character of Spetses in terms of scale, form, materials, and stylistic elements, contributing to the re-establishment of visual and typological continuity within the urban landscape. Furthermore, the increasing awareness of the cultural and economic value of the island's heritage encouraged efforts to restore and reuse existing structures, aligning tourism development with broader principles of sustainability and cultural preservation.

Overall, the evolution of tourism in Spetses reflects the broader tensions and challenges faced by many insular Mediterranean settlements — oscillating between the pressures of tourism-induced growth and the imperatives of heritage conservation. The two distinct periods of tourism development on the island illustrate different approaches to managing these dynamics: an initial phase marked by rapid and largely unregulated expansion, followed by a more recent phase characterized by regulatory control and the selective integration of heritage values into tourism-oriented development practices.

Figure 15. Typical building of the 20th century that contributes to the morphological and typological deterioration of the traditional settlement

https://www.google.gr/maps/@37.2588317,23.1641189,3a,75y,270.18h,86.39t/data=!3m7!1e1!3m5!1sAF1QipPgplo4qg26nPbwFuPfUrQO8ldSPOc9MaK3A!2e10!3e12!7i5120!8i2560?entry=ttu (retrieved on 24/6/2023)

Figure 16. New buildings designed to blend in the traditional residential landscape

https://www.andro.gr/zoi/speteses-potiropoulos/ (retrieved on 28/9/2024)

Contemporary efforts for the preservation and enhancement of the traditional qualities of Spetses continue to operate within the framework of the existing regulatory restrictions, which have been institutionalized over recent decades. Despite the enduring intensity of tourism activity, this regulatory framework — coupled with increased social awareness of the value of local heritage — has been relatively successful in mitigating the risk of further degradation or irreversible alteration of the island's historical and architectural character.

Architectural typologies, morphological features, and the traditional spatial organization of the settlement remain largely intact. A significant proportion of tourism-related functions are accommodated within preserved or restored existing structures, contributing to the safeguarding of the built heritage while simultaneously enhancing the visual identity and aesthetic appeal of the settlement. In this regard, tourism development, particularly when oriented towards the adaptive reuse of heritage buildings, functions not only as an economic driver but also as a mechanism for cultural preservation and landscape valorization.

Moreover, the spatial distribution of public uses reflects a conscious attempt to reconcile the imperatives of heritage protection with the functional demands of contemporary urban life. The restriction of non-residential uses within the core of the settlement supports the maintenance of a predominantly residential character in certain areas, limiting the pressures of commercialization and promoting the continuity of local everyday practices.

At the same time, however, tourism represents the principal vector for economic revitalization, providing employment opportunities for the permanent population and sustaining the viability of commercial and service sectors. The emergence and expansion of residential, dining, and hospitality services have reinforced the island's economic base, particularly in view of the limited alternatives available within the insular context.

Nevertheless, the presence of tourism is not without its ambivalent or adverse consequences. The most pronounced negative impact is associated with the intensification of seasonal fluctuation, which has given rise to a 'two-speed' town. This phenomenon manifests in the spatial and functional polarization of Spetses between the densely populated summer months and the significantly quieter off-season period. The needs and everyday life of the permanent population — which numbers 3,748 residents (2021 Census, ELSTAT) — are often subordinated to the demands of seasonal visitors, leading to a mismatch between the infrastructure provided and the actual needs of year-round inhabitants.

Additionally, problems within the mobility and transport network further complicate daily life. While restrictions on vehicle circulation have been implemented to preserve the traditional urban fabric, they have simultaneously resulted in the extensive use of motorized two-wheelers by both residents and tourists. This widespread reliance on motorcycles generates notable environmental and noise pollution, contributing to the deterioration of the urban landscape and adversely affecting the quality of life.

Moreover, the scarcity of available public spaces and urban squares is exacerbated during the summer months, as many of these areas are informally appropriated for parking purposes. This practice undermines the social function of public space, preventing its use as a locus of social interaction, leisure, and community life. The gradual alienation and functional erosion of public space reflect deeper processes of social transformation, marked by the estrangement of community members and the growing predominance of temporary residents or visitors within the settlement.

Neighborhood gathering spaces, traditionally vital for fostering social cohesion and everyday interaction, are progressively disappearing. The urban structure of Spetses is increasingly oriented towards the coastal front, which operates as the primary space of collective life, catering simultaneously to local residents and the supra-local demands of tourism (Figure 17). This spatial concentration of social activity reinforces patterns of fragmentation within the urban fabric, diminishing the vibrancy and resilience of the town's interior spaces and neighborhood networks.

Figure 17. The seafront esplanade summer

https://www.dreamingreece.com/greek-islands/argosaronic/spetses-island (retrieved on 24/6/2023)

In conclusion, tourism exerts a multifaceted influence on both the traditional settlement and the social fabric of Spetses, generating a complex interplay of positive and negative effects. To a considerable extent, the regulatory framework implemented in previous decades has proved effective in safeguarding the distinctive character of the urban fabric, ensuring that new development occurs in harmony with the island's architectural and cultural heritage. However, while the preservation of traditional identity remains a central objective, it is equally imperative to acknowledge and address the evolving needs and aspirations of contemporary society.

Accordingly, the future development of Spetses requires a dynamic and adaptive approach to spatial planning and heritage management — one that maintains a delicate balance between the protection of cultural and architectural values and the enhancement of the island's functional capacity and quality of life. Continuous monitoring, the periodic revision of existing regulations, and the incorporation of contemporary sustainability principles are essential in fostering a resilient and vibrant community.

For tourism-dependent destinations such as Spetses, the challenge lies in moving beyond a static model of heritage conservation towards a more integrated and context-sensitive approach. Effective planning and policy interventions must aim not only to preserve the unique characteristics of the traditional settlement but also to support the vitality of everyday life for permanent residents. Ultimately, the goal should be to avoid the risk of transforming these historic settlements into mere open-air museums catering exclusively to visitors, and instead to promote their evolution as living, dynamic environments that simultaneously accommodate tourism and sustain local community life.

REFERENCES

Delenda, E. (2013). *The impacts of tourism on small communities: The case of Spetses* (Student project). Retrieved from: http://digilib.teiemt.gr/jspui

Mpia, N. (2018). *Socio-economic transformations on the island of Spetses and their impacts on the natural environment: A diachronic investigation* (Master's thesis). Retrieved from: https://apothesis.eap.gr/archive/item/155521

Vassi, A., & Christodoulopoulou, G. (2014). Planning for sustainability at the destination level through limiting accessibility on the island of Spetses, Greece. WIT Transactions on Ecology and The Environment. *Sustainable Tourism, VI*, 187–198. DOI: 10.2495/ST140151

Charitatos, P. (2004). *Unexplored Spetses*. Road.

Filindra, M. "Spetses," in D. Philippidis (Ed.), *Greek Traditional Architecture, Volume 4: Peloponnese A* (pp. 245-248). Athens: Melissa, 1987.

Galettas, I., & Mpouzoumpardi, M. (2004). *Spetses: History - Ethnography* (Vol. I). Spetsiot Union.

Chapter 10
Examining Touristification in the EU Regions through a Composite Indicator Methodology

Kostas Gourzis
https://orcid.org/0000-0001-7434-9270
University of the Aegean, Mytilene, Greece

Dimitris Psarologos
https://orcid.org/0009-0003-9440-0338
School of Social Sciences, Mytilene, Greece

George Sykas
https://orcid.org/0009-0004-9947-6398
School of Social Sciences, Mytilene, Greece

Stelios Gialis
https://orcid.org/0000-0001-5204-6009
School of Social Sciences, Mytilene, Greece

ABSTRACT

Over the past few decades, tourism has become a key economic activity, with the crises of the last 15 years making many peripheral economies in Europe and beyond increasingly dependent on it. However, empirical research on touristification at the regional level remains limited. To address this gap, we introduce a composite index that measures the territorial and social pressures from tourism's supply and demand,

DOI: 10.4018/979-8-3373-1554-6.ch010

as well as regions' dependence on tourism in terms of employment and output. Our analysis covers all EU NUTS-2 regions from 2009 to 2022 and identifies three main zones of touristification: the sunlust insular/seaside EU South, the Italian and Austrian Alps, and various urban destinations across Central and Northern EU. Ultimately, two key asymmetries emerge: one between tourism-oriented Southern EU and production-oriented Northern EU, and within the former, another between tourism-dependent Greece, Portugal, and Croatia, and the economically diverse Spain and Italy.

INTRODUCTION

Tourism has proven to be a crucial avenue for capitalist accumulation in a period otherwise characterised by successive crises (Bianchi, 2018). In the EU, when the combined forces of tariffs' flexibilisation, commodity and financial markets' unification and liberalisation, increasing cultural interdependence, and deindustrialisation intertwined with the effect of the Global Crisis of 2008/2009, they reshaped power dynamics within and across member countries and highlighted new sectoral priorities. Specifically, under intense recessive pressures, most Southern countries saw their industrial structures being dismantled, all the while tourism was exhibiting a notable resilience (Herod et al., 2022). This led them to develop a heavy dependence upon the latter, actively lowering entry barriers for institutional investors and global capital into local housing and tourism markets. Specifically, these strategies were initially spearheaded by the introduction of residence-by-investment programs (i.e., Golden Visas) and subsequently by a series of policies facilitating the short-term rental market of via online peer-to-peer platforms (e.g., Airbnb; Gourzis et al., 2019).

Against this backdrop, the study at hand conducts an empirical analysis of the geography of touristification on a pan-EU level based on an original composite indicator, covering the tumultuous period of the past 15 years, which has been characterised by successive crises. To frame touristification as our research problem and distinguish it from overtourism, we have drawn from Critical Urban Studies (Estevens et al., 2023). Moreover, to bolster our approach theoretically, we have considered the role of investment capital in constructing tourism spaces, drawing from Geographical Political Economy (Bianchi, 2018), and we have approach tourism areas' status within the frame of dynamic evolutionary trajectories, drawing from Evolutionary Economic Geography (Ma & Hassink, 2013).

Specifically, adhering to a rigorous CI methodology (OECD/EU/JRC, 2008), we develop the "Tourism Activity Intensity, Density and Dependence Composite Index" (TAIDD CI onwards), which comprises three dimensions: tourism's (i) social and (ii) territorial impact and (iii) the dependence that regions develop upon tourism.

Drawing from commonly adopted approaches within Tourism Studies in measuring demand and supply (see for instance Peeters et al., 2018), the first two dimensions refer to visitor numbers and bed places adjusted by population and area respectively. The third dimension comprises variables relevant to employment and output. The CI is calculated and mapped for (nearly) all EU NUTS 2 regions from 2009 to 2022.

In what follows, we first frame our research by operationalising the concept of touristification. Then, we lay out our methodology in detail and present our analysis results. We close with a brief discussion of our main findings.

DEFINITIONAL AND METHODOLOGICAL CONSIDERATIONS

While touristification has recently been mostly associated with the urban context—in particular, the uncontrollable proliferation of platform-mediated short-term rentals (Estevens et al., 2023)—it originates from earlier geographical work on the reorientation of economies towards tourism (Ojeda & Kieffer, 2020). Here, we suggest an operationalisation of touristification as a key concept for an empirical analysis, proposing four conditions that articulate it.

The first condition involves a notable increase in the supply of the tourism product, primarily achieved by expanding accommodation capacity—including converting long-term accommodation into short-term rentals—and aligning recreational offerings with tourists' preferences and aesthetics (Estevens et al., 2023). The second condition is a significant increase in the demand for the tourism product, encompassing both accommodation and recreation. This occurs because touristified areas are already at an advanced stage of their evolutionary path, having developed complex agglomeration economies comprising a diverse array of businesses (Papatheodorou, 2004). Given that touristification is characterised by both increased tourist traffic and the proliferation of tourism-related land uses (Gourzis et al., 2022), a third important condition is that there must be a relative balance between supply and demand. The final, and most important, condition is the deepening dependence that the economic base of touristified areas develops on tourism, manifesting as the dominance of tourism-related output and labour. Based on the above, we structured the original TAIDD CI as shown in *Table 1*. The index adjusts tourism's supply and demand according to population and area to account for both social and territorial pressures. In this way, it considers a variety of tourism types and, consequently, markets; for instance, urban tourism (e.g., heritage, conference/convention-related, shopping), which exhibits higher density compared to rural tourism (e.g., alpine, sun-sea-sand), which demonstrates much higher intensity. The third dimension has been given a heavier weight due to its relatively higher importance. All data used for the TAIDD CI have been retrieved from Eurostat except for tourism's direct GDP. The latter was drawn from the World Bank[1] due to data scarcity, which stems from inherent difficulties in framing tourism activity and collecting such data (Eurostat, 2023; UNWTO, 2022). Both arrivals and bed places refer to all categories of tourist accommodation, namely NACE codes I55.1-3. However, National Statistical Authorities can exclude establishments with less than ten bed places from observation[2], which can omit part of the short-term rental market. Arrivals were preferred over the nights spent and net occupancy rate to filter out longer stays that are not necessarily related to the type of touristification investigated here (e.g., digital nomadism; Estevens et al., 2023).

The temporal scope of the analysis covers the years from 2009 to 2022, a tumultuous period characterised by the 2008/2009 Global Crisis, the Great Recession, which peaked around 2013/2014, an anaemic recovery during the second half of the 2010s, and the COVID-19-induced recession from 2020 to 2022 (Herod et al., 2022).

Table 1. TAIDD CI dimensions, variables, calculations, and weights

Dimension	Variable	Calculation	Weight	
Intensity	BEDS/POP	Bed places per inhabitant	0.3	0.15
	TOUR/POP	Tourist arrivals per inhabitant		0.15
Density	BEDS/AREA	Bed places per square kilometer	0.3	0.15
	TOUR+POP/ AREA	Tourist arrivals plus inhabitants per square kilometer		0.15
Dependence	EMPL%	Share of NACE I (accommodation and catering) in total employment	0.4	0.20
	TDGDP%	Share of tourism's direct contribution to GDP		0.20

The geographical scope encompasses all EU27 NUTS 2 regions except for those being geographically irrelevant (i.e., French overseas regions, Spanish autonomous cities in northern Africa) and FI20 (i.e., Åland in Finland) that constantly exhibits missing values. The 2009 NUTS 2 nomenclature was followed throughout the study period, meaning that changes in a given country's regions and/or their borders during 2009-2023 have been ignored. Specifically, in Croatia the initial HR04 is kept by summing up HR02, HR05 and HR06 (which appeared from 2019 onwards); in Poland, the initial PL12 is kept by summing up PL91 and PL92 (appearing in 2013); in Lithuania, the initial LT00 is kept by summing up LT01 and LT02 (appearing in 2013); in Hungary, the initial HU10 is kept by summing up HU11 and HU12 (appearing in 2013). Ireland, where new regions occurred by moving borders and dividing previous ones, is treated as a single-region country (similarly to Kelly et al., 2022).

In cases of missing data regarding arrivals and bed places, alternative official sources were utilised. Data for DE14, NL31, and NL32 in 2009, as well as DE50, DEG0, and PT20 in 2014, were retrieved from the respective National Statistical

Authorities. However, for Ireland, the missing data in 2022 were calculated by multiplying the 2021 values by the change from 2021 to 2022 for the EU. Similarly, regarding employment, the missing data for FRM0 in 2009 were estimated by multiplying the 2010 values by the national change from 2009 to 2010. The missing data for DEC0, PL52, PL72, and PL84 in 2022 were estimated by multiplying the 2019 values (we avoided 2020 and 2021 due to the pandemic) by their respective national changes from 2019 to 2022.

The data were normalised using z-scores; although this method is not preferable when outliers distort the results (OECD/EU/JRC, 2008), our analysis seeks to highlight these outliers, as it focuses on touristification. To check for multicollinearity issues, Pearson's R was calculated between all variables (see *Table 3* in the Appendix). The matrix there reveals strong correlations between supply and demand-related variables, primarily between BEDS/POP and ARRIV/POP, as well as BEDS/AREA and ARRIV+POP/AREA. However, of the 15 combinations examined, only 2 exhibit a strong correlation and another 2 a very strong correlation in most years. In fact, these correlations indicate cause-and-effect links that should not be overlooked in our analysis, given that touristification is defined as a relative balance between intense supply and demand. Considering also that, despite their correlations, these variables refer to crucial parameters of the phenomenon under study (OECD/EU/JRC, 2008), all of them were retained. Subsequently, to check for redundant dimensions and variables, Principal Components Analysis was conducted, showing that no redundancies were present, as almost all variables had a single strong component loading (i.e., above 0.6 in absolute value; see *Table 4* in the Appendix). Lastly, adjusted according to their attributed weights (see above), the TAIDD CI was calculated by linearly aggregating all variables. Linear aggregation was chosen due to the complementarity of the index's dimensions (Mendola & Volo, 2017); specifically, regions with lower values in the density dimension than in the intensity dimension (e.g., rural tourist areas) can be as touristified as those with the opposite balance (e.g., urban tourist areas).

Critical steps in the OECD/EU/JRC (2008) guidelines related to the decomposition of a CI into its original variables, the calculation of alternative versions to assess robustness (or sensitivity to specific variables), and the correlation between CI scores and other metrics have been undertaken but are not included here due to space limitations. All calculations are available upon request.

EMPIRICAL ANALYSIS OF THE TAIDD CI

In what follows, we scrutinise the results for each year and the changes in the index over the course of the study period. At this point, we must emphasise that the

TAIDD CI provides a relative picture of touristification; due to data normalisation, the average value of all scores is 0 in any given year. Additionally, it is important to note that changes in scores do not necessarily indicate an absolute increase in the extent of a region's touristification but rather reflect an enhanced significance relative to other regions' touristification.

With this in mind, *Figure 1* reveals three main zones of touristification that do not shift significantly throughout the study period. The first zone encompasses the EU South, which consistently constitutes the most touristified part of the EU. Specifically, the TAIDD CI exhibits exceptionally high values mainly in regions with significant sunlust destinations: insular Greece (the South Aegean, the Ionian Islands, and Crete), Malta, Adriatic Croatia, insular Spain (the Balearic and Canary Islands), and insular, southern, and metropolitan Portugal (Madeira, Algarve, Lisbon). Trailing behind with slightly lower index values are Catalonia and Cantabria in Spain, Corsica in France, Liguria, Veneto, and Tuscany in Italy, as well as Central Macedonia and the Peloponnese in Greece. Additionally, it is important to note that significant sunlust destinations which rank high among all regions in 2009, such as eastern and southern Spain (Valencia, Andalucía) and southern France (Côte d'Azur and Languedoc-Roussillon), gradually decline in significance. The second zone of touristification encompasses Alpine regions in Italy (Bolzano and Aosta Valley) and Austria (mainly Tirol and Salzburg). Following with lower scores are Italian region of Trento and the Austrian regions of Kärnten and Vorarlberg, while Eastern Austria (Steiermark and Burgenland), albeit initially exhibiting notable scores, gradually recedes in significance. Notably, the French Rhône-Alpes and Oberbayern in the Bavarian Alps (as well as the adjacent mountainous region of Niederbayern) did not particularly stand out during the study period to be considered part of this zone. The third category is not a zone per se; rather, certain metropolitan regions with global and/or heritage cities in Central and Northern Europe act as islands of touristification. Specifically, these cities include Brussels in Belgium, Vienna in Austria, Prague in the Czech Republic, Berlin and Hamburg in Germany, and Amsterdam (Noord-Holland) in the Netherlands. Meanwhile, Paris (Île-de-France), which initially exhibited high TAIDD CI scores, is gradually receding in significance. Some urban heritage destinations in the EU South, such as Barcelona (Catalonia), Athens (Attica), Florence (Tuscany), and Lisbon, could also be part of this zone. Malta constitutes a unique case, as, apart from a sun-and-sand destination, it is one of the most densely populated areas in the world, with urban heritage (including architecture and museums) playing a major role in its tourism product (Borg, 2023). Nevertheless, it was preferred that all latter destinations to be considered part of the EU South cluster.

Apart from the above, *Figure 1* shows three secondary touristification zones comprising peripheral destinations (Papatheodorou, 2004). The first includes certain northern European seaside regions, specifically Zeeland in the Netherlands and Schleswig-Holstein and Mecklenburg-Vorpommern in Germany. The second peripheral zone comprises a diverse range of mixed urban heritage regions in Central and Northern Europe, such as Trier and Bremen in Germany, Limburg in the Netherlands, and Bruges in Belgium (West Flanders). In this category we should also consider Copenhagen (Hovedstaden) in Denmark, which in fact exhibits one of the most notable score increments during the study period (see *Table 2*). Lastly, peripheral sun-and-sand destinations can be found in the Bulgarian regions along the Black Sea (Severoiztochen and Yugoiztochen), although their significance diminishes over time.

In contrast to the above primary and secondary zones of touristification, the northern and eastern parts of the EU periphery consistently exhibit the lowest TAIDD CI scores, encompassing almost all Eastern EU countries (Poland, Romania, Bulgaria, and Czechia, except for Prague), as well as the Baltic States (Lithuania, Latvia, Estonia) and Nordic countries (Sweden, Finland, and Denmark).

Remarkably, the Greek insular regions of South Aegean and the Ionian Islands consistently rank first in all years for which the index is calculated, with a notable gap separating them from the trailing Balearic Islands in Spain, Brussels in Belgium, Adriatic Croatia, and Malta. *Table 2* displays the top-ranking regions only for 2022; nevertheless, throughout the study period the top 20 positions in terms of TAIDD CI scores are occupied by nearly the same regions. Similarly, the bottom 20 positions steadily feature regions mainly from Poland, Romania, and Lithuania. This indicates that the Great Recession did not subvert spatial pattens of touristification but rather reinforced them. Still, important changes occur during the study period (2009-2022). Specifically, most of the initially high-ranking Alpine regions exhibit great index losses. Moreover, significant urban destinations such as Berlin, Prague, and Paris follow a similar course. Apart from those, many of the German mixed urban secondary destinations display score decreases.

Figure 1. TAIDD CI results for 2009 (left) and 2022 (right)

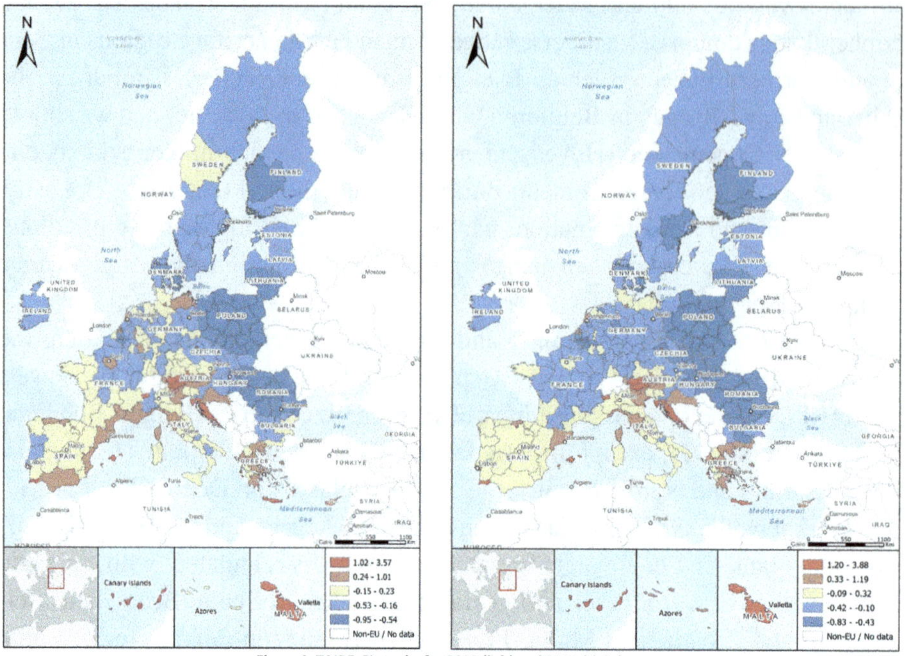

Figure 1: TAIDD CI results for 2009 (left) and 2022 (right)

Source: Eurostat, World Bank, authors' own elaboration

Source: Eurostat, World Bank, authors' own elaboration

Table 2. Top and bottom 30 regions in terms of the TAIDD CI score and its changes, 2009, 2022, 2009-2022

rank	Highest index scores, 2022			Greatest index increments, 2009-2022			Lowest index scores, 2022			Greatest index losses, 2009-2022		
1	EL42	Notio Aigaio	3.88	HR03	Jadranska Hrvatska	0.85	PL52	Opolskie	-0.83	AT33	Tirol	-0.57
2	EL62	Ionia Nisia	3.58	EL62	Ionia Nisia	0.72	PL84	Podlaskie	-0.82	AT32	Salzburg	-0.49
3	HR03	Jadranska Hrvatska	2.63	PT20	Azores	0.59	PL72	Swietokrzyskie	-0.81	AT21	Kärnten	-0.46
4	BE10	Bruxelles	2.26	PT30	Madeira	0.49	RO41	Sud-Vest Oltenia	-0.79	ES53	Illes Balears	-0.40
5	MT00	Malta	2.21	PT11	Norte	0.41	RO31	Sud - Muntenia	-0.76	AT11	Burgenland	-0.36
6	ES53	Illes Balears	2.14	FRM0	Corse	0.40	RO42	Vest	-0.74	DE80	Mecklenburg-Vorpommern	-0.34
7	AT13	Wien	2.11	PT16	Centro (PT)	0.38	RO21	Nord-Est	-0.73	DEC0	Saarland	-0.31
8	ITH1	Bolzano	1.87	PT17	Lisbon	0.36	PL81	Lubelskie	-0.71	AT34	Vorarlberg	-0.29
9	PT15	Algarve	1.81	EL42	Notio Aigaio	0.31	PL41	Wielkopolskie	-0.70	DEB2	Trier	-0.27
10	EL43	Kriti	1.77	EL41	Voreio Aigaio	0.30	RO11	Nord-Vest	-0.70	MT00	Malta	-0.27

continued on following page

Table 2. Continued

rank	Highest index scores, 2022			Greatest index increments, 2009-2022			Lowest index scores, 2022			Greatest index losses, 2009-2022		
11	CZ01	Praha	1.55	PT18	Alentejo	0.29	PL71	Lódzkie	-0.68	DEB1	Koblenz	-0.25
12	ES70	Canarias	1.51	EL43	Kriti	0.29	PL61	Kujawsko-Pomorskie	-0.66	CZ01	Praha	-0.25
13	DE30	Berlin	1.45	DK01	Hovedstaden	0.24	PL43	Lubuskie	-0.66	DE22	Niederbayern	-0.24
14	PT30	Madeira	1.45	EL54	Ipeiros	0.23	PL82	Podkarpackie	-0.66	FRL0	Côte d'Azur	-0.22
15	AT33	Tirol	1.43	PT15	Algarve	0.22	RO12	Centru	-0.64	BE10	Brussels	-0.22
16	ITC2	Valle d'Aosta	1.37	ITF4	Puglia	0.22	PL22	Slaskie	-0.64	DE73	Kassel	-0.20
17	AT32	Salzburg	1.19	RO21	Nord-Est	0.20	PL62	Warminsko-Mazurskie	-0.64	AT31	Oberösterreich	-0.19
18	DE60	Hamburg	0.95	EL51	Eastern Macedonia and Thrace	0.20	PL12	Mazowieckie	-0.63	FR10	Île de France	-0.19
19	FRM0	Corse	0.95	EL65	Peloponnisos	0.19	RO22	Sud-Est	-0.62	DE91	Braunschweig	-0.19
20	ITH2	Trento	0.94	EL52	Kentriki Makedonia	0.18	BE31	Prov. Brabant wallon	-0.59	DE30	Berlin	-0.19
21	EL41	Voreio Aigaio	0.92	ITI3	Marche	0.18	LT00	Lithuania	-0.59	FRD1	Basse-Normandie	-0.18
22	NL34	Zeeland	0.86	RO11	Nord-Vest	0.18	SK02	Západné Slovensko	-0.56	DE27	Schwaben	-0.18
23	EL65	Peloponnisos	0.67	ITG2	Sardegna	0.18	HU32	Észak-Alföld	-0.55	DE23	Oberpfalz	-0.17
24	EL54	Ipeiros	0.65	NL32	Noord-Holland	0.17	BE32	Prov. Hainaut	-0.55	DE13	Freiburg	-0.17
25	PT17	Lisbon	0.62	ITF5	Basilicata	0.16	PL51	Dolnoslaskie	-0.55	ES12	Principado de Asturias	-0.17
26	CY00	Kypros	0.55	RO41	Sud-Vest Oltenia	0.16	BG31	Severozapaden	-0.55	DED2	Dresden	-0.17
27	AT21	Kärnten	0.54	PL42	Zachodniopomorskie	0.16	BG32	Severen tsentralen	-0.54	FRF2	Champagne-Ardenne	-0.16
28	EL52	Kentriki Makedonia	0.53	EL61	Thessalia	0.15	HU33	Dél-Alföld	-0.53	DE40	Brandenburg	-0.16
29	AT34	Vorarlberg	0.49	RO22	Sud-Est	0.15	PL63	Pomorskie	-0.53	ES70	Canarias	-0.16
30	NL32	Noord-Holland	0.48	BE32	Prov. Hainaut	0.15	PL21	Malopolskie	-0.52	LU00	Luxembourg	-0.16

Source: Eurostat, World Bank, authors' own elaboration

Overall, the above reflect Germany's and Austria's generalised "de-touristification", at least in comparison to the rest of the EU. Interestingly, as noted above, similar trends are observed in multiple sun-and-sand regions in Spain and France, while even Malta gradually recedes in terms of index scores. All the above leave touristification intensifying further in certain parts of the EU South, namely the Greek, Portuguese and Croatian regions. Indicatively, 15 regions from these countries exhibit some of the highest index increments during 2009-2022 (see *Table 2*). Among them are not only traditionally touristy regions like Algarve, Madeira, Lisbon, South Aegean, Crete, and Adriatic Croatia, but also formerly peripheral destinations such as the Peloponnese, North Aegean, and Epirus in Greece, or Azores and Norte in Portugal, exhibit some of the largest increments.

As a result of the above, dualisms emerge across various scales. The first dualism pertains to the entire EU, where the South contains vastly more touristified regions. The second dualism relates to the interior of the EU South, with Spain and Italy, on one hand, being less touristified than Greece, Portugal, and Croatia. In fact, albeit omitted from in the tables or figures presented here, both dualisms reached their peak just before the COVID-19 pandemic in 2019, a year characterised by widespread polarisation. Furthermore, it is worth noting that the regional patterns of touristification were established during the deep recession years (2009-2014), with many regions that exhibited notable increases during the Great Recession also standing out in 2022.

DISCUSSION ON TOURISTIFICATION IN THE FACE OF DEEPENING ASYMMETRIES

The study above found three primary zones of touristification: i) the sunlust destinations of the EU South regions, ii) the Alpine resorts of Italy and Austria, and iii) a mix of urban hotspots across Central and Northern EU. Furthermore, the study pointed to secondary zones such as seaside Northern or Eastern European areas and a blend of mixed urban regions in Germany, the Netherlands, and Belgium. Importantly, our analysis highlighted two gradually deepening polarisations: one between the touristifying South and the "detouristifying" North, and one within the former, with Greece, Portugal, and Croatia exhibiting a hyper-touristification. From the above analysis, two key points emerge.

Firstly, the geography of touristification remained relatively stable throughout the study period, confined to three clearly defined zones of core destinations. This indicates that initial conditions such as climate and landscape play a key role in the emergence of tourist areas. However, for core destinations to be distinguished from secondary ones, the consolidation of large-scale infrastructure in transportation and accommodation is equally important. This is particularly evident in the EU South, where pre-crisis construction-driven growth models, largely responsible for the development of large hotels, airports, and road networks, were swiftly replaced by tourism-dependent models when the former were dismantled during the Great Recession (Gourzis & Gialis, 2022). In fact, as such infrastructure takes a long time to consolidate, an "agglomeration hysteresis" prevents peripheral destinations from developing into core ones (Papatheodorou, 2004).

Secondly, the significant differentiation observed among countries within the same EU South cluster highlights the key role of regional and national path dependencies in the development of tourist areas (Ma & Hassink, 2013). Specifically, certain moments during the past 15 years have served as critical points of inflection, steering the trajectories of regions and entire countries. These points have arisen from the varied responses of countries to recessive pressures (James et al., 2023). For instance, Greece and Portugal, as a counterbalance to the austerity measures implemented under Economic Adjustment Programmes, aggressively pursued foreign direct investment (FDI) by formulating favourable Golden Visa programmes and subsequently pushing for touristification through a series of legislative changes that facilitate the operation of the short-term rental market (Gourzis et al., 2019). In contrast, Italy and Spain, as larger economies with significant industrial output that have avoided the intervention of international institutions during the EU's debt crisis (Herod et al., 2022), sought to diversify their tourism products.

Juxtaposing our results with the relevant literature, it is important to note that, despite overcommercialisation and lack of innovation in their tourism product (see Ma & Hassink, 2013), Southern EU has displayed exponential tourism development in recent years. That said, this growth may prove to be short-lived, especially given the shift away from the increasing temperatures of the South (Gibson, 2021) and the rise of 'coolcation'. Another critical point is that, despite the dominance of the urban context in recent touristification literature, we found that rural destinations have retained their past significance. This could be partially attributed to the level of analysis employed (NUTS 2) or the lack of targeted data on short-term rentals, which have mostly reshaped and boosted urban tourism. Nevertheless, our results strongly resonate with key tourism studies in recent literature using Big Data, which have found that the majority of total nights spent in the EU are concentrated in southern sunlust destinations (Batista e Silva et al., 2021). Furthermore, recent overtourism trends have elicited strong responses in both urban and rural areas (Peeters et al., 2018). Therefore, we conclude this study by advocating for a strengthening of the literature on rural touristification and on the relationship between touristification and the escalating housing crisis.

REFERENCES

Batista e Silva, F., Barranco, R., Proietti, P., Pigaiani, C., & Lavalle, C. (2021). A new European regional tourism typology based on hotel location patterns and geographical criteria. *Annals of Tourism Research, 89*, 103077. DOI: 10.1016/j.annals.2020.103077

Bianchi, R. (2018). The political economy of tourism development: A critical review. *Annals of Tourism Research, 70*, 88–102. DOI: 10.1016/j.annals.2017.08.005

Borg, N. (2023, August 18) Fact-check: Does Malta have the highest population density in Europe? *Times of Malta.* Available at: https://timesofmalta.com/article/factcheck-does-malta-highest-population-density-europe.1049809

Estevens, A., Cocola-Gant, A., López-Gay, A., & Pavel, F. (2023). The role of the state in the touristification of Lisbon. *Cities (London, England), 137*, 104275. DOI: 10.1016/j.cities.2023.104275

Eurostat (2023) Tourism Satellite Accounts in Europe – 2023 edition. *Statistical Reports.* Available at: https://ec.europa.eu/eurostat/web/products-statistical-reports/w/ks-ft-22-011

Gibson, C. (2021). Critical tourism studies: New directions for volatile times. *Tourism Geographies, 23*(4), 659–677. DOI: 10.1080/14616688.2019.1647453

Gourzis, K., Alexandridis, G., Gialis, S., & Caridakis, G. (2019). Studying the spatialities of short-term rentals' sprawl in the urban fabric: The case of Airbnb in Athens, Greece. In *Artificial Intelligence Applications and Innovations: AIAI 2019 IFIP WG 12.5 International Workshops: MHDW and 5G-PINE 2019, Hersonissos, Crete, Greece, May 24–26, 2019* [Springer International Publishing.]. *Proceedings, 15*, 196–207.

Gourzis, K., Herod, A., Chorianopoulos, I., & Gialis, S. (2022). On the recursive relationship between gentrification and labour market precarisation: Evidence from two neighbourhoods in Athens, Greece. *Urban Studies (Edinburgh, Scotland), 59*(12), 2545–2564. DOI: 10.1177/00420980211031775

Herod, A., Gialis, S., Psifis, S., Gourzis, K., & Mavroudeas, S. (2022). The impact of the COVID-19 pandemic upon employment and inequality in the Mediterranean EU: An early look from a Labour Geography perspective. *European Urban and Regional Studies, 29*(1), 3–20. DOI: 10.1177/09697764211037126

James, L., Halkier, H., Sanz-Ibáñez, C., & Wilson, J. (2023). Advancing evolutionary economic geographies of tourism: Trigger events, transformative moments and destination path shaping. *Tourism Geographies, 25*(8), 1819–1832. DOI: 10.1080/14616688.2023.2285324

Kelly, E., Tuda, D., Whelan, A., Gourzis, K., Avagianou, A., Krommyda, V., . . . Judge, C. (2022) *Baseline Study on Impact of Youth Employment Policies.* Cowork4YOUTH Report. Accessed at: https://www.cowork4youth.org/wp-content/uploads/2022/11/Cowork4YOUTH_Output-3_Baseline-Study_Final-1_compressed.pdf

Ma, M., & Hassink, R. (2013). An evolutionary perspective on tourism area development. *Annals of Tourism Research, 41*, 89–109. DOI: 10.1016/j.annals.2012.12.004

Mendola, D., & Volo, S. (2017). Building composite indicators in tourism studies: Measurements and applications in tourism destination competitiveness. *Tourism Management, 59*, 541–553. DOI: 10.1016/j.tourman.2016.08.011

OECD/European Union/EC-JRC. (2008). *Handbook on Constructing Composite Indicators: Methodology and User Guide.* OECD Publishing., DOI: 10.1787/9789264043466-

Ojeda, A. B., & Kieffer, M. (2020). Touristification. Empty concept or element of analysis in tourism geography? *Geoforum, 115*, 143–145. DOI: 10.1016/j.geoforum.2020.06.021

Papatheodorou, A. (2004). Exploring the evolution of tourism resorts. *Annals of Tourism Research, 31*(1), 219–237. DOI: 10.1016/j.annals.2003.10.004

Peeters, P., Gössling, S., Klijs, J., Milano, C., Novelli, M., Dijkmans, C., . . . Postma, A. (2018) Overtourism: impact and possible policy responses. *Research for TRAN Committee.* Available at: https://www.europarl.europa.eu/RegData/etudes/STUD/2018/629184/IPOL_STU(2018)629184_EN.pdf

UNWTO. (2022) Measuring the Sustainability of Tourism (MST) Estimating Tourism Direct GDP with limited data. *UNWTO Reports.* Available at: https://webunwto.s3.eu-west-1.amazonaws.com/s3fs-public/2022-12/Measuring_TDGDP.pdf

ENDNOTES

1 Accessed at: https://tcdata360.worldbank.org/indicators/tot.direct.gdp?country=SVK&indicator=24648&countries= SVN&viz=line_chart&years=2008,2022

2 Eurostat Metadata. Occupancy of tourist accommodation establishments (tour_occ). Available at: https://ec.europa.eu/eurostat/cache/metadata/en/tour_occ_esms.htm

APPENDIX

Table 3. Pearsons' R between TAIDD CI variables

2009	**BEDS/POP**	**ARRIV/ POP**	**BEDS/ AREA**	**ARRIV+POP/ AREA**	**EMPL %**	**TDGDP**
BEDS/POP	1.000					
ARRIV/POP	0.883	1.000				
BEDS/AREA	0.267	0.375	1.000			
ARRIV+POP/AREA	-0.041	0.124	0.871	1.000		
EMPL %	0.717	0.732	0.347	0.108	1.000	
TDGDP	0.292	0.338	0.164	0.056	0.505	1.000
2014	BEDS/POP	ARRIV/POP	BEDS/ AREA	ARRIV+POP/ AREA	EMPL %	TDGDP
BEDS/POP	1.000					
ARRIV/POP	0.868	1.000				
BEDS/AREA	0.247	0.393	1.000			
ARRIV+POP/AREA	-0.035	0.159	0.905	1.000		
EMPL %	0.732	0.701	0.310	0.087	1.000	
TDGDP	0.346	0.353	0.141	0.034	0.596	1.000
2019	BEDS/POP	ARRIV/POP	BEDS/ AREA	ARRIV+POP/ AREA	EMPL %	TDGDP
BEDS/POP	1.000					
ARRIV/POP	0.912	1.000				
BEDS/AREA	0.237	0.366	1.000			
ARRIV+POP/AREA	-0.023	0.141	0.920	1.000		
EMPL %	0.756	0.755	0.295	0.079	1.000	
TDGDP	0.372	0.378	0.143	0.027	0.619	1.000
2022	BEDS/POP	ARRIV/POP	BEDS/ AREA	ARRIV+POP/ AREA	EMPL %	TDGDP
BEDS/POP	1.000					
ARRIV/POP	0.928	1.000				
BEDS/AREA	0.220	0.319	1.000			
ARRIV+POP/AREA	-0.015	0.106	0.927	1.000		
EMPL %	0.761	0.752	0.270	0.063	1.000	

continued on following page

Table 3. Continued

2009	BEDS/POP	ARRIV/POP	BEDS/AREA	ARRIV+POP/AREA	EMPL %	TDGDP
TDGDP	0.382	0.371	0.150	0.033	0.584	1.000
	Legend					
Very Weak	$0.0 \leq \lvert R \rvert < 0.2$					
Weak	$0.2 \leq \lvert R \rvert < 0.4$					
Moderate	$0.4 \leq \lvert R \rvert < 0.6$					
Strong	$0.6 \leq \lvert R \rvert < 0.8$					
Very Strong	$0.8 \leq \lvert R \rvert < 1.0$					

Table 4. Principal Component Analysis, TAIDD CI variables

2009	Variance	Proportion	Cum. Proportion	BEDS/POP	ARRIV/POP	BEDS/AREA	ARRIV+POP/AREA	EMPL %	TDGDP
PC(1)	3,012	50,202	50,202	0,477	0,511	0,360	0,213	0,488	0,312
PC(2)	1,664	27,736	77,938	-0,289	-0,180	0,581	**0,704**	-0,173	-0,146
PC(3)	0,829	13,813	91,751	-0,355	-0,295	-0,053	0,034	0,119	**0,877**
PC(4)	0,307	5,124	96,874	0,278	0,310	0,033	0,027	**-0,845**	0,333
PC(5)	0,120	1,999	98,873	0,532	**-0,655**	0,434	-0,309	-0,044	0,039
PC(6)	0,068	1,127	100,000	0,450	-0,306	-0,584	0,601	0,039	0,016
2014	**Variance**	**Proportion**	**Cum. Proportion**	**BEDS/POP**	**ARRIV/POP**	**BEDS/AREA**	**ARRIV+POP/AREA**	**EMPL %**	**TDGDP**
PC(1)	3,074	0,512	0,512	0,479	0,505	0,339	0,205	0,496	0,335
PC(2)	1,726	0,288	0,800	0,260	0,118	-0,597	**-0,698**	0,195	0,193
PC(3)	0,795	0,133	0,933	-0,381	-0,358	-0,017	0,084	0,146	**0,836**
PC(4)	0,239	0,040	0,972	0,145	0,412	-0,014	0,013	**-0,814**	0,383
PC(5)	0,115	0,019	0,992	**-0,675**	**0,650**	-0,266	0,114	0,179	-0,077
PC(6)	0,051	0,008	1,000	0,283	-0,106	**-0,676**	**0,671**	0,019	0,000
2019	**Variance**	**Proportion**	**Cum. Proportion**	**BEDS/POP**	**ARRIV/POP**	**BEDS/AREA**	**ARRIV+POP/AREA**	**EMPL %**	**TDGDP**
PC(1)	3,141	0,523	0,523	0,487	0,511	0,313	0,186	0,501	0,343
PC(2)	1,764	0,294	0,817	0,232	0,117	**-0,615**	**-0,701**	0,178	0,176
PC(3)	0,772	0,129	0,946	-0,378	-0,356	-0,008	0,067	0,133	**0,842**

continued on following page

Table 4. Continued

PC(4)	0,202	0,034	0,980	0,258	0,309	0,011	0,010	**-0,834**	0,378
PC(5)	0,080	0,013	0,993	**-0,630**	**0,686**	-0,310	0,186	0,050	-0,018
PC(6)	0,041	0,007	1,000	0,322	-0,181	**-0,654**	**0,659**	0,034	0,004

2022	**Variance**	**Proportion**	**Cum. Proportion**	**BEDS/ POP**	**ARRIV/ POP**	**BEDS/ AREA**	**ARRIV+POP/ AREA**	**EMPL %**	**TDGDP**
PC(1)	3,105	0,518	0,518	0,499	0,513	0,295	0,173	0,502	0,344
PC(2)	1,790	0,298	0,816	-0,219	-0,131	**0,628**	**0,703**	-0,164	-0,140
PC(3)	0,769	0,128	0,944	-0,345	-0,355	-0,013	0,036	0,111	**0,861**
PC(4)	0,227	0,038	0,982	0,278	0,311	0,006	0,012	**-0,840**	0,347
PC(5)	0,067	0,011	0,993	**0,648**	**-0,684**	0,277	-0,184	-0,041	-0,005
PC(6)	0,042	0,007	1,000	0,294	-0,171	**-0,665**	**0,664**	0,041	0,005

Chapter 11
The Legislative Framework for Tourism Structures in the Greek Coastal Zones

Despina Dimelli
https://orcid.org/0000-0002-0234-1293
Technical University of Crete

ABSTRACT

This chapter examines the role of tourism in the development of coastal zones, focusing on the legislative frameworks shaping tourism-related infrastructure. Since tourism depends on quality infrastructure such as accommodation, transport, and recreational facilities, the chapter analyzes laws that either support or restrict such development in coastal areas. A comparative approach is applied, reviewing frameworks from various Mediterranean countries to highlight similarities, differences, and the presence or absence of environmental safeguards. Special emphasis is given to Greece, where coastal zones are both ecologically sensitive and economically strategic for tourism. The analysis assesses the Greek legal framework's ability to balance environmental protection with tourism growth, examining how current laws and planning instruments influence the scale, location, and type of infrastructure. Overall, the chapter contributes to understanding how tourism infrastructure can align with sustainability and spatial justice goals.

DOI: 10.4018/979-8-3373-1554-6.ch011

INTRODUCTION

Tourism is fundamentally an activity driven by human desires and motivations, involving the temporary movement of individuals from their usual place of residence to a different geographical location for reasons beyond regular work-related commuting. Netto (2009) identifies several key components that define tourism, including the physical act of travel itself, the specific motivations that drive such travel—typically linked to leisure, recreation, or business—and active engagement in a wide range of activities once the traveler reaches their destination. This definition highlights tourism as a multifaceted socio-economic phenomenon, shaped by the behaviors of individual travelers, as well as by broader structural and societal conditions that influence travel patterns on a local, national, and even global scale.

The World Tourism Organization (UNWTO), which provides an internationally accepted framework for defining tourism, describes a tourist as any visitor who stays at a destination for more than 24 hours, and whose primary purpose for the visit is either leisure or professional-related activities. This expansive definition underscores the diverse and multi-dimensional nature of tourism, capturing a broad spectrum of relationships, activities, and phenomena that are inherently linked to travel, regardless of the specific underlying motivations that prompt the journey. In its 2024 report, the UNWTO noted a substantial rise in international tourism, with approximately 1.286 billion international overnight visitors recorded in 2023. This figure represents a remarkable 34% increase from 2022, equivalent to an additional 325 million tourists worldwide, which further highlights the dynamic and growing nature of the global tourism sector.

In recent decades, the nature of modern tourism has evolved significantly, responding to increasingly diversified and specialized patterns of consumer demand. Tourists today exhibit more complex preferences and expectations when it comes to the quality, uniqueness, and personalization of services and experiences offered by destinations. This evolution in consumer behavior has contributed to the emergence of niche markets within the broader tourism sector, such as ecotourism, cultural tourism, and adventure tourism, which cater to travelers seeking specific types of experiences. As a result, the tourism industry has become increasingly segmented and dynamic, creating new opportunities and challenges for both demand-side and supply-side stakeholders. Furthermore, tourism development is inherently spatially differentiated, shaped by the unique geographical, cultural, economic, and environmental factors that define each destination, influencing the ways in which tourism is developed and managed in different locations around the world.

The relationship between tourism and local or regional development is particularly important in areas that are rich in environmental and cultural value. Traditionally, mass tourism has been associated with the "3S" model—sun, sea, and sightseeing—

especially in coastal regions, where these natural attractions are abundant. Coastal zones have long been recognized as prime destinations for tourism, due to their breathtaking landscapes, biodiversity-rich ecosystems, and the wide range of recreational activities they offer. These regions thus play a vital role in driving both national and local economies, contributing significantly to employment opportunities, revenue generation, and broader regional development. In this context, tourism serves as both an economic engine and a driver of social and infrastructural development, fostering greater interaction between local communities and visitors from around the globe.

Coastal zones, however, are conceptualized and understood in a variety of ways, depending on the perspective from which they are being examined. These perspectives may be institutional, ecological, or geographical in nature. The United Nations Environment Programme (UNEP) defines coastal zones as dynamic and ever-changing interface areas where terrestrial and marine environments come into contact. These zones encompass a variety of features, such as shorelines, estuaries, wetlands, and other transitional ecosystems, all of which are shaped not only by natural forces, such as tides, waves, and erosion but also by anthropogenic influences, including urban expansion, agricultural activities, and industrial development (Post & Lundin, 1996). Similarly, the European Union's Integrated Coastal Zone Management (ICZM) Recommendation (2002/413/EC) provides a definition of coastal zones as regions that include both marine and terrestrial components. These zones extend from the outer boundary of territorial waters to the landward boundary, defined by areas where human activities significantly impact coastal ecosystems and environments.

Tourism in coastal areas represents one of the most significant and rapidly expanding sectors of the global tourism industry, largely due to the intrinsic appeal of these environments and their ecological value. Coastal regions are consistently ranked among the most visited destinations in the world, drawing tourists with their pristine beaches, luxury resorts, maritime activities, and diverse habitats, all of which appeal to a wide range of visitors. However, the increasing intensity of tourism in these regions has led to a variety of pressing environmental, socio-economic, and cultural challenges. These challenges include habitat degradation, pollution, resource overuse, the loss of biodiversity, social displacement of local communities, and growing pressure on infrastructure and services. The increasing concentration of tourists in coastal regions has exacerbated these issues, putting significant strain on local ecosystems and the communities that depend on them for their livelihoods.

As a result, there is an urgent need for the development and implementation of sustainable tourism strategies that are specifically tailored to the unique environmental, economic, and social conditions of coastal zones. Sustainable coastal tourism should be designed to balance the goals of environmental conservation, economic viability, and social equity, aiming to promote long-term resilience, adaptability, and sustainability in these regions. To achieve this, policy frameworks need to integrate a range of tools, including environmental regulations, spatial planning mechanisms, and community-based governance structures, all of which are essential for mitigating the negative impacts of tourism and ensuring the protection of valuable coastal ecosystems. Furthermore, effective stakeholder engagement, cross-sectoral coordination, and the active involvement of local communities in the decision-making process are all critical to ensuring that tourism contributes positively to regional development. In this way, it can enhance the local economy and provide benefits to local residents without compromising the ecological integrity and sustainability of coastal areas.

In conclusion, the rapid growth of tourism in coastal regions has placed significant pressures on local environments and communities, necessitating the adoption of sustainable tourism practices. By recognizing the complex relationships between tourism, development, and environmental conservation, it is possible to create policy frameworks that not only support the growth of the tourism industry but also protect and preserve the unique natural and cultural assets of coastal zones for future generations. Through a balanced approach that includes effective management, stakeholder involvement, and an emphasis on sustainability, coastal tourism can contribute to both the economic prosperity and ecological well-being of these vital regions.

LEGISLATION FOR TOURISM DEVELOPMENT IN MEDITERRANEAN COASTAL ZONES

This chapter offers a critical examination of the legislative framework governing the development of tourism infrastructure in Greece's coastal zones, with a particular focus on the presence or absence of regulatory restrictions and the implications these have for both spatial planning and environmental dynamics. Coastal regions, particularly those with high ecological sensitivity, face complex challenges in balancing economic growth through tourism with the need to preserve fragile ecosystems. Effective governance in these areas is essential, as tourism infrastructure development must be managed through legal frameworks that ensure environmental sustainability while promoting economic growth. Given the multiple and often competing land uses—such as urban expansion, agriculture, and tourism infrastructure—regulations must be in place to safeguard the environmental integrity of Greece's coastal zones, without compromising the socio-economic needs of local populations and industries.

Coastal areas are often characterized by intense human activity coupled with natural vulnerability, making effective legislation critical to mitigating the environmental impacts of tourism. These regions are highly attractive to both international tourists and local communities, which increases the pressure on their resources and ecosystems. Understanding the legal context surrounding tourism development is vital, as the dynamics of coastal tourism require a multifaceted approach. Effective coastal tourism governance must consider both environmental protection and the socio-economic demands of the region. These dual objectives make the regulation of tourism in coastal zones a uniquely complex and challenging task, demanding nuanced legal frameworks that take into account the various pressures exerted on these regions.

In an effort to enhance the analysis of coastal tourism regulation in Greece, this chapter also incorporates a comparative perspective by evaluating the legislative frameworks of other Mediterranean countries. This comparative approach will focus on countries that have large tourism industries and extensive coastal zones, such as Spain. By examining these legal frameworks, the chapter aims to identify best practices, regulatory gaps, and potential policy solutions that could inform Greece's approach to coastal tourism development, ultimately leading to more sustainable spatial planning.

Spain, with its approximately 7,270 kilometers of coastline, serves as a relevant and insightful case study for this analysis. Spain's coastal regions are home to diverse ecosystems, which are important both economically and ecologically. Approximately 25% of Spain's coastline consists of sandy beaches, which are vital to tourism, and the population density along the coast reaches 6,380 inhabitants per kilometer. This reflects the significant urban and tourism-related pressures placed on Spain's coastal areas. Over the past five decades, tourism development has played a pivotal role in shaping Spain's national economy, contributing roughly 12% to the Gross Domestic Product (GDP). Coastal zones attract around 5% of the total international tourists visiting Spain each year, further underscoring the economic significance of coastal tourism.

However, the rapid growth of tourism infrastructure has resulted in substantial land-use changes, particularly in areas traditionally dedicated to agriculture and fishing. In many coastal regions, tourism-related services and energy infrastructure, such as fuel storage facilities, have replaced these traditional industries. This transformation is especially noticeable around Spain's 530 port facilities (Pellach & Carmon, 2016), where tourism infrastructure expansion has significantly altered land use. The competition between economic growth through tourism and the need to preserve natural habitats and protect the environment illustrates the complex challenges faced by Mediterranean coastal regions.

The example of Spain highlights the need for a well-designed and integrated legislative framework capable of managing both the economic development and environmental protection of coastal regions. Spain's experience in balancing these two objectives provides valuable insights for Greece, enabling the country to assess whether its existing legal instruments are effectively addressing the challenges posed by the expansion of tourism and urbanization along its coastline. The comparative analysis underscores the importance of regulatory frameworks that can adapt to changing conditions, including the pressures from rising tourist numbers and the need for environmental conservation in coastal zones.

Despite the strides Greece has made in developing regulatory frameworks for coastal zone management, the experience of Spain offers lessons in achieving a more balanced and sustainable approach to coastal tourism development. Spain's approach emphasizes the importance of integrating economic development with environmental protection, offering potential strategies for Greece to improve its own legal framework. As both countries face similar challenges, it is clear that their regulatory frameworks must evolve to account for emerging issues, such as climate change, increased tourist numbers, and the growing need for biodiversity conservation in coastal areas.

The increasing impacts of climate change, such as rising sea levels and coastal erosion, as well as the ongoing pressure from tourism and urbanization, make it essential for Greece's legal framework to adapt. The comparative analysis with Spain serves as a useful reference, suggesting that an integrated approach that balances tourism development with environmental protection is not only necessary but urgent. In the long term, both Greece and Spain must refine their coastal tourism legislation to ensure that it supports the sustainable development of their coastal areas while safeguarding the ecological and economic benefits they provide.

Figure 1. The Costa Brava area

(Pellach & Carmon, 2016)

PUBLIC AND THE ENVIRONMENT

The legal framework governing tourism development along Spain's coastal zones has undergone substantial changes over the years, evolving in response to shifting economic, environmental, and socio-political conditions. Spain's journey toward formalizing tourism and real estate development along its coastlines began in 1959, with the introduction of groundbreaking legislation aimed at promoting both real estate development and tourism as integral components of a broader national economic strategy. This initiative gained significant momentum with the passage of Law 28/1969 during the Francoist regime, which was specifically designed to attract foreign investment and stimulate the tourism sector as a key aspect of the country's post-war economic recovery. The motivation behind this legislation was rooted in a broader desire to modernize Spain's economy, increase the national GDP, and enhance Spain's position in the rapidly expanding international tourism market (Fernando, Rosete Verges, & Larruga, 2024). One particularly crucial piece of legislation, Law 197/1963, played a pivotal role by establishing Centers and Areas of National Tourist Interest. This law was specifically aimed at organizing the development of tourism in space, designating particular zones for development, and ensuring that tourism expansion aligned with national financial priorities, thus enabling the structured and systematic economic exploitation of coastal areas (Ley 197/1963, 1963).

However, by 1978, after decades of fast-paced and often unregulated tourism development, Spain's Constitution formally recognized the coastal zone as a public domain, highlighting both its social and ecological significance. This shift in constitutional policy marked a critical turning point in the country's approach to coastal zone management and environmental protection. In subsequent years, the enactment of the Coastal Law (Ley de Costas) in 1988 became a major legislative milestone, designed to address three key threats to coastal areas: rampant overdevelopment, the privatization of public land, and worsening environmental degradation. The Coastal Law aimed to limit unchecked development and safeguard Spain's coastal ecosystems, ensuring that tourism expansion did not cause irreversible damage to these essential natural resources. However, the law faced strong resistance, particularly from the tourism and real estate sectors. Critics of the legislation contended that the stringent environmental protections embedded in the law would obstruct the sector's ability to grow and compete in an increasingly globalized tourism market. Despite the law's ambitious environmental goals, illegal tourism-related constructions, especially along Spain's picturesque coastlines, continued to increase, undermining the effectiveness of these legal safeguards and raising concerns about the law's ability to prevent environmental harm.

In response to the economic challenges presented by the 2007–2008 financial crisis, the Spanish government sought to amend the Coastal Law as part of an effort to boost the real estate and tourism sectors. These amendments, which aimed to ease regulatory restrictions on coastal development, were part of a broader economic strategy to revitalize the country's economy during the downturn. However, the subsequent recession and downturn in the construction industry limited the success of these amendments, particularly the modifications introduced in 2013, which were designed to provide temporary relief to the real estate sector (Martínez Cordero, 2013). More recently, new legislative measures, such as the Climate Change and Energy Transition Law, have incorporated more sustainability-focused provisions into coastal development policies, including a maximum duration of 75 years for coastal concessions. This shift reflects a growing recognition of the need to balance economic development with long-term environmental sustainability. Furthermore, amendments to the Coastal Law in 2022, which eliminated technical criteria for defining the inland boundary of the seashore, have further complicated the governance and development of coastal areas, presenting new challenges for spatial planning and environmental protection (Fernando, Rosete Verges, & Larruga, 2024).

Looking forward, future policy measures aimed at improving the efficiency and effectiveness of Spain's legal framework for coastal tourism development must prioritize simplifying legal processes, ensuring regulatory coherence, and fully integrating sustainability principles into decision-making processes. Strengthening the enforcement of regulations, improving transparency in governmental decision-making, and proactively addressing emerging climate-related risks, such as rising sea levels and increased storm activity, will be essential for ensuring the long-term resilience of Spain's coastal regions. By implementing these reforms, Spain could achieve a more balanced approach, ensuring that the undeniable economic benefits of tourism do not come at the expense of coastal ecosystems. This would allow future developments to contribute positively to both regional prosperity and the preservation of natural habitats.

In comparison, Malta, with its relatively short coastline of only 198 kilometers, presents a different yet equally instructive case. Malta's coastal zone, while geographically small, faces significant challenges due to its high population density—2,170 inhabitants per kilometer of coastline—and the presence of several culturally and environmentally important sites. Since the 1950s, a large portion of Malta's population has migrated from inland areas to the coastal regions, leading to intensified urbanization and mounting tourism pressures along the coastline. Despite this rapid development, Malta's coastal zone today also includes a number of abandoned hotels, which are remnants of earlier phases of tourism expansion. In 1990, Malta enacted its first comprehensive coastal zone management plan, recognizing the vulnerability of the coastal zone due to the significant pressures exerted by tourism infrastructure. The 1990 law prohibited the construction of new permanent structures in the coastal zone and mandated the demolition of any existing structures. The law also designated the coastal zone as a public good, ensuring that it would be owned and protected by the state for the benefit of the general public (Zammit Pace, Bray, Baily, & Potts, 2018).

Despite the existence of this legal framework, unauthorized construction continues to be a persistent issue in Malta's coastal zone. Although these unauthorized structures are technically prohibited, many of them have been legitimized through the imposition of fines on their owners. Additionally, buildings constructed before 1967 are exempt from demolition requirements, which further complicates efforts to manage coastal zone development. During the summer months, these areas, including those designated as protected sites, experience heightened pressure from both tourism and local population activities, exacerbating environmental degradation. However, the political will to enforce these legal provisions and demolish illegal structures has been lacking, with authorities often failing to fully implement the laws designed to protect the coastal zone (Malta Ministry of Tourism, n.d.). This ongoing issue highlights the complexities of coastal zone management and underscores the need for stronger political commitment and more robust enforcement mechanisms to safeguard Malta's coastal heritage.

Israel, with its coastline extending over 205 kilometers, faces unique challenges in balancing tourism and recreational development with the imperative to protect its coastal resources. Israel's population density along its coastline is among the highest in the world, with 40,000 inhabitants per kilometer, exerting immense pressure on the country's coastal ecosystems. Israel's legislative framework permits tourism and recreational activities along the coast, provided these activities do not conflict with the broader objectives of environmental protection. However, ambiguities within the legislative framework have allowed for the exploitation of coastal areas, leading to developments that sometimes contradict the central goals of environmental preservation.

In 2004, Israel introduced a set of new regulations aimed at addressing coastal degradation. These regulations included the establishment of a special committee to evaluate proposed developments, with particular emphasis on prioritizing projects that serve the public interest over those driven by private profit. This initiative was supported by the enactment of the Protection of the Coastal Environment Law in 2004, which sought to protect Israel's beaches and coastal areas by recognizing them as invaluable public resources that should be safeguarded for public use and enjoyment. The law established a 300-meter buffer zone inland from the shoreline, designating this area as a public resource to be protected from overdevelopment and degradation (Ministry of Environmental Protection, 2020).

Despite these legal protections, Israel's coastal zone management framework remains complicated by the involvement of multiple regulatory bodies, each with its own set of directives and priorities. As a result, coastal development projects often fail to align with the core principles of environmental protection. Recognizing the need for reform, Israel is in the process of revising its coastal zone development framework to better integrate environmental protection principles and ensure that future interventions are in line with the long-term interests of both the public and the environment (Pellach & Carmon, 2016). This ongoing revision of the legal framework reflects Israel's commitment to refining its approach to coastal management, ensuring the health and sustainability of its coastal ecosystems for future generations.

In conclusion, the comparative analysis of Spain, Malta, and Israel's coastal tourism management frameworks illustrates the complexities involved in balancing tourism development with environmental preservation. Despite differences in geographical, political, and economic contexts, all three countries face similar challenges in ensuring that tourism growth does not come at the expense of their coastal ecosystems. Each country's legal framework has evolved in response to both internal and external pressures, and continued efforts to refine these frameworks will be crucial for addressing emerging environmental and socio-political challenges. By further improving these frameworks, these nations can better ensure that their coastal areas remain viable, sustainable, and beneficial for future generations.

Figure 2. The Hof Ha-Carmel hotel in Israel

THE LEGISLATIVE FRAMEWORK FOR TOURISM DEVELOPMENT IN GREEK COASTAL ZONES

Greece is home to an expansive and diverse coastline that spans approximately 15,150 kilometers, making it one of the largest coastlines in Europe. With a population density of 720 inhabitants per kilometer along its coastal areas—double the national average—the country faces unique challenges and opportunities in managing its coastal resources. Tourism plays an integral role in Greece's economy, contributing significantly to the nation's Gross Domestic Product (GDP), with tourism accounting for roughly 20% of the country's total GDP. This underscores the importance of coastal zones, as they are primarily utilized for tourism-related activities, which dominate over other uses such as fishing or maritime transport infrastructure (Pellach & Carmon, 2016). Over time, Greek legislation governing the development and use of these coastal zones has evolved, with a notable shift from a focus on land ownership and land use regulation to a more integrated, environmentally conscious approach that reflects both national priorities and the growing influence of European Union (EU) policies. This legal evolution has resulted in the development of a sophisticated and well-defined regulatory framework designed to promote sustainable development and safeguard Greece's valuable coastal ecosystems from further degradation.

Initially, Greek coastal legislation was largely concerned with issues of land ownership and controlling unauthorized construction, with minimal focus on environmental sustainability or public access. However, as environmental awareness grew, particularly in the latter half of the 20th century, and as EU policies gained influence, Greek coastal zone legislation began to incorporate broader concerns, including the protection of coastal ecosystems, public access, and the need to adapt to climate change. These shifts resulted in the establishment of a legal framework that integrates national laws, EU directives, and international conventions, all of which focus on protecting and managing Greece's marine and coastal environments in a way that prioritizes both environmental conservation and sustainable development.

The history of coastal zone management in Greece spans several centuries and reflects the country's evolving understanding of the value of its coastal areas, which are essential for both the economy and the environment. Coastal management practices in Greece can be traced back to the 19th century, with the first significant legal measures being introduced in 1837. The aim was to prevent the privatization of coastal land, which was recognized as a crucial public resource. Two key legal texts, the 1837 Decree and the 1940 Law No. 2344, defined the seashore as public property, marking the beginning of legal attempts to ensure that coastal areas remained publicly accessible and protected. As Greece entered the mid-20th century, however, coastal regulations were often fragmented and insufficient to address the increasing pressures from tourism development. This issue became particularly pronounced in the 1960s and 1970s, a period of rapid tourism expansion that resulted in substantial coastal development that frequently neglected the environmental consequences of such growth.

A significant turning point in Greece's approach to coastal zone management came in 1985, with the passage of the Coastal Zone Planning Law. This law was designed to provide a more structured and organized framework for managing coastal land, with an emphasis on regulating land use and protecting coastal ecosystems (Levantis, 2001). The law marked the beginning of a more sophisticated approach to coastal management that would continue to evolve over the following decades. Additionally, the Greek Constitution, amended in 1975, explicitly reinforced the obligation of the state to preserve the natural environment, including coastal areas, and to manage them in a sustainable manner. This constitutional commitment was complemented by the enactment of Law 1650/1986, which was Greece's first comprehensive environmental protection legislation. This law provided a legal foundation for the protection of ecosystems, including coastal and marine environments, and laid the groundwork for the modern coastal zone management framework.

Another important milestone in the evolution of Greek coastal legislation came with the introduction of Law 2971/2001. This law expanded upon the provisions of earlier laws and incorporated more contemporary values, including the need for environmental conservation, public access to coastal areas, and the prevention of illegal construction. One of the key provisions of this law was the designation of the seashore and beaches as public property, thus making illegal constructions along the coast subject to demolition. This law also introduced a more systematic approach to issuing temporary permits for tourism and business activities on the coast, further reinforcing the importance of regulating coastal land use. Additionally, stricter enforcement measures were introduced to curb illegal constructions, marking a significant step toward improving the overall effectiveness of coastal management in Greece.

As the 21st century progressed, climate change began to emerge as a critical issue for coastal zone management, with rising sea levels and increased coastal erosion posing substantial risks to both infrastructure and ecosystems. In response to these growing challenges, the Greek government enacted Law 4414/2016, which placed a stronger emphasis on climate change adaptation within the context of coastal management. This law, along with subsequent policies, introduced measures to address the risks of erosion and flooding, both of which threaten coastal infrastructure and ecosystems. Additionally, the National Climate Adaptation Strategy of 2016 outlined specific actions for managing coastal areas in light of climate change, including the implementation of maritime spatial planning (MSP). This strategy aims to balance the needs of various industries, including tourism, shipping, and fisheries, while ensuring the long-term protection of coastal ecosystems.

The integration of EU directives into Greek coastal management has played an essential role in modernizing the country's approach to coastal protection, ensuring that its policies align with international standards. This integration has allowed

Greece to better safeguard its extensive coastline, ensuring the resilience of its coastal ecosystems amid growing environmental challenges. Despite the advancements in Greece's legal frameworks for coastal zone management, challenges remain in the enforcement and implementation of these regulations. In particular, economic pressures, illegal construction, and competing land use interests continue to undermine the achievement of environmental goals in certain regions. Consequently, while Greece's legal framework for coastal zone management is comprehensive, further adaptation of these laws to address emerging challenges, such as climate change, along with the strengthening of enforcement mechanisms, will be crucial in ensuring the long-term sustainability of the country's coastal zones.

THE COASTAL ZONES IN GREEK TOURISM SPATIAL PLANNING

Tourism growth in Greece has historically been accompanied by significant infrastructure development, much of which remains largely underutilized during the off-peak seasons. While there are various legal measures intended to protect the environment, these infrastructural developments often contribute to the unsustainable and disproportionate expansion of tourism-related facilities along the country's coastline. In many cases, these developments remain idle for long periods, especially during the off-peak months, resulting in inefficient use of resources and exacerbating the environmental impact caused by overdevelopment. These underused developments create unnecessary waste and contribute to the degradation of natural resources, all while failing to meet the actual seasonal demands of the tourism sector. As a result, valuable resources are wasted, and ecosystems are subjected to further harm due to the excess infrastructure built to cater to tourism, rather than responding to real needs.

In response to these issues, Greece's tourism spatial planning policies have become increasingly focused on ensuring that tourism infrastructure development is sustainable and aligned with broader environmental objectives. One of the primary strategies employed involves promoting the development of low-density accommodations, encouraging energy-efficient construction practices, and integrating responsible waste management systems into tourism projects. These measures aim to mitigate the environmental impact of tourism infrastructure while still supporting economic growth. Another key focus is diversifying the types of tourism promoted, including forms such as agro-tourism, cultural tourism, and eco-tourism. By nurturing a variety of tourism sectors, the Greek government seeks to distribute tourism's impact more evenly across the country, thus alleviating the heavy concentration of tourism activities in already highly developed coastal regions. The implementation of frameworks like the Special Spatial Planning Framework for Tourism (SSPFT) represents a significant step in managing the growth of tourism infrastructure. This framework establishes clear guidelines for developers, ensuring that new tourism projects, especially those located near ecologically sensitive areas, align with environmental conservation goals and do not compromise the integrity of these fragile ecosystems.

The Greek system of spatial planning designates areas for tourism development based on a range of factors, including environmental sensitivity, the potential for urban expansion, and the proximity to cultural or historical landmarks. This classification of areas is designed to guide the sustainable growth of tourism infrastructure while safeguarding the country's natural and cultural heritage. However, while the primary goal of the legal framework is to foster sustainable tourism growth, challenges continue to arise in its implementation. One of the key obstacles is the frequent modification and revision of the laws that govern tourism development, especially in coastal areas. These frequent changes often create unpredictability and inconsistency in the planning and regulatory processes, undermining long-term sustainability objectives. The constant revisions result in a climate of uncertainty, with developers often taking advantage of temporary legal loopholes or shifts in regulations to push forward projects that might otherwise be restricted or denied. This lack of predictability complicates efforts to create a stable and effective planning environment for sustainable tourism development in the country.

A particularly concerning issue is the retroactive legalization of previously illegal structures, often facilitated by what is referred to as a "building amnesty." Under this system, developers can pay fines to legalize structures that were built without the necessary permits or that violate zoning laws. While this approach might be viewed as a solution to the problem of illegal construction, it actually exacerbates the situation by encouraging developers to bypass established legal frameworks. The existence of such amnesty promotes a disorganized and haphazard approach to tourism development. Rather than preventing illegal construction, this policy sends a message that regulations can be circumvented for a fee, thus undermining the integrity of zoning laws and further jeopardizing the protection of vulnerable coastal ecosystems. This creates a cycle of unchecked growth, where the legal framework is disregarded, and the environment continues to suffer as a result.

Another significant challenge in Greece's coastal tourism development framework is the fragmentation of authority and responsibility across multiple government agencies. Ministries such as the Ministry of Environment, the Ministry of Tourism, and various local municipal authorities all play a role in the decision-making processes regarding coastal tourism projects. However, the division of responsibilities among these entities often leads to confusion, inefficiency, and sometimes conflicting decisions. The boundaries of coastal zones and the scope of inland regulations are often unclear or applied inconsistently. This lack of clarity in the delineation of regulatory authority provides opportunities for developers to exploit gaps in the system, bypassing regulations or engaging in practices that would otherwise be prohibited. In some cases, developers may even exert pressure on local authorities to expedite the approval process for certain projects, especially if the projects are perceived to offer significant economic benefits to the area. The lack of coherence among various governmental bodies weakens the effectiveness of the overall legal framework and makes it more difficult to enforce policies that prioritize long-term sustainability.

Additionally, local communities and stakeholders often have limited opportunities to participate in decision-making processes regarding tourism development in their coastal areas. The decision-making process is frequently centralized, with major decisions being made at the national or regional level, leaving little room for input from residents or small business owners who may be most directly affected by these developments. This top-down approach to planning often results in projects that are misaligned with the needs and concerns of local communities. Consequently, the interests of large investors, such as multinational hotel chains or large real estate developers, tend to dominate the decision-making process, while smaller, locally owned businesses are often excluded from the benefits of tourism growth. The prioritization of large-scale investments over the inclusion of local enterprises reflects a broader issue within Greece's tourism sector, where small and medium-sized businesses struggle to compete with multinational corporations that are better positioned to dominate the market.

The disproportionate focus on large-scale tourism developments, as opposed to small, locally owned businesses, not only has detrimental effects on the local economy but also erodes efforts to foster a more equitable and sustainable approach to tourism development. Small businesses, with their deeper understanding of local traditions, cultures, and the unique characteristics of their communities, are typically more adept at aligning their operations with sustainable practices. They are better positioned to operate in ways that honor and preserve local customs and protect the environment. In contrast, large-scale, corporate-driven tourism ventures often prioritize profit over these critical considerations, leading to the commercialization and homogenization of local culture, which may result in a loss of the very essence that attracts tourists in the first place.

When the focus is placed solely on large-scale developments, Greece risks sacrificing the authentic charm and distinctiveness of its coastal regions. Such projects often bring with them environmental degradation, social displacement, and a disruption of the local way of life. These impacts are especially pronounced in areas that have historically relied on tourism as a primary economic driver. Local communities that have traditionally depended on tourism for their livelihoods are often marginalized by the influx of large investors, who may not have the same long-term commitment to the community or the environment. This alienation of local populations not only affects the social fabric but also undermines the sense of ownership and pride that local businesses and residents have in their regions.

Furthermore, the dominance of large investors in the tourism sector leads to a centralization of economic power, often concentrating profits in the hands of a few multinational corporations. This reduces the potential for wealth distribution within local communities, which are instead left to contend with rising living costs, overcrowding, and a lack of access to the benefits of tourism. Small, locally owned ventures, in contrast, have a more direct link to the community and are more likely to reinvest in local infrastructure and services, ensuring that the economic benefits of tourism are more widely shared. As a result, the local population's ability to engage meaningfully in and benefit from tourism is diminished.

Ultimately, the emphasis on large-scale tourism developments over locally owned businesses undermines efforts to create a balanced, sustainable tourism model that benefits all sectors of society. A more inclusive approach that fosters the growth of smaller, locally driven enterprises is crucial for ensuring that the tourism industry in Greece can thrive without sacrificing the environmental integrity, cultural richness, and social cohesion of its coastal regions. Only by striking a balance between large-scale developments and local entrepreneurship can Greece achieve a more sustainable, equitable, and community-oriented tourism industry.

In summary, while the legal framework for tourism infrastructure development in Greece's coastal areas was designed with the intention of fostering sustainable growth and protecting the country's natural and cultural assets, several significant challenges continue to hinder its effective implementation. These challenges include legal inconsistencies, inadequate coordination among government agencies, limited local participation in decision-making processes, and the dominance of large investors over smaller, locally owned enterprises. To overcome these obstacles and ensure a more equitable and sustainable tourism development process in Greece's coastal areas, there is a clear need for stronger consistency in the application of regulations, better coordination among agencies, greater local community engagement, and a more inclusive approach to tourism planning that actively involves small businesses. Only by addressing these issues can Greece hope to create a tourism sector that is not only economically successful but also environmentally responsible and socially inclusive.

CONCLUSION

Despite Greece's recognition of the importance of promoting sustainable tourism practices, the existing legal framework has not yet fully incorporated comprehensive and well-coordinated spatial planning regulations that are specifically tailored to the country's coastal zones. This significant gap in the legal and regulatory structure has resulted in a variety of inefficiencies and shortcomings in the country's coastal tourism development legislation. These inefficiencies can be traced to a combination of bureaucratic delays, insufficient enforcement mechanisms, and inconsistencies in zoning regulations and environmental standards that currently govern coastal zone management. The development and management of tourism infrastructure in Greece's coastal regions are governed by a complex, multilayered system of laws and regulations, which span across different levels of governance, including national, regional, and local authorities. The fragmentation of these regulatory frameworks often leads to conflicting requirements and overlaps in regulations, which complicates the compliance process for developers and severely hampers effective governance and management practices in these sensitive areas.

The issue of regulatory fragmentation and inefficiency is not unique to Greece. Many other Mediterranean countries that rely heavily on tourism, such as Spain, Malta, and Israel, also face similar challenges. These nations, too, have experienced rapid and often unchecked tourism growth, which has led to various environmental degradation issues, social conflicts, and the deterioration of fragile coastal ecosystems. These problems represent long-term threats to the sustainability of the coastal tourism industry in these regions. For Greece, the challenges that it shares with other Mediterranean countries only serve to amplify the vulnerability of its coastal regions, further increasing the uncertainty surrounding the future of coastal tourism, which is increasingly threatened by environmental degradation. The effects of such degradation may significantly undermine the country's ability to maintain its competitive edge in the global tourism market, especially if its coastal ecosystems continue to suffer irreversible damage.

A central issue contributing to these challenges is the prevailing emphasis on maximizing the immediate economic benefits derived from tourism development, often at the expense of long-term environmental sustainability. This economic-driven approach, which tends to prioritize the construction and expansion of tourism-related infrastructure, often fails to account for the substantial long-term environmental costs that come with such developments. These environmental costs include habitat destruction, depletion of biodiversity, soil erosion, water resource depletion, and the degradation of essential coastal ecosystems that are crucial to the balance and health of these regions. While there are various European Union regulations and directives designed to promote more sustainable practices in tourism development, Greece's adaptation and integration of these frameworks have been slow and inconsistent. This delay in effectively aligning with EU regulations has led to significant gaps in compliance, enforcement, and implementation at the national level, which only further exacerbates the challenges of achieving truly sustainable coastal tourism development in the country.

The combination of these various factors has created a situation where the objectives of economic growth through tourism are often prioritized over environmental considerations. While this may yield short-term economic gains, it leads to long-term environmental losses that could jeopardize the stability and health of Greece's coastal ecosystems, as well as the communities that rely on these ecosystems for their livelihoods. To overcome these challenges and create a more sustainable and balanced approach to tourism development, there is an urgent need for a fundamental redefinition and restructuring of the legal framework governing coastal tourism in Greece. This restructuring should focus on simplifying and streamlining bureaucratic processes, improving the consistency and rigor of regulations, and, above all, prioritizing sustainable development as the core objective of tourism infrastructure projects. By shifting the focus toward long-term benefits derived from environmental protection, social well-being, and sustainable economic development, the legal framework can be better aligned with the overall goal of fostering tourism growth that is both economically viable and ecologically sustainable in the long run.

In order to protect Greece's coastal ecosystems and ensure that they remain intact for future generations, it is essential to reconsider and reorient the country's approach to tourism development. A comprehensive reevaluation and reorientation of Greece's legal, regulatory, and planning frameworks are necessary to foster a balance between tourism growth and environmental preservation. The new framework should place a strong emphasis on environmental sustainability while simultaneously supporting the economic importance of tourism to Greece's overall economy. By ensuring that tourism infrastructure is developed in a manner that minimizes environmental harm, integrates ecological concerns into planning processes, and promotes sustainable land-use practices, Greece can establish a more sustainable and balanced tourism model that benefits all stakeholders involved.

Furthermore, a comprehensive overhaul of Greece's coastal zone legislation should adopt a more holistic and integrated approach to land-use planning. This integrated approach should recognize the interconnectedness of economic, environmental, and social factors that influence the development and use of coastal areas. By embracing such an integrated framework, Greece can better position itself to address current and future challenges associated with coastal tourism development. This approach would allow the country to preserve its coastal areas as valuable resources for both ecological and economic purposes, ensuring their continued viability as tourism destinations. In this way, Greece can ensure that its coastal regions maintain their unique character and provide long-term economic, ecological, and social value, all of which are essential for the country's sustainable development in the future.

Achieving these objectives will require coordinated and concerted efforts across multiple governmental departments, stakeholders, and local communities to ensure that Greece's coastal areas are preserved as public resources for the benefit of both current and future generations. The management of these areas must be approached in a way that ensures that tourism development is done with the utmost respect for the carrying capacity of these ecosystems. A comprehensive, well-integrated approach to coastal zone management is essential for fostering sustainability, ensuring the viability of the tourism industry, and preserving Greece's natural coastal environments for future generations. Only by adopting such an approach can Greece's tourism sector thrive while simultaneously protecting and preserving the unique and irreplaceable coastal ecosystems that are vital to the country's identity and long-term well-being.

REFERENCE

EU. (2002). *European Parliament and Council Recommendation on ICZM (2002/413/ EC)*. European Commission.

Fernando, A., Rosete Verges, F., & Larruga, J. (2024). Coastal Laws evolution in Spain: An analysis from the public policy cycle. *Marine Policy, 162*, 106052. Advance online publication. DOI: 10.1016/j.marpol.2024.106052

Levantis, E. (2001). Greek National Report. *Revue juridique de l'Environnement Année 2001 H-S pp.*, pp. 239-254.

Ley 197/1963. (1963). *Ley 197/1963 sobre Centros y Zonas de Inter´es Turístico Nacional.* Boletín Oficial del Estado, 313,.

Malta Ministry of Tourism. (n.d.). *National Tourism Policy 2015-2020.* Retrieved from https://tourism.gov.mt/wp-content/uploads/2023/04/National-Tourism-POlicy-2015-2020.pdf

Martínez Cordero, J. (2013). Modificaciones conceptuales de la ribera del mar,en la reforma de la Ley de costas. In E. Perez Galvez, & A. Aleman Monterreal, *Costas y urbanismo. El litoral tras la Ley 2/2013, de protecci'on y uso sostenible del litoral y de modificaci'on de la Ley de Costas* (pp. 173-224). Wolters Kluwer Espana.

Ministry of Environmental Protection. (2020, 08 17). *Ministry of Environmental Protection.* Retrieved from Protecting Israel's Coast: https://www.gov.il/en/pages/coastal_environment_preservation

Netto, A. (2009). What is tourism? Definitions, theoretical phases and principles. In J. Tribe, *Philosophical Issues in Tourism* (pp. 43-61). Channel View Publications. DOI: 10.21832/9781845410988-004

Pellach, C., & Carmon, D. (2016). *Mare Nostrum Project-Final Report: Legal-Institutional Instruments for Integrated Coastal Zone Management (ICZM) in the Mediterranean.* Mare Nostrum Project.

Post, J., & Lundin, C. (1996). *Guidelines for integrated coastal zone management.* The International Bank for Reconstruction and Development/The World Bank. DOI: 10.1596/0-8213-3735-1

UNWTO-World Tourism Organization. (2024). *World Tourism Barometer 2024-Volume 22 • Issue 1.* https://www.unwto.org/publications. Retrieved from https://webunwto.s3.eu-west-1.amazonaws.com/s3fs-public/2024-01/UNWTO_Barom24_01_January_Excerpt.pdf?VersionId=IWu1BaPwtlJt66kRIw9WxM9L.y7h5.d1

Zammit Pace, M., Bray, M., Baily, B., & Potts, J. (2018). Beach management review of the Maltese Islands. *Proceedings of the Thirtneeth International MEDCOAST Congress on coastal and marine sciences, engineering, management and conservation*, (pp. 203-214).

Chapter 12
Natural Hazards Geography Assessment in South Lesvos Island, Greece

Kostas D. Kalabokidis
https://orcid.org/0000-0002-4942-802X
University of the Aegean, Mytilene, Greece

ABSTRACT

There is an increasing need to develop new holistic approaches that address the underlying causes, behavior and impacts of natural disasters that appear in rural and residential areas, threatening people, property and heritage, as well as other values at risk. An example of integrated risk management and socioecological perspective is thus offered as a case study in south Lesvos Island, Greece. Two successive natural disasters "hurt" and drastically limited the local environmental, residential, productive, social, economic and tourist capital of the area. It was considered of utmost importance to generally monitor and analyze the post-disaster situation, in order for authorities and public services to plan a series of actions for the implementation of the necessary short-term and medium-to-the-long term rehabilitation measures, and overall reconstruction and development of the affected areas. All in all, we focused on the socioeconomic and biophysical causes and effects of the pathology of natural hazards (i.e. natural hazards geography) rather than trying to treat the symptoms.

DOI: 10.4018/979-8-3373-1554-6.ch012

INTRODUCTION

The recurring catastrophic events of large disasters (e.g. wildfires, storms, floods, landslides, erosion, etc.) are now evolving into major environmental, economic and social threats and crises amid climate change in Greece and throughout the world (Darques et al. 2024). Nowadays, there is an increasing recognition of the need to develop new holistic approaches that address the underlying causes, behavior and impacts of disaster events that develop in rural and residential areas, threatening human lives and property, monuments of natural and cultural heritage, as well as other values at risk (EADO 2023). For example, many attribute the "tsunami" of wildfires to the climate crisis, but this cannot be an excuse for not planning an integrated risk management approach (i.e. with research, risk reduction, readiness, response and recovery planning). Especially in cases of mixing forests and settlements, as happened in south Lesvos Island, Greece, where a July 23, 2022, wildfire started and spread from the nearby semi-mountainous forest and agricultural areas (Figure 1) and ended up in the unmanaged peri-urban forest of the Vatera seaside resort (without fortunately lamenting human casualties in this case). Indeed, the exceptional high aesthetic and protective values of the natural and anthropogenic environment of the Vrissa-Vatera area on the island of Lesvos suffered significant damages from the wildfire, which took place five years after the great destruction of a powerful earthquake (Figure 2) of June 12, 2017 (magnitude 6.3 on the Richter scale) (Chatzis et al. 2019, Vlachakis et al. 2020). These successive natural disasters "hurt" and drastically limited the local environmental, residential, productive, social, economic and tourist capital of the area.

Responding to the extremely emergency conditions prevailing in the earthquake-traumatized area of Vrissa and Vatera in Lesvos, followed by the devastating fire of approximately 2,500 hectares burned on 23-29/7/2022 (Figure 3), a pilot application report was prepared with the objective of the inventory, research, analysis and mapping of the wider area, as a "donation" contribution to the local community[1] (Kalabokidis 2024). It is considered of utmost importance to generally inventory and monitor the post-disaster situation, in order for authorities and public services to plan a series of actions for the implementation of the necessary short-term and medium-/long-term rehabilitation measures, and overall reconstruction and development of the affected areas. Public services do not have at large the overall know-how, scientific training and computational complexity that such processes require to achieve multiple resource objectives, but with scientific assistance authorities can acquire and appropriately utilize socioecological and spatial data, models and technologies.

Figure 1. Development of the wildfire and the huge column of smoke (i.e. plume-dominated) in the semi-mountainous forest and agricultural areas of the wider area in Vrissa and Vatera of Lesvos Island, Greece.

Figure 2. Damages in Vrissa, Lesvos Island, Greece, in the aftermath of the earthquake of June 12, 2017 (magnitude 6.3 on the Richter scale).

Figure 3. Mapping of the burned area of Vrissa and Vatera wildfire with low, medium and high severity levels in south Lesvos Island, Greece

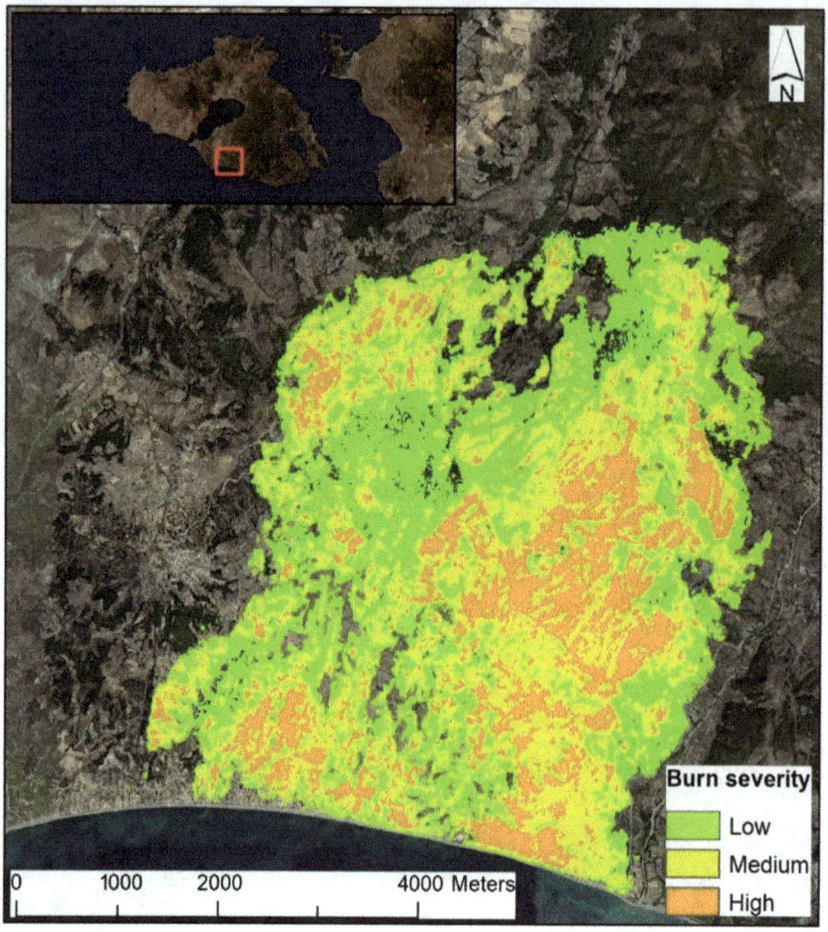

(after Vasilakos and Verykios 2024).

Ground, aerial, satellite, spatial and geo-information techniques (Kavroudakis 2024, Soulakellis et al. 2024 Vasilakos and Verykios 2024) were used to determine and analyze fire behavior and effects in forest, peri-urban and rural ecosystems of the affected area (Dalezios et al. 2018, Arianoutsou and Kazanis 2024, Kalabokidis et al. 2024, Palaiologou and Kalabokidis 2024, Xanthopoulos 2024). Also, a multitude of critical issues were analyzed spatially and thematically, such as rural development and economy, post-fire soil erosion and water resources management, urban and spatial development planning, promotion of natural and cultural heritage, as well as local development (Anagnostou et al. 2024, Gaganis et al. 2024, Kavroudakis 2024, Nagopoulos et al. 2024, Palaiopanis et al. 2024, Tzoraki et al. 2024, Zouros et al. 2024). All in all, we focused on the socioeconomic and biophysical causes and effects of the pathology of natural hazards (i.e. *Natural Hazards Geography*) rather than trying to treat the symptoms (Fischer et al. 2016). In addition, impact assessment in disaster-prone areas of southern Lesvos Island, Greece, and proposals for their environmental and socioeconomic reconstruction and sustainable development after recent catastrophic events were discussed, intended mainly for stakeholders and end-users.

INTEGRATED RISK MANAGEMENT

An Integrated Risk Management strategy was proposed to exploit the synergies between Prevention-Preparation, Detection-Response and Recovery-Adaptation, with a framework of modern technologies and social-ecological-economic solutions to enhance the resilience of forest, agricultural and human ecosystems in a changing climate. Individual objectives were:

- Addressing the phenomena of destructive wildfires and increasing the fire resilience of forest ecosystems in a changing climate.
- The development, exploitation and promotion of innovative solutions at the administrative, social, economic, technological, ecological and health protection levels.
- The integration and local implementation of a specialized level of holistic solutions throughout the affected area and beyond, strengthening cooperation between the public sector, the scientific community, local society, citizen groups and individuals at the various stages of land management.
- The strengthening of social awareness regarding extreme forest fire incidents and other disasters, and the transfer of the necessary knowledge and experience (know-how) to interested parties.

Clearing and thinning of pine forests, reduction of fuels, creation of fire protection zones and management of the wildland-urban interface by experts are necessary to reduce the risk of a fire outbreak and spread, and the safety of those living or operating in the area (Kalabokidis 1996, Bachantourian et al. 2023). Scientific investigation showed the need to open new fire protection zones approximately 20 kilometers long, 50 meters wide and with a total area of 100 hectares in order to achieve the minimum protection of settlements and forests in the wider study area of Polichnitos, Vrissa, Vatera and Stavros in south Lesvos Island (Palaiologou and Kalabokidis 2024). The opening of these zones will also protect the olive groves of the area, but also the particularly important tourist development and recreation areas of both Vatera and Nyfida in southern Lesvos. In addition, property owners that accept these fire protection measures will be protected consequently. The protection and economic development of the area is in everyone's interest. The proposed fire protection measures that include biomass removal cannot be permanent, since any measure implemented will have a lifespan of 3 to 10 years, after which the fuel is expected to recover depending on the type of vegetation that will prevail afterwards. This requires maintenance of the new firebreaks and other fuel reduction projects whenever necessary, so that they maintain their ability to prevent future wildfires (Palaiologou and Kalabokidis 2024).

Identification of negative or positive elements resulting from the vegetation and existence or absence of its fragmentation, existence of zones of mixing settlements with forest or agricultural lands along with population (and its way of reacting) constitute important elements for proposals of improvements in the future. Any "obstacles" to the spread of forest fires with the existence of specific types of land cover and/or vegetation are acceptable. The biophysical and other conditions that contributed to the burning or not of the settlements in the specific area, analyzed in relation to the firefighting efforts, along with the rational management of the vegetation, replacement of plant species, as well as the creation of sheltered fire zones in the affected area should also be under serious consideration (Xanthopoulos 2024). The exposure of settlements with Wildland-Urban Interface (WUI) elements to risk is possible, and in order to limit this possibility and improve the safety of residents and visitors, it is necessary to make serious efforts to prepare all dangerous and vulnerable settlements in Lesvos for the event of a wildfire. This includes Vatera as vegetation will recover, houses will be repaired and maintained, and new ones will be created. In this direction, it is proposed to assess the probability of exposure to a high-intensity fire and the vulnerability of each settlement and categorize it, as well as to take measures to improve fire safety in residential areas (Kalabokidis 1996, Menemenlis et al. 2023, Xanthopoulos 2024). This also includes measures regarding residents at the local level (e.g. information about the risk, ways to prepare themselves and their homes, how to react in the event of a fire, cultivating volunteerism, a Community Wildfire Protection Plan) (Xanthopoulos 2024). The use of stochastic fire simulation models is particularly helpful in the coordination and implementation of the most effective measures for the prevention, suppression and monitoring of forest fires, considering the operational readiness of the services involved, mainly in matters of early intervention and ground action (Kalabokidis et al. 2014). Additionally, the continuous study of weather variables helps in the systematic assessment and understanding of meteorological variability in space and time (Kalabokidis et al. 2013), as well as in the more general planning (prevention, simulation and management) of fire risk (Ager et al. 2017), always aiming at assessing the risk of occurrence and management of natural disasters for the information and education of the authorities and the general public at large (Figure 4).

Figure 4. Meteorological conditions via a web platform

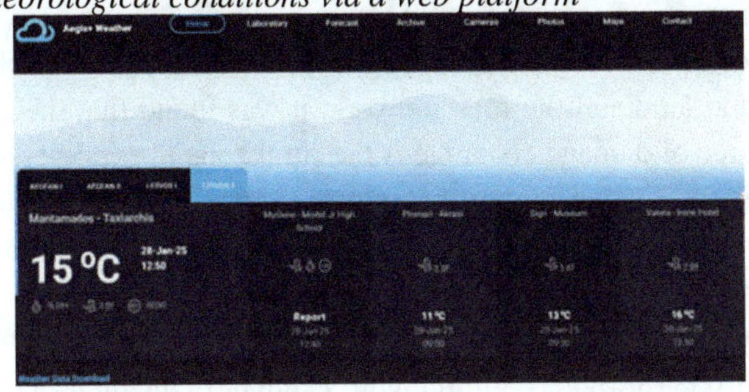

(i.e. http://meteo.aegean.gr/) that allows the display and retrieval of measurements in graphs and tables for Greece, including Lesvos Island.

Regarding the geo-hazard protection of the area, it should be mentioned that the Lesvos Forestry Directorate has worked towards both the soil erosion (Gaganis et al. 2024) and flood (Tzoraki et al. 2024) control of the wider affected area, and an expensive corrective project in the burned areas has already completed as of 2022-2023 with the construction of simple technical works (e.g. log bundles, log dams, branch nets and wire mesh boxes), aiming to mitigate the impacts of flooding and erosion phenomena (Figure 5). As additional measures to these, where the flood risk appears higher due to the extent of the potentially flooded areas and the types of land uses located within them, the construction of embankments along the sections of the local Almyropotamos river that requires special attention (e.g. after the settlement of Vrissa, as well as along the downstream section of the watercourse in Vatera) and other nature-based solutions are indicatively needed (Tzoraki et al. 2024). The appropriate configuration and widening of the watercourse bed could be examined in places, so that in each case their drainage capacity is increased, especially where the new biological sewage treatment system of the region will join Almyropotamos river's watercourse in the near future. Regular cleaning of the watercourse from debris is also considered important, as well as checking for possible points where the flow in the watercourse may be obstructed (Tzoraki et al. 2024).

According to model simulations, almost all of the areas flooded by waters are predicted to be agricultural areas, with the residential area in the area being affected to a less extent. As for the three adjacent streams along the Vatera coastline, the picture regarding the size of the flooded areas differs significantly compared to that of the Almyropotamos river one, as in the three branches under study the simulated flood limits extend a short distance on either side of their bed, covering small areas of land (Tzoraki et al. 2024). For these three watercourses (streams) in Vatera considerably high flow velocities are observed (especially in their upstream section) and a smaller free surface width, which is due to the more pronounced relief of the three adjacent basins for the section of the watercourses and suburbanization coverage downstream (Tzoraki et al. 2024).

Based on economic analyses of floodplains for the land uses of the residential zone, agricultural land and the road network, it was found that the category with the highest degree of damage coverage is the rural zone (Tzoraki et al. 2024). The road network is the second highest category in terms of area affected in the potential flood zone. The urban zone category is affected the least according to the simulated potential flood maps, with floodplains increasing by up to 25% from the 5-year return period to the 100-year period. Regarding the economic damages due to flooding return periods, agricultural lands again have the highest estimates. The damage values for this specific zone differ by three orders of magnitude, maintaining these in all three return periods (i.e. 5-, 50- and 100-year floods). The residential areas have a greater economic loss compared to the road network that has a larger affected area (Tzoraki et al. 2024).

Figure 5. Log bundles (above), log dams (below left) and wire mesh boxes (below right) for erosion and flood protection in the affected areas

(photos by Lesvos Forestry Directorate, Greece)

The wildfire in the wider Vrissa area caused significant social and economic consequences with long-term effects. The reconstruction of the area is a major challenge for eliminating the socioeconomic problems that are created as well as halting the population decline. In addition to the damage to the pine forests of the area, the wildfire destroyed a significant part of the Vrissa olive grove with varying degrees of severity (Figure 6) (Vasilakos and Verykios 2024). Almost three years later, the absence of a coherent plan for the reconstruction of the olive grove is a clear example of the inability to understand the long-term effects of the disaster and the imperative need to eliminate them as quickly as possible from the day after (Palaiopanis et al. 2024). The time and costs of rehabilitation after the catastrophic event highlight the importance of prevention and strengthening rehabilitation mechanisms, for the faster processing of bureaucratic, institutional and financial issues. However, the one-dimensional aspect focusing on the temporary compensation of those affected is not recommended as it will be a disastrous option for the wider region in the long run (Palaiopanis et al. 2024).

Figure 6. Firefighting operations in the burned olive groves of Vrissa, Lesvos Island, Greece

(after Palaiopanis et al. 2024)

The wildfire highlighted serious issues related to the physical safety of olive groves (Palaiopanis et al. 2024). Olive cultivation occupies approximately 40% of the island of Lesvos and is directly or indirectly linked to the entire local population and demographics. The olive-growing areas do not constitute a single entity but are spread across the entire island, showing high fragmentation both in geographical space and in the distribution of plots and ownerships. Furthermore, they show heterogeneity in the cultivated varieties, soils, altitudes, morphological conditions, etc. (Palaiopanis et al. 2024). The juxtaposition of these areas to forests, shrublands and grasslands, and their development around urban areas and settlements increases the risk of causing considerable damages or destruction by fire or flooding. The changing climatic conditions and the intense anthropogenic interventions are creating a new reality in our era for risk mitigation and management. The protection of the olive groves of Lesvos against natural hazards should constitute a special chapter of research and application in connection with the other areas of the island, treating it as one of the largest wealth-producing resources of the island (Palaiopanis et al. 2024).

SOCIOECOLOGICAL PERSPECTIVES

Olive groves, pine forests, riverside vegetation, water resources, beach fronts, flora, fauna, livestock, thermal springs, archaeological and historical sites, traditional heritage, rural-urban landscapes, properties, infrastructure and human resources constitute an entire local capital, and only a holistic approach to rehabilitation will bring direct and multiplying benefits for the region. The lack of a culture of prevention, the inability to quickly adapt to a constantly changing and multi-hazard environment, the difficulty of changing priorities and redirecting resources are elements that must be reformed in this era of climate, environmental and socioeconomic crises. The Vrissa-Vatera wildfire can serve as an example for raising awareness and education of the population through knowledge mining, with the organization of space and resources according to the principles of Mother Nature (Nagopoulos et al. 2024, Palaiopanis et al. 2024). The enhancement of social awareness regarding extreme forest fire incidents and the transfer of the necessary knowledge and experience (know-how) to the interested user bodies should be implemented through appropriate and targeted information and training initiatives (Nagopoulos et al. 2024).

Mediterranean-type ecosystems are well adapted to frequent wildfires, which means that if the natural succession is not disturbed by human interventions for at least 3-5 years after the fire in the area, the system will probably return to its original stage (Kalabokidis and Palaiologou 2019, Arianoutsou and Kazanis 2024). An in-situ inventory in various locations confirmed that in general the regeneration of pine trees was progressing at a satisfactory rate, except for the sites that had been burned with high fire intensities (Kazanis et al. 2023). The eventual density of pine trees will indicate sites where regeneration support may be needed in order to maintain the forest character of the area. Land use planning to reduce the future exposure of forests, crops and settlements to fire, and the increased focus on the prevention and deterrence of human-caused disasters in the midst of climate change are two approaches with important priorities and prospects (EADO 2023). The importance of prevention, information, and education of the bodies involved should be emphasized in order to improve the organization and coordination in the management of natural disasters (Nagopoulos et al. 2024).

Climate change affects all spheres of social actions and has significant impacts on today's societies of risks and disasters worldwide (Beck 2013). In this light, climate change is not only an environmental problem that requires technical and management solutions (EADO 2023), but also a social one that has significant impacts on human communities; and therefore, it requires a holistic approach towards risk mitigation and resilience improvement to ecosystems and societies, along with restoration planning after disaster events (EADO 2023, Nagopoulos et al. 2024). It is highly recommended to continue monitoring and strictly adhere to the above principles (Kalabokidis 2024). At the same time, there is even now a lack of environmental awareness at a social level, resulting in this phenomena being treated not as an "anthropogenic climate change", and neglecting also the interplay of socio-psychological factors in managing the impacts and addressing the risks in the modern era (Beck 2013, Nagopoulos et al. 2024).

Social research in the affected area reveals the need to establish an organizational environment that will contribute to strengthening feelings of community, cooperative disposition among the residents of the area and expanding local social capital for the protection of the natural environment as a key factor of local development, social cohesion and geographical planning (Kalabokidis et al. 2021, Nagopoulos et al. 2024). An organizational environment is needed that can shield local communities against risks and strengthen awareness of the value of active participation of citizens in a common action plan to prevent threats and risks due to natural disasters (Kalabokidis et al. 2008). The degree of preparedness in dealing with future emergencies is inextricably linked to the establishment of a common action plan, the result of networking and cooperation between the State, local government, social organizations and citizens (Nagopoulos et al. 2024). The risks that threaten human societies may not be completely avoidable; however, it is important to identify, develop and implement ways to mitigate their impacts through a preventive governance of the risks before they occur (Morehouse et al. 2011). It is therefore important to diagnose forms of risk early, i.e. a responsible and integrated strategy linked to the role of institutions, organizations and human behaviors in the face of uncertainties and potential risks, which in order to be effectively addressed should be further examined (Nagopoulos et al. 2024).

Furthermore, addressing urban and spatial planning issues of the Vatera-Vrissa settlements, as well as the exploitation of their traditional cultural heritage, can contribute to attracting a form of alternative, low-disturbance tourism to the surrounding area (Anagnostou et al. 2024, Kavroudakis 2024). The mild tourist development will add a quite basic economic resource for the residents at this difficult turning point in the history of the region, complementing the agricultural, livestock and fishing employment of the residents and it will function as a lever for socioeconomic and cultural development, improving the quality of life of the residents and at the same time ensuring the tradition, cultural heritage and history of the place (Figure 7) (Anagnostou et al. 2019).

With the aim of developing alternative forms of tourism in the wider area, it is proposed to highlight its wealth through the design of walking trails and other routes with the aim of connecting sites and other points of interest that include geological monuments, hot springs, waterfalls, sandy and rocky coasts, wetlands, habitats and observatories of wild bird fauna, salt pans, sites of ecological and paleontological interest (Figure 8), and cultural monuments, museums, collections and visitor information areas (Zouros 2007). The most important of these sites have already been verified as geo-sites in the official recognition of Lesvos as a UNESCO Global Geopark (Zouros et al. 2024). In the wider area, within pyroclastic formations, significant fossils have been identified, and in particular fossilized tree trunks with complete petrification provide the possibility of determining the species of fossilized trees. It is also known that significant fossils of terrestrial vertebrate animals approximately 2 million years old have been discovered in the area. All of these remain unexploited and await actions to highlight them with possible excavations and their proper promotion by competent local and scientific bodies (Zouros et al 2024). Such a perspective could work towards enriching the tourist product of the area and with the development of educational activities and internships and/or summer schools for students, as well as for other scientific meetings and conferences with the creation of appropriate tourist infrastructure (Zouros et al. 2024).

Figure 7. Traditional cultural heritage of the wider rural and residential landscape of south Lesvos Island, Greece

(after Anagnostou et al. 2024)

Figure 8. Coastal geotopes and rocky beach fronts along the Vatera coastline in Lesvos Island, Greece

(after Zouros et al. 2024)

Natural disasters and catastrophes are an integral part of human and everyday life and therefore must be given due attention. The area of south Lesvos Island has recently suffered a major earthquake with great destructions mainly in the town of Vrissa (Figure 9). Some physical and human geography aspects of this catastrophe still remain unchanged and are worth highlighting. Issues of proper disaster response, efficient civil protection, effective rehabilitation, exemplary volunteerism and humanitarian support for the earthquake-stricken region were serious milestones from the beginning of the incidence. Critical questions of how we could prepare to prevent the disaster, to reduce the risk and to react properly if a disaster occurs are of relevance, in addition to reducing the impact of a disaster on human lives, material damages and environmental consequences. Thus, some pertinent responses include rescuing trapped people during the event, planning for relief of the population in the first life-threatening hours and recovery of the affected region in the next critical months after the event. Some elements may be summarized as below (Nikellis 2017):

- Removing the risk of the area, whereas in the case of the 2017 earthquake there has been a significant delay for almost eight years now; and still without any official removal of the hazard of the town of Vrissa, despite given promises by the authorities.
- Restoring the opening of the roads by removing debris etc. to allow access to all points of the settlement.
- Taking measures to avoid possible flood damages (collapses, household goods, etc.) to buildings and removing flammable and toxic materials.

• Finalizing the results of geological and geophysical studies before reconstruction decisions are made and on what settlement is required to rebuild, with the participation of directly interested residents and consultation with the competent State authorities.
• Creating open and public common spaces and constructing public infrastructure developments and utilities in the new settlement.
• Providing adequate funding with transparency, control and accountability.

Figure 9. An aerial view of the destroyed town of Vrissa, Lesvos Island, Greece, from a drone coverage

(photo by Ermioni Papadopoulou).

The volunteer activity is of enormous value in the times of disaster both for management of the event and the people's relief as well. Besides the economic and technical matters of a natural disaster, the human factors in terms of psychological support have been also of tremendous importance for the suffering population. The collapse of the houses, the injury of persons, the eviction, the violent changes (for the worse) in daily habits affect children and adults to a greater or lesser extent (Figure 10). For all those affected, support from psychologists has been considered absolutely necessary. However, there are other actions that could alleviate the problems to some extent (Nikellis 2017):

• The immediate treatment in the first instances of the consequences of the disaster (e.g. securing shelter and food), the removal of uncertainty both about the evolution of the phenomenon and about the next day. In Vrissa, it took a month and a half for the relevant Joint Ministerial Decree of rebuilding to be issued and there are plenty of uncertainties about it even now, almost eight years after the event.

• An important role in encouraging those affected is also played by the feeling of security that solidarity inspires. For this reason (and of course not only) in Vrissa there was an action to utilize some money given directly to the local council by distributing foods that are perishable and ready-made meals to the elderly or sick people. With this action, the local community and networks wanted to say "we are by your side" to the helpless and the weak.

• A good injection of self-confidence and optimism was also the existence of various activities that life continues and will soon fully regain its rhythm (e.g. in Vrissa the local Library's cultural center was operating and has already given these signals with the even increasing activities mainly in Vatera till nowadays).

Figure 10. A view of the deserted main-activity (Platanus) square and streets in Vrissa, Lesvos Island, Greece, in the aftermaths of the 2017 earthquake.

CONCLUSION

The purpose of this study was to conduct applied research concerning support for the overall reconstruction and development of the affected areas of Vrissa-Vatera, Lesvos, Greece, after the wildfire of 23/07/2022 (and following the devastating earthquake of June 2017). The fire incident, which led to the destruction of forest, agricultural and peri-urban areas, was characterized as an extreme event of an erratic wildfire due to the rapid spread, the development of a large vertical combustion plume (smoke column) and the unpredictable behavior of the fire, amidst intense dry/ thermal and topographic conditions (Kalabokidis et al. 2024). The scope of the project was achieved through a series of research and scientific analyses and works, numerous field visits for data collection, discussions, workshops, meetings, personal communication, interviews, outreach, educational seminars, conferences, graduate and undergraduate theses (Lahouri et al. 2024). Thus, conclusions and proposals emerged as discussed and reported in this chapter. Through these, the following strategic objectives were pursued (Kalabokidis 2024):

- Promoting sustainable economic development
- Proposals for the establishment of public utility services, and
- Implementation of rational environmental policies.

All of the above constitute priorities with the aim of revitalizing the wider area of Vatera and Vrissa in Lesvos Island, Greece, through a modern plan (with conditions, criteria, rules, possibilities, timetables, implementation resources and effectiveness) on issues of rational sustainable development (e.g. farming, agriculture, agroforestry, tourism, culture, etc.) and the location of activities (i.e. economy, society and environment). Emphasis should be placed on the necessity of improving and developing proper tourist infrastructure on the occasion of the summer period, when the area exceeds its capacity many times over, a fact that would benefit the place in the rest of the period and is understandable by the people and authorities (Anagnostou et al. 2019, 2024). Of course, all of this presupposes the will of a synchronized State, which should finance/ coordinate/ implement/ control in a timely and valid manner the overall socioeconomic and environmental reconstruction and support of the place with its competent services, but also the local people themselves (society and local bodies) in the role of protagonists and not spectators. It should also be understood that the resort town of Vatera has been the "vehicle" and an especially important "asset" of the earthquake-prone and fire-stricken Vrissa throughout time, in the face of the need and vision of salvation and regeneration of the wider region (Anagnostou et al. 2019, 2024).

After all, we do talk about a wider "considered isolated" Greek region of enormous national, geopolitical, environmental, agricultural, geological, cultural, thermal, tourist and socioeconomic importance and value on the borders of the Aegean Sea, Greece and Europe. Any interventions to address the problems should yield tangible results with a future and perspective, and not simply temporary, pointless and demagogic interventions without continuity. Finally, the socioeconomic support and rehabilitation of these nationally and multi-faceted sensitive island regions of Greece is absolutely necessary and crucial for the survival, retention and future attraction and reintegration of permanent residents into these places. The flee of young, skilled workers has a further negative impact on the aging of the population, the renewal of generations and rural-urban development. If the increasing population "desertification" continues with low birth rates and abandonment of the region, then the current demographic changes and imbalances can have a significant impact on various aspects of human lives, both in society and the economy as well as in culture and the environment. All of these constitute a serious existential challenge for Greece at a national and geopolitical level (Anagnostou et al. 2019).

REFERENCES

Ager, A., Finney, M., Kalabokidis, K., & Moore, P. (2017). Wildland fires. In Dalezios, N. R. (Ed.), *Environmental Hazards Methodologies for Risk Assessment and Management* (pp. 305–336). International Water Association Publishing. DOI: 10.2166/9781780407135_0307

Anagnostou, S., Kalabokidis, K., & Sidiropoulos, G. (2019, Fall). Vatera of Lesvos in the next decade: The need for a plan for regeneration, reconstruction and development. [In Greek]. *Echo of Vrissa*, (66), 19–22.

Anagnostou, S., Tataris, G., Sidiropoulos, G., Pelekou, E., & Gougoulios, N. 2024. Spatial development planning. In: Kalabokidis K. (Ed.). Impact study and expertise for addressing problems in the affected area of Vrissa-Vatera-Stavros in Lesvos Island, Greece, after the fire of July 23, 2022. University of the Aegean. ISBN: 978-618-86408-1-8. Pp. 206-223. (In Greek).

Arianoutsou, M., & Kazanis, D. 2024. Assessment of the potential for natural regeneration of Pinus brutia forests in the burned area of Vrissa-Vatera-Stavros in southern Lesvos. In: Kalabokidis K. (Ed.). Impact study and expertise for addressing problems in the affected area of Vrissa-Vatera-Stavros in Lesvos Island, Greece, after the fire of July 23, 2022. University of the Aegean. ISBN: 978-618-86408-1-8. Pp. 93-104. (In Greek).

Bachantourian, M., Kalabokidis, K., Palaiologou, P., & Chaleplis, K. (2023). Optimizing fuel treatments allocation to protect the wildland–urban Interface from large scale wildfires in Greece. *Fire (Basel, Switzerland)*, *6*(2), 75. DOI: 10.3390/fire6020075

Beck, U. (2013). *World at Risk*. Polity Press.

Chatzis, N., Kkallas, C., Papazachos, C., Anthymidis, M., Rovithis, E., Karakostas, C., & Papaioannou, C. 2019. Stochastic simulation of seismic motion and site-effects studies of ambient noise and seismic data: The case of the Vrisa settlement and the 2017 M = 6.3 Lesvos earthquake. 15th International Congress of the Geological Society of Greece, Athens, 22-24 May 2019 | Harokopio University of Athens, Greece. Bulletin of the Geological Society of Greece, Sp. Pub. 7 Ext. Abs. GSG2019-250. Pp. 24–25.

Dalezios, N., Kalabokidis, K., Koutsias, N., & Vasilakos, C. 2018. Chapter 10. Wildfires and Remote Sensing: An Overview. In: Remote Sensing of Hydrometeorological Hazards, G.P. Petropoulos and T. Islam (eds.). CRC Press, Taylor & Francis Group, Boca Raton, Florida, US. Pp. 211-236.

Darques, R., Sidiropoulos, G., & Kalabokidis, K. (Eds.). (2024). *The Geography of Greece - Managing Crises and Building Resilience. World Regional Geography Book Series*. Springer. DOI: 10.1007/978-3-031-29819-6

EADO. (2023). *Resilience of the Greek Forest Ecosystems to Climate Change. Committee for the Resilience of the Greek Forest Ecosystems to Climate Change (EADO)* [M. Arianoutsou, C. Zerefos, K. Kalabokidis, A. Poupkou, F. Aravanopoulos (Eds.)]. Academy of Athens. (In Greek)

Fischer, A. P., Spies, T. A., Steelman, T. A., Moseley, C., Johnson, B. R., Bailey, J. D., Ager, A. A., Bourgeron, P., Charnley, S., Collins, B. M., Kline, J. D., Leahy, J. E., Littell, J. S., Millington, J. D. A., Nielsen-Pincus, M., Olsen, C. S., Paveglio, T. B., Roos, C. I., Steen-Adams, M. M., & Bowman, D. M. J. S. (2016). Wildfire risk as a socioecological pathology. *Frontiers in Ecology and the Environment, 14*(5), 276–284. DOI: 10.1002/fee.1283

Gaganis, P., Kontos, T., & Anastasiou, M. 2024. Soil erosion risk assessment of the Vatera area of Lesvos Island as a consequence of the disastrous fire of July 2022. In: Kalabokidis K. (Ed.). Impact study and expertise for addressing problems in the affected area of Vrissa-Vatera-Stavros in Lesvos Island, Greece, after the fire of July 23, 2022. University of the Aegean. ISBN: 978-618-86408-1-8. Pp. 106-120. (In Greek).

Kalabokidis, K. 1996. Wildland fire safety in residential areas. In *Proceedings International Fire Safety Conference*, 24-25 May 1996, Santorini, Greece. Department of Civil Engineering, Aristotle University of Thessaloniki, Greece. Pp. 267-278.

Kalabokidis, K. (Ed.). 2024. Impact study and expertise for addressing problems in the affected area of Vrissa-Vatera-Stavros in Lesvos Island, Greece, after the fire of July 23, 2022. University of the Aegean. ISBN: 978-618-86408-1-8. 376 p. (In Greek).

Kalabokidis, K., Iliopoulos, N., & Gliglinos, D. (2013). *Fire Meteorology and Forest Fire Behavior in a Changing Climate*. ION Publications. (In Greek)

Kalabokidis, K., Iosifides, T., Henderson, M., & Morehouse, B. (2008). Wildfire policy and use of science in the context of human–environment system on the Aegean Archipelago. *Environmental Science & Policy, 11*(5), 408–421. DOI: 10.1016/j.envsci.2008.01.006

Kalabokidis, K., Korres, G., Soulakellis, N., & Feidas, H. (Eds.). (2021). *Social Sciences and Geography: Theory, Methods and Techniques of Spatial Analysis. In memoriam to Professor Pavlos Kanaroglou*. University of the Aegean, Department of Geography. (In Greek)

Kalabokidis, K., & Palaiologou, P. (2019). Mediterranean Forest Fuels. In Manzello, S. (Ed.), *Encyclopedia of Wildfires and Wildland-Urban Interface (WUI) Fires*. Springer. DOI: 10.1007/978-3-319-51727-8_29-1

Kalabokidis, K., Palaiologou, P., & Finney, M. 2014. Fire behavior simulation in Mediterranean forests using the Minimum Travel Time algorithm. In: Wade DD & Fox RL (Eds.), Robinson ML (Comp.). Proceedings of 4th Fire Behavior and Fuels Conference, 18-22 February 2013, Raleigh, NC, USA and 1-4 July 2013, St. Petersburg, Russia. Published by the International Association of Wildland Fire: Missoula, MT, USA. Pp. 468-492.

Kalabokidis, K., Roussou, O., Vasilakos, C., Cristaki, M., Zikoulis, A., Palaiologou, P., Kypriotellis, P., & Gougoulios, N. 2024. Fire ecology and management. In: Kalabokidis K. (Ed.). Impact study and expertise for addressing problems in the affected area of Vrissa-Vatera-Stavros in Lesvos Island, Greece, after the fire of July 23, 2022. University of the Aegean. ISBN: 978-618-86408-1-8. Pp. 18-70. (In Greek).

Kavroudakis, D. 2024. Spatial analysis. In: Kalabokidis K. (Ed.). Impact study and expertise for addressing problems in the affected area of Vrissa-Vatera-Stavros in Lesvos Island, Greece, after the fire of July 23, 2022. University of the Aegean. ISBN: 978-618-86408-1-8. Pp. 194-203. (In Greek).

Kazanis, D., Roussou, O., Bazos, I., Kalabokidis, K., & Arianoutsou, M. 2023. Estimating the post-fire regeneration potential of Pinus brutia forests at south Lesvos island, Greece. Poster paper in Proceedings of the 11th HELECOS Conference. Patras, Greece, 4-7 October 2023. Pp. 191-192.

Lahouri, D., Kalloniatis, C., Kalabokidis, K., Tzoraki, O., Roussou, O., Dimou, I., Kopsacheilis, V., Andreadis, O., Zikoulis, A., & Krallis, F. 2024. Communication and promotion issues. In: Kalabokidis K. (Ed.). Impact study and expertise for addressing problems in the affected area of Vrissa-Vatera-Stavros in Lesvos Island, Greece, after the fire of July 23, 2022. University of the Aegean. ISBN: 978-618-86408-1-8. Pp. 311-340. (In Greek).

Menemenlis, D., Palaiologou, P., & Kalabokidis, K. (2023). Wildfire-residential risk analysis using building characteristics and simulations to enhance structural fire resistance in Greece. *Fire (Basel, Switzerland)*, *6*(10), 403. DOI: 10.3390/fire6100403

Morehouse, B. J., Henderson, M., Kalabokidis, K., & Iosifides, T. (2011). Wildland fire governance: Perspectives from Greece. *Journal of Environmental Policy and Planning*, *13*(4), 349–371. DOI: 10.1080/1523908X.2011.611678

Nagopoulos, N., Paraskevopoulos, D., Maragoudakis, M., Kouroutzas, C., ... 2024. Qualitative social research on the protection of the natural environment as a key factor of local development and social cohesion. In: Kalabokidis K. (Ed.). Impact study and expertise for addressing problems in the affected area of Vrissa-Vatera-Stavros in Lesvos Island, Greece, after the fire of July 23, 2022. University of the Aegean. ISBN: 978-618-86408-1-8. Pp. 242-285. (In Greek).

Nikellis, E. 2017. Personal communication on local civil protection issues in the aftermath of the Vrissa 2017 earthquake. Vrissa-Vatera resident and administrator of a local webpage (i.e. https://vatera.gr/platanos/, in Greek), Lesvos Island, Greece.

Palaiologou, P., & Kalabokidis, K. 2024. Fuel breaks and estimated burning conditions with the aim of breaking the continuity of forest vegetation in the wider area of Vatera, Vrissa, Stavros and Polichnitos. In: Kalabokidis K. (Ed.). Impact study and expertise for addressing problems in the affected area of Vrissa-Vatera-Stavros in Lesvos Island, Greece, after the fire of July 23, 2022. University of the Aegean. ISBN: 978-618-86408-1-8. Pp. 71-82. (In Greek).

Palaiopanis, R., Pappas, I., Kizos, A., & Tscheulin, T. 2024. Agricultural issues in the affected area of Vrissa- Vatera-Stavros of Lesvos by the fire of 23 July 2022. In: Kalabokidis K. (Ed.). Impact study and expertise for addressing problems in the affected area of Vrissa-Vatera-Stavros in Lesvos Island, Greece, after the fire of July 23, 2022. University of the Aegean. ISBN: 978-618-86408-1-8. Pp. 170-192. (In Greek).

Soulakellis, N., Vasilakos, C., Tataris, G., Papadopoulou, E., Proestakis, S., Hatzivassiliou, T., & Chroni, A. 2024. Web cartography. In: Kalabokidis K. (Ed.). Impact study and expertise for addressing problems in the affected area of Vrissa-Vatera-Stavros in Lesvos Island, Greece, after the fire of July 23, 2022. University of the Aegean. ISBN: 978-618-86408-1-8. Pp. 287-309. (In Greek).

Tzoraki, O., Siarkos, E., Hatzivassiliou, T., & Tsoulibrk, A. 2024. Assessment of flood risk and expected cost of damage. In: Kalabokidis K. (Ed.). Impact study and expertise for addressing problems in the affected area of Vrissa-Vatera-Stavros in Lesvos Island, Greece, after the fire of July 23, 2022. University of the Aegean. ISBN: 978-618-86408-1-8. Pp. 121-167. (In Greek).

Vasilakos, C., & Verykios, V. (2024). Burned olive trees identification with a deep learning approach in Unmanned Aerial Vehicle images. *Remote Sensing (Basel), 16*(23), 4531. DOI: 10.3390/rs16234531

Vlachakis, G., Vlachaki, E., & Lourenço, P. B. (2020). Learning from failure: Damage and failure of masonry structures, after the 2017 Lesvos earthquake (Greece). *Engineering Failure Analysis*, *117*, 104803. DOI: 10.1016/j.engfailanal.2020.104803

Xanthopoulos, G. 2024. Wildland-urban interface area. In: Kalabokidis K. (Ed.). Impact study and expertise for addressing problems in the affected area of Vrissa-Vatera-Stavros in Lesvos Island, Greece, after the fire of July 23, 2022. University of the Aegean. ISBN: 978-618-86408-1-8. Pp. 83-92. (In Greek).

Zouros, N. (2007). Geomorphosite assessment and management in protected areas of Greece. Case study of the Lesvos Island – coastal geomorphosites. *Geographica Helvetica*, *62*(3), 169–180. DOI: 10.5194/gh-62-169-2007

Zouros, N., Valiakos, E., Lambrakopoulos, A., & Antonakis, E. 2024. Display of geological – natural heritage of Vrisa area. In: Kalabokidis K. (Ed.). Impact study and expertise for addressing problems in the affected area of Vrissa-Vatera-Stavros in Lesvos Island, Greece, after the fire of July 23, 2022. University of the Aegean. ISBN: 978-618-86408-1-8. Pp. 226-241. (In Greek).

ENDNOTES

1 The full report is available and can be downloaded from the following link (in Greek): http://aegisplusrisk.aegean.gr/ekthesi-vrisiis/

Chapter 13
Exploring the Role of the Ecomuseum Model as a Catalyst for Territorial Development:
The Case of the Non-Profit Civil Company "Messolonghi By Locals"

Evangelos S. Pavlis
https://orcid.org/0000-0001-8210-541X
Agricultural University of Athens, Greece

ABSTRACT

This article examines territorial development as a strategic approach for sustainable regional growth, with a focus on ecomuseums as catalysts for socio-economic and cultural advancement. Territorial development is a place-based model emphasizing local governance, participatory management, and resource optimization, reinforced by territorialization, which strengthens local identity through community-led initiatives. Using the "Messolonghi By Locals" (MbL) case, the article demonstrates how ecomuseum-inspired initiatives promote regional identity, public engagement, and sustainable tourism. Findings highlight the impact of community-driven cultural initiatives in fostering social, cultural, and economic benefits, underscoring ecomuseums' role in enhancing community resilience and informing territorial policy.

DOI: 10.4018/979-8-3373-1554-6.ch013

1. INTRODUCTION: THE MEANING AND CHARACTERISTICS OF TERRITORIAL DEVELOPMENT

Territorial development refers to the strategic planning and implementation of policies aimed at enhancing the economic, social, and environmental conditions within a specific geographical region. It encompasses multiple dimensions, including economic, social, cultural, environmental, and institutional development, addressing issues such as governance, environmental conservation, community engagement, and the promotion of sustainable practices across these domains (Magnaghi 2005). This approach is grounded in the participatory management of shared natural and cultural resources, promoting their distinctive characteristics and value, with a particular focus on local governance (Pavlis 2024). While territorial development emphasizes localized and immediate concerns, it contrasts with the broader and long-term orientation of sustainable development (Goussios, Gaki, & Faraslis 2023). Territorial development encompasses activities such as agriculture, folk art, craftsmanship, industry, thematic tourism, sports, and cultural engagement, including educational initiatives. It is characterized by an experiential rather than observational approach (Wylie 2007). Additionally, it seeks to restore the relationships between agriculture and food production, producers and consumers, local and global dynamics, and humans and the landscape, as a means of addressing increasing global competition (Pavlis 2024).

Territorialization is achieved by establishing control over the relationship between local society, space, and its resources, based on a socio-political agreement. This process involves building a cultural identity that forms the foundation for constructing and defining the territorial area (Campagne and Pecqueur 2014; Anthopoulou and Goussios, 2007; Goussios, Gaki, and Faraslis 2023). It fosters forms of self-organization and self-governance, which can lead to the emancipation of rural areas from state control, enabling more effective collaborating with state structures (Goussios, 2013). This concept complements territorial development by emphasizing the importance of local control and cultural identity in shaping the development and governance of specific regions.

Territorial potential builds upon the foundational concepts of territorial development and territorialization by emphasizing the set of cooperative, coordinated, and interdependent relationships that collectives or groups of individuals maintain with their space. This potential is expressed through the reconnection of consumers with producers, the revival of food cultures, rural traditions, and local identities, and the adoption of environmentally friendly management practices. It also involves the re-exploitation of local resources within the broader political and economic framework of the emerging "new rurality" which centers on a common territorial plan (Goussios 2013). By harnessing territorial potential, communities strengthen their capacity for self-organization and self-governance. This reconnection fosters a deeper relationship between rural areas and their inhabitants, reinvigorating local economies and reinforcing cultural heritage. As a result, territorial potential plays a critical role in addressing contemporary challenges such as global competition, environmental sustainability, and social cohesion, while aligning with the broader goals of territorial development, which seek to enhance the well-being of specific regions through participatory management and the strategic use of natural and cultural resources (Pavlis 2024). In addition, the concept of territorial capital encompasses human, social, natural, infrastructure, physical/material and financial capital and integrates a range of development and sustainability concepts in a comprehensive manner, acknowledging their potential to enhance the long-term well-being of the community (Morreta 2021).

The scope of the article is to explore the role of the ecomuseum model in fostering territorial development by utilizing existing local resources, such as cultural heritage, landscapes, and community assets, to promote sustainable growth and identity-building. It aims to investigate how the ecomuseum's bottom-up approach, which emphasizes community involvement, networking, and the preservation of local traditions, contributes to socio-economic development without the need for extensive new infrastructure. By analyzing the case of "Messolonghi By Locals" the article will provide a practical example of how participatory cultural initiatives can enhance local engagement, strengthen regional identity, and promote sustainable tourism, demonstrating the potential of ecomuseums as catalysts for regional transformation.

2. THE CHARACTERISTICS OF THE "ECOMUSEUM" MODEL

The concept of the ecomuseum was introduced by museologists Hugues de Varine and Georges H. Riviere in 1971. The term "ECO" derives from the Greek word 'οίκος', meaning natural environment, and "MUSEUM" refers to a place or temple dedicated to the Nine Muses of Greek mythology, the goddesses of various arts.

An ecomuseum is essentially a "museum without walls" integrating culture, nature, and history by associating exhibits with their broader natural and human environments, conditions, and functions. It operates as a community museum, placing the local population at the heart of its management and interpretation (de Varine 1996). Unlike traditional museums, ecomuseums adopt a holistic approach to interpreting cultural landscapes, which allows them to capture the unique characteristics of a place that conventional museums often cannot.

Ecomuseums emphasize the unique characteristics of local areas, while linking key aspects such as health, food, and environmental awareness with the surrounding landscape. By promoting short supply chains, fair trade, and innovation for new job opportunities, ecomuseums help revitalize local economies and ensure that development is grounded in the specific context of each territory (Dal Santo et al. 2017; Navajas Corral and Fernández 2022). Ecomuseums also strengthen local participation in both the management and decision-making processes related to their regions, fostering a sense of ownership and collaboration within communities (Dal Santo et al. 2017). Moreover, ecomuseums can contribute significantly to the repopulation of marginalized or rural areas by enhancing their connection to public spaces and offering opportunities for meaningful engagement with local heritage and the economy. By encouraging sustainable practices, innovation, and active participation, ecomuseums ensure that development is both locally driven and culturally embedded, reinforcing the socio-political agreements and governance structures essential for long-term territorial sustainability (Corsane et al. 2007; Dogan and Timothy 2019; Johnston and Marwood 2017; Wuisang et al., 2018).

Heritage related to ecology, ethnography, folklore, history, and archaeology is exhibited throughout the community in various settings such as gardens, roads, neighborhoods, public buildings, and cultural centers. This method of display ensures that heritage remains connected to its environment, as "living evidence" of an era should not be separated from its surroundings—otherwise, it risks becoming merely a dead object (Davis 2011). Ecomuseums are based on structured narratives that focus on selected topics, particularly the relationship between humans and their place within the landscape, providing a deeper, context-rich interpretation of cultural heritage.

According to Davis (2011), the ecomuseum serves as a platform for local development by utilizing common cultural and natural assets, with a focus on quality and the promotion of living heritage. A typical ecomuseum's goal is to provide an integrated and immersive experience that connects people with the heritage of their region. Additionally, an ecomuseum can function as an operational tool for policy implementation, particularly in lagging or underdeveloped areas. It offers a framework for fostering local development through the preservation and active engagement with cultural and natural resources. Accordingly, the primary objective of a successful ecomuseum is to act as a platform for the activation, organization, and promotion of local assets. Its core aims include: (a) improving quality of life and strengthening the sense of community, (b) preserving common heritage assets through sustainable management, and (c) empowering local capital (cultural, social, economic, and ecological).

In terms of improving quality of life and strengthening community bonds, successful ecomuseums emphasize local empowerment and the activation of civil society (Dogan and Timothy 2019). This is achieved through a bottom-up approach, where the local community takes the lead in deciding which aspects of their cultural and natural heritage are significant, addressing the values attached to landscapes and other local features. The process is grounded in a social contract that involves shared responsibilities and cooperation among local actors, administrative bodies, scientists, NGOs, and citizens (Pappalardo 2020). Decisions are made collectively through general stakeholder meetings, ensuring that the management and planning are done "by the local community for the local community".

The protection and sustainable management of common heritage assets are also central goals. Ecomuseums aim to reframe the local development model by reviving local ethics, customs, traditions, and celebrations. This approach encourages residents to build their own sense of identity by fostering a deeper connection between nature and culture. Furthermore, ecomuseums work to raise awareness of environmental, cultural, and landscape issues, promoting the development of a strong environmental and landscape conscience (Pavlis and Terkenli 2023; Pavlis 2017). This is accompanied by efforts to protect, manage, and plan the landscape in a sustainable manner.

Empowering local capital is another critical focus of ecomuseums. Local capital is understood as a combination of cultural, natural, social, and human resources. Ecomuseums support the rejuvenation of traditional professions, such as stonemasonry, tailoring, needlework, soapmaking, carpentry, and basketry, alongside others. They also foster the development of new products, services, and ideas, and promote local products and cuisine. Additionally, ecomuseums encourage the growth of alternative forms of tourism, including agro-tourism, ecotourism, cultural tourism, gastronomic tourism, and science tourism, all of which contribute to local economic resilience and cultural vitality (Dal Santo et al. 2017; Dogan 2015).

3. THE CASE OF "MESSOLONGHI BY LOCALS"

3.1. Non-Profit Civil Company's Profile: Messolonghi by Locals

"Messolonghi by Locals" (MbL) is a Non-Profit Civil Company (AMKE, the Greek initials) founded in 2018 by a group of young individuals committed to their rural town of Messolonghi during a period of significant social and economic crisis. Non-Profit Civil Companies (AMKE) in Greece focus on social benefit, reinvesting any surplus into their objectives and often collaborating with public or private entities for funding. Their flexible structure supports a wide range of cultural, social, and environmental activities. MbL was established as part of the "START: Create Cultural Change" program, the organization aims to foster a sense of community and strengthen collective action among the residents of Messolonghi.

The mission of MbL is centered on the development of a participatory mapping initiative that captures the city's rich cultural identity. Through this mapping process

(Picture 1), the organization seeks to actively engage community members and encourage their involvement in various cultural and natural activities. By organizing and promoting these activities, MbL not only enriches the local cultural landscape but also fosters a deeper connection among people and place. The organization focuses on studying, promoting, and preserving the local cultural heritage and environment. By prioritizing the unique traditions and natural beauty of Messolonghi, MbL plays a crucial role in ensuring that these elements are recognized and valued, by organizing cultural tourism and ecotourism experiences for both locals and visitors. Through its various initiatives, MbL is dedicated to building a vibrant and sustainable future for the town, highlighting the importance of community involvement, cultural preservation and landscape conservation.

Figure 1. Participatory mapping

(source: MbL webpage)

3.2. The Case Study Area: Messolonghi

Messolonghi, a rural town in western Greece, holds significant historical and ecological importance, particularly due to its proximity to the Messolonghi lagoon, which is part of a complex wetland ecosystem designated under the Ramsar Convention for its rich biodiversity (Picture 2). This lagoon serves as a critical habitat for numerous bird species, including migratory populations, thereby playing an essential role in maintaining ecological balance in the region. Historically, Messolonghi is renowned for its pivotal involvement in the Greek War of Independence, notably during the heroic "Exodus of Messolonghi" in 1826, an event that has inspired extensive historical and sociopolitical studies. The town's fortifications and memorial sites, such as the Garden of Heroes, provide valuable insights into 19th-century warfare and the social resilience of besieged communities. Additionally, Messolonghi has a longstanding tradition of salt production, intricately linked to the lagoon's natural conditions, which shapes both the local economy and culture. This interplay of ecological, historical, and cultural elements makes Messolonghi an intriguing case study for examining human-environment interactions, sustainable natural resource management, and cultural heritage preservation. Consequently, the town serves as a multidisciplinary platform for research in fields such as environmental science, geography, history, and anthropology.

Figure 2. Messolonghi lagoon

(source: author)

3.3. The Method

On-site research for this study was conducted using semi-structured, in-depth interviews with two members of the MbL on 19 July 2024, complemented by participatory observation. The remaining data collection involved secondary sources, such as online resources, informational leaflets, and other available published materials. Additionally, qualitative data was gathered from a convenience sample of 69 reviews on Tripadvisor covering the period from July 2020 to June 2024, encompassing all available reviews related to MbL. The integration of in-depth interviews, participatory observation, and secondary data analysis—including a convenience sample of user-generated Tripadvisor reviews—provides a robust methodological framework. This mixed-methods approach combines first-hand qualitative insight with secondary textual data analysis, allowing for a well-rounded perspective on participants' experiences and attitudes. Such triangulation enhances data richness and validity, offering the potential to uncover nuanced themes and insights, especially valuable when the aim is to explore specific perspectives and lived experiences within a defined community context.

4. THE RESULTS

4.1. The Actions of "Messolonghi by Locals"

4.1.1. The Continuously Developing Network of Thematic Cultural Routes

The network of thematic routes encompasses diverse areas, including nature routes (such as the sealake of Klisova, the sealakes of Tholi and Paliopotamos, and Mount Arakinthos, Picture 3), antiquity routes (including the ancient settlement of Oiniades, archaeological sites of Kalidona and Plevrona, and ancient theaters), tradition routes (highlighting the feasts of Aitolia and Saint Simios), architecture routes, arts routes, cycling routes, and local products routes.

Figure 3. Thematic routes on sealakes

(source: MbL FB page)

4.1.2. Participatory Mapping and Recording

Participatory mapping and recording is a dynamic process that focuses on developing thematic cultural routes. This process involves extensive research, experiential learning, and documentation through various methods such as photography, video, and painting (Image 1). Additionally, this participatory mapping process fosters knowledge sharing and interaction with the local community, encouraging collaborative planning efforts. The ultimate aim is to provide locals with the opportunity to deepen their understanding of their place while also offering travelers valuable insights into the cultural and environmental wealth of Messolonghi.

An educational initiative (Educational programme "Mapping our place") focused on participatory mapping aims to explore the cultural identity of Messolonghi and its surrounding municipality by involving students from regional schools. Each school develops cultural routes and generates maps showcasing students' unique interpretations of local heritage. Events are organized for students to present their mapped cultural routes to the local community. Students are exposed to experiential learning techniques and diverse stimuli that enrich their sense and understanding/ appreciation of the cultural heritage, history and traditions that shape their communities. The program seeks to immerse students in democratic decision-making processes, research methodologies, project implementation, collaboration, teamwork, and networking with peers from other educational institutions. The initiative enhances students' creativity and critical thinking, preparing them for active participation in their communities.

Figure 4. Participatory mapping and recording

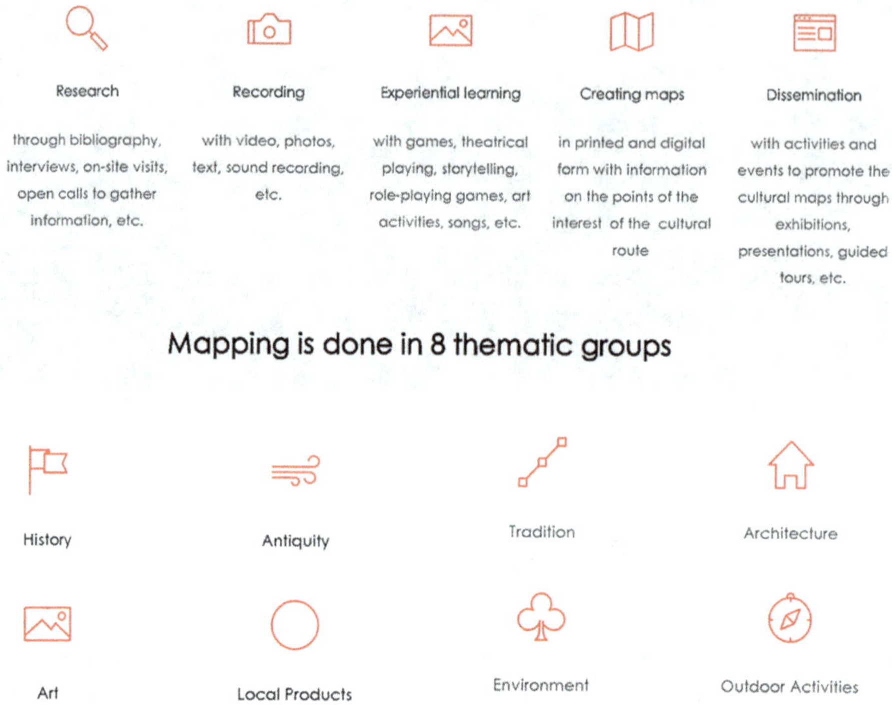

(source: MbL webpage)

4.1.3. Organizing Various Events and Promoting Local Products and Artifacts

MbL has established and currently administers the "Local Hub" a social and cultural center serving as a dynamic venue for expression, networking, learning, co-creation, and collaboration, facilitating a diverse range of activities that enrich the community (Picture 4). Interdisciplinary events take place every month, featuring screenings, presentations, discussions, and other activities designed to engage the community and stimulate dialogue on a variety of topics. Among the events hosted at the Local Hub are community meetings, exhibitions, workshops, seminars, screenings, and coworking opportunities, along with a reading room. MbL encourages active participation through experiential workshops and skill development seminars, but also free discussions and debates for issues arising from their interests or the challenges facing the local community, fostering an environment of collaboration, expression and innovation. Additionally, the center hosts children's art competitions, nurturing creativity and providing young residents with opportunities to showcase their talents. The Local Hub also includes a souvenir shop that features local products and handcrafted items, promoting the unique cultural identity of Messolonghi.

Figure 5. Activities outside the Hub

(source: MbL webpage)

4.1.4. Organizing the Annual Festival

An annual three-day festival, which started in 2019, is deeply rooted in the cultural identity of Messolonghi and serves as a vital platform for disseminating data gathered through the organization's "participatory mapping" initiative. During this festival, participants share the mapped cultural routes that have been developed through community-based research, providing valuable insights into the various landscapes. This exchange of knowledge enhances the understanding of Messolonghi's cultural tapestry and its ecological significance. The festival's program is diverse and engaging, including guided urban walks, hiking expeditions, sports activities, exhibitions, film screenings, concerts, and presentations of local gastronomy and products (Picture 5).

Figure 6. Activities during the annual Festival

(source: MbL FB page)

4.2. The In-Depth Interviews

The key question "How can we enhance and promote our community?" introduced by the representative of MbL, encapsulates the central aim of this inquiry into community development and engagement. Another statement highlights a critical challenge faced by many communities, particularly in rural or less populated areas: "the migration of young people poses a significant challenge to our local area". This phenomenon can lead to a decline in local culture, economic vitality, and community cohesion. By emphasizing the necessity of creating opportunities for youth within their local area, both the members suggested a strategic focus on local empowerment and economic development as solutions to mitigate the impacts of youth outmigration.

Internally, MbL is supported by five full-time partners, alongside 4-5 external collaborators, such as tour guides and artists. This structure promotes a diverse range of skills and perspectives, crucial for fostering innovative solutions and initiatives. Externally, the extensive network comprising 30-40 individuals, including producers, graphic designers, and educators, reflects a broad engagement with various stakeholders. The inclusion of 54 volunteers in the annual festival, along with partnerships with primary and secondary schools and local cultural organizations, suggests a commitment to community involvement and capacity building. The basic funding mechanisms mentioned are the National Strategic Reference Framework, ticket sales from guided tours, and donations.

According to their words, they decided to embark on a "journey" of discovery through Messolonghi, exploring its hidden corners, rich history, traditions, and the small details that often go unnoticed in everyday life". This signifies an active engagement with the local context, highlighting the value of place-based initiatives. The exploration of hidden corners, history, traditions, and often-overlooked details aligns with principles of cultural heritage management and emphasizes the significance of local narratives in fostering community pride and identity. They claimed that "by actively engaging with our surroundings, we aim to foster an engaged community that builds a strong sense of belonging among its members, ultimately enhancing and promoting the vibrancy of our local area", because "the solution lies in creating opportunities for youth within their own community, allowing them to thrive where they are rooted"... The emphasis on belonging is crucial in countering the effects of youth migration, as fostering strong social ties can encourage young people to remain in or return to their communities, thereby contributing to local vitality.

The analysis of these components reveals a strategic framework for enhancing and promoting community engagement. By addressing the challenges of youth migration through collaboration, resource utilization, and active participation in local discovery, the community can cultivate a vibrant and resilient environment. This approach not only seeks to enhance the local economy but also strengthens social bonds, ensuring that the community remains a dynamic and appealing place for current and future residents.

4.3. The Tripadvisor Reviews' Research

The research conducted on Tripadvisor reviews of MbL provides valuable insights into the perceptions and experiences of individuals who participated in the activities offered by the organization. Analyzing the comments (textual analysis and codification, Image 2) written between July 21, 2020, and June 6, 2024, reveals a total of 69 reviews, predominantly in Greek (51) and English (18). Notably, the overall rating for the activities was remarkably high, with 98.5% of the reviews achieving a perfect score of 5 out of 5. This overwhelmingly positive feedback suggests a strong appreciation for the experiences provided by "Messolonghi by Locals," indicating a high level of satisfaction among participants.

Qualitative analysis of the reviews highlighted several key categories that emerged from the comments, reflecting the aspects of the experiences that resonated most with participants. The most frequently mentioned category was the "boat tour of the lagoon" with 25 mentions. The "educational aspect" received 22 mentions. The category labeled "experience" was also prominent, with 21 mentions, which likely encompasses the overall enjoyment and engagement of participants during the activities. Other noteworthy categories included "hospitality" (13 mentions) and "organizers' love for the region" (13 mentions), both of which emphasize the personal touch and commitment of the staff. There have been 12 mentions for "organization" and "history" categories and 11 mentions for the "vision/tourism model". Other aspects such as "landscape" (9 mentions), "seminars & exhibitions" (7 mentions), and "local products and hub" (6 mentions) were also highlighted by some visitors.

Figure 7. The textual analysis of the TripAdvisor reviews

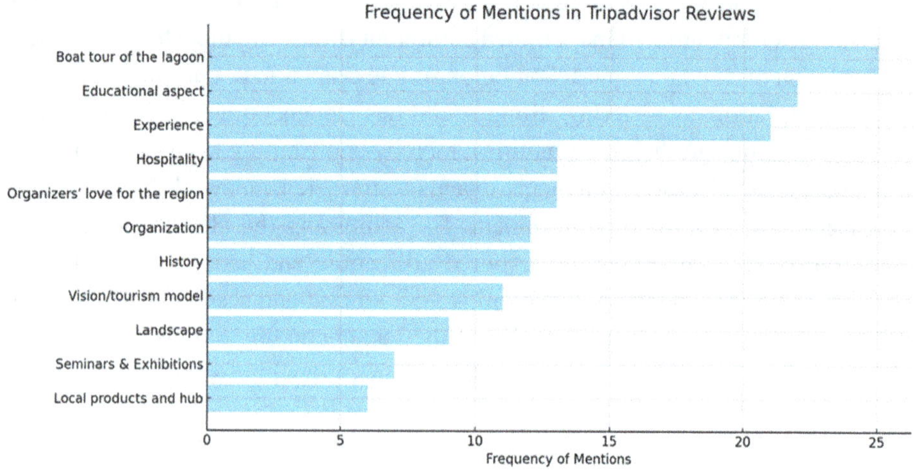

5. DISCUSSION

5.1. Community Engagement

5.1.1. Community Engagement Through Diverse Events

The varied events and the annual festival play a crucial role in enhancing community engagement with Messolonghi's cultural and environmental heritage. By focusing on underexplored cultural and environmental elements, these events create spaces for community members to discover, reflect upon, and celebrate their heritage. Through activities such as interactive workshops, guided tours, map creation and hands-on experiences, the community gains a richer understanding of the local environment and history and the community members can be empowered to discover and document their values and resources, transforming passive observation into active participation (Fox and Le Dantec, 2014; Thrift, 2007). These events generate interest in local heritage by prioritizing in-depth, culturally rich experiences that attract both residents and culturally inclined visitors who seek meaningful cultural engagement, positioning the local population at the core of resource management (Stroja, 2018; de Varine, 1996). Such engagement not only strengthens local pride but also establishes an emotional connection with Messolonghi's heritage, potentially fostering future generations of community stewards. However, as Wallace (2022) suggests, heritage management approaches that emphasize community empowerment through the celebration of local traditions and awareness-raising do not, on their own, necessarily lead to enhanced active engagement of local communities with heritage sites; consequently, additional measures are needed to foster deeper involvement.

5.1.2. Participatory Management

The emphasis on participatory management further deepens community engagement by inviting residents to take an active role in the stewardship of local resources. This framework goes beyond mere participation in events, encouraging self-organization and local governance in managing shared assets (Pavlis, 2024). By positioning the community as central to decision-making processes, it empowers individuals and groups to contribute to shaping the future of their landscape and underscores the importance of local empowerment and the engagement of civil society (Dogan and Timothy, 2019). This participatory approach allows for a more nuanced management of heritage that reflects the unique perspectives and needs of the community itself. Heritage values, place attachment, and community ideals are significant drivers of participation, while tourism perceptions play a lesser role (Dragouni and Fouseki, 2017). Additionally, participatory management builds social capital within the community, reinforcing trust, cooperation, and resilience as residents collaborate to protect their heritage in meaningful ways.

5.1.3. Impact on Community and Reputation

MbL has successfully positioned itself as a key promoter of Messolonghi's heritage, garnering a positive reputation both locally and beyond. The organization's work has had a tangible impact on the community, as evidenced by increased local involvement and enhanced pride in regional identity. By integrating local and visitor experiences, MbL creates a feedback loop: the appreciation of cultural assets by visitors often reinforces the community's own sense of value in their heritage. Thus, community-based heritage management can strengthen local identity (Beal et al, 2019). This recognition also benefits Messolonghi's reputation as a destination that respects and elevates its cultural roots, attracting visitors interested in sustainable, meaningful tourism experiences. Through this approach, MbL contributes to both the preservation of cultural heritage and the town's socioeconomic well-being, providing a model of how community-driven cultural programming can support identity formation, promote sustainability, and foster local pride. Local participation enhances the sense of identity, fosters social capital, and contributes to the socioeconomic well-being of destinations (Béal et al., 2019; Russo, 2005).

5.2. Collaborating and Networking

5.2.1. Engaging a Diverse Audience Through a Variety of Activities

MbL's collaborative approach is exemplified by its involvement in a range of initiatives/ activities encompassing agriculture, folk art, thematic tourism, sports, and education. These initiatives reflect a commitment to addressing multiple dimensions—economic, social, cultural, and ecological. By engaging with different aspects of local life, MbL fosters a holistic understanding of the region's heritage, appealing to a broader audience with varied interests. Additionally, this diversity in initiatives facilitates partnerships with a wide range of collaborators, from local artisans to environmental groups, educators, and tourism operators. This networked approach enhances MbL's capacity to promote sustainable development and community resilience, as it leverages the expertise and resources of multiple stakeholders.

MbL demonstrates a robust approach to community engagement by designing activities that appeal to a wide range of participants. Through well-structured and thoughtfully organized events, MbL reaches diverse demographic groups—locals and visitors alike—thereby fostering broad-based community involvement. This approach to engagement is essential for effective heritage preservation and sustainable tourism, as it ensures that a variety of voices and perspectives contribute to the interpretation and celebration of Messolonghi's cultural and environmental assets (Li and Hunter, 2015). By tailoring these activities to be accessible and meaningful to different audiences, MbL not only strengthens local identity but also raises awareness about the region's resources among outsiders, thus expanding its network of support and advocacy for heritage conservation.

5.2.2. Hospitality and Organizer Passion as Catalysts for Community Building

A notable feature of MbL's impact lies in the recurring emphasis on the hospitality and enthusiasm of the organizers, which creates a welcoming and inclusive atmosphere for participants. This personal touch, derived from the organizers' deep appreciation for Messolonghi, cultivates a sense of community among participants and enhances the quality of engagement. Studies have shown that emotional investment by organizers can be a decisive factor in successful community-based projects, as it builds trust and rapport among participants, leading to a more cohesive and committed community (Bostock et al, 2016). Practices that foster emotional engagement contribute to building trust in partnerships involving multiple stakeholders (Sloan and Oliver, 2013). The organizers' evident passion for the region adds authenticity to MbL's initiatives, making activities feel genuine and purpose-driven, which further attracts participants who value meaningful engagement with local culture (Kim and Lee, 2022). This sense of community fosters deeper emotional connections among attendees, contributing to the long-term sustainability of MbL's efforts by nurturing a loyal and supportive network.

5.2.3. Effective Network of Collaborators

MbL uses an internal and an external network of collaborators, a fact that underlines the importance of a collaborative approach in community enhancement. The collaborative network of internal, external collaborators and volunteers not only strengthens community ties but also enhances the resource base for cultural and educational initiatives, thereby fostering a sense of shared responsibility and collective identity. Collaboration across diverse sectors enriches MbL's programming and also reinforces its role as an integrative force within the community. By engaging with collaborators/ stakeholders from various backgrounds, MbL cultivates a network that supports knowledge exchange, resource sharing, and joint problem-solving. This structure allows for adaptive and responsive programming, ensuring that MbL's activities remain relevant to community needs and aligned with sustainable development goals. Collaborative, community-based monitoring promotes shared ecological understanding, builds trust, encourages social learning, and supports adaptive management (Fernandez-Gimenez et al, 2008). Moreover, a broad network of collaborators strengthens MbL's credibility, as the organization becomes a central hub for cultural and environmental initiatives in Messolonghi, further enhancing its reputation and impact. The role of stakeholder engagement and partnerships in promoting sustainable development and community resilience has been emphasized by other researchers (Mitchell, 2019; Koščak, 2016).

5.3. Integrating Culture, Education, Ecology, and Utilization of Local Resources

5.3.1. Integration of Cultural and Natural Assets Through Spatially-Informed Design

MbL employs a spatially-informed approach that seamlessly integrates the region's cultural and natural heritage. This strategy involves the careful design of cultural and ecological routes that not only highlight Messolonghi's natural beauty but also elevate its cultural identity. By recognizing and showcasing the ecological and aesthetic value of these landscapes, MbL promotes a comprehensive view of heritage that links cultural landmarks with natural surroundings (Keramitsoglou, et al, 2023; Kyvelou and Gourgiotis, 2019). This integration underscores the interconnectedness of place and identity, fostering a holistic sense of regional character that resonates with both locals and visitors. The spatial strategy aligns with sustainable tourism principles by encouraging visitors to engage with the environment in ways that respect and enhance local resources, thus supporting both conservation and community pride. In this line, Loulanski and Loulanski (2011) in their meta-analysis of 483 studies propose fifteen factors as essential elements for creating a more effective theoretical framework and evidence-based policy in cultural heritage and tourism, with the goal of promoting sustainability.

5.3.2. Ecomuseum Principles as a Framework for Heritage Interpretation

Although MbL does not officially identify as an ecomuseum, its approach closely mirrors ecomuseum principles by highlighting the dynamic interplay between heritage and environment. Ecomuseums typically emphasize a decentralized, community-centered model of heritage interpretation, where sites are not confined to single locations but rather spread across a "territory" (Davies, 2011). MbL's focus on thematic, narrative-driven routes reflects this concept, presenting heritage as a "living evidence" of the evolving relationship between people and their environment. This framework encourages participants to view Messolonghi's landscapes not merely as backdrops but as integral parts of local heritage that are continuously shaped by human-environment interactions (Pavlis, 2017). By promoting heritage as contextually embedded and evolving, MbL positions Messolonghi as a site of ongoing cultural expression, deepening participants' understanding and appreciation of local identity.

5.3.3. Educational Value and Experiential Learning

A key component of MbL's mission is the educational impact of its activities, which aim to promote cultural understanding through experiential learning. Participants consistently express appreciation for the educational aspects of MbL's programming, which aligns with the organization's emphasis on meaningful engagement with cultural heritage. By incorporating experiential learning, MbL enables participants to actively engage with history, local traditions, and ecological knowledge, rather than passively observing them. This educational dimension supports a deeper, more lasting connection with local heritage, as it combines recreation with learning to create a holistic experience, enables practical wisdom and social change incorporating experiential learning and critical thinking to address sustainability challenges (Jamal et al., 2011). Experiential learning fosters critical thinking and a personal connection to the heritage landscape, which may inspire participants to become advocates for cultural and environmental preservation, further supporting the long-term sustainability of MbL's efforts and promoting the development of environmental and landscape conscience (Pavlis and Terkenli 2023; Pavlis 2017).

5.3.4. Activation of Territorial Capital for Community Development

MbL's strategy effectively aligns with territorial development principles by leveraging local resources—cultural assets, natural landscapes, and community traditions—as part of a broader approach to enhance community well-being. Through activities that reconnect the community with its traditions, environment, and landscape, MbL activates territorial capital, the unique resources inherent to a particular place. It does so, by successfully developing strategies that utilize local resources in a manner that is not only economically efficient but also socially, culturally, and environmentally sustainable (Morreta 2021; Campagne and Pecqueur 2014). This activation fosters a renewed sense of pride and belonging among community members, promoting socio-economic resilience and cultural continuity. By developing strategies that utilize local resources, MbL strengthens the community's capacity to shape its development trajectory, supporting sustainable practices that are locally grounded and culturally relevant. MbL's actions seem to enhance territorial potential by reconnecting the community with its traditions, resources, and environment (Goussios 2013). The organization's approach demonstrates how territorial development can be achieved by drawing on a region's distinctive characteristics, thus reinforcing community identity while creating economic opportunities through cultural tourism and heritage conservation tailored to the distinct needs, resources, and cultural identity of the region acknowledging their potential to enhance the long-term well-being of the community (Morreta 2021).

5.4. Territorializing Tourism

5.4.1. Integrated Embodied Visitor Experience Through More-Than-Representational Tourism

MbL employs an approach that transcends conventional representational tourism by emphasizing embodied, multi-sensory experiences for visitors. Similarly, Waterton (2014) promotes an understanding of heritage through embodied, sensory, and emotional engagement beyond just visual consumption and Eshuis and Ripoll González (2024) propose a new definition of place brands as embodied, multi-sensory experiences that go beyond just symbolic or cognitive representation. Additionally, Wang (2023), in his study, emphasizes the crucial role of unexpected bodily experiences and participation in celebrations and events for fostering a sense of belonging. He highlights the allure of spontaneous, unplanned moments and the significant impact of the more-than-human rural environment. Such trends/ strategies align with the principles of "more-than-representational" tourism, which promotes active engagement with place beyond visual consumption, involving visitors in experiences that evoke sensory and emotional responses tied to local culture and landscape. Over the past two decades, embodiment theory in tourism studies has shifted from traditional visual-centered analyses to examining the roles of taste, smell, touch, and sound in shaping tourist experiences (Everingham et al., 2021). MbL's events and activities, which range from agricultural experiences to folk art demonstrations, provide participants with hands-on interactions that foster a deep connection with the region's heritage. The success of this approach is evident in the diversity of participant responses, which reflect the personal and immersive quality of their experiences. This embodied engagement encourages visitors to perceive Messolonghi not merely as a destination but as an evolving cultural landscape, fostering an appreciation for the area's unique characteristics and enhancing the memorability of their visit.

5.4.2. Contextual Development and Local Identity Preservation

A core tenet of MbL's territorial tourism approach is the grounding of development within the context of local production and culture, ensuring that consumption remains closely tied to place. This contextual development model is essential for preserving local cultural heritage and identity, as it prevents the detachment of products and experiences from their origins. By promoting local production and consumption patterns, MbL enhances the 'authenticity' of tourism experiences, providing visitors with a genuine insight into the area's cultural practices. Simultaneously, the destination creates potential tourism products adjusted to fit the demands of niche markets (Koščak, 2016). This grounding approach not only supports the local economy but also strengthens community pride and cohesion, as residents see their culture and resources being valued and protected. Additionally, it mitigates the risks of cultural commodification, where local practices are altered to cater to external demands, thus maintaining the integrity of Messolonghi's heritage. MbL's emphasis on keeping development embedded within local contexts represents a sustainable strategy that reinforces tourism as a positive force within the community, preserving both cultural identity and environmental integrity (Navajas Corral and Fernández 2022).

5.4.3. Sustainable Tourism Beyond Entertainment

MbL's tourism model consciously moves beyond entertainment-driven tourism by integrating cultural preservation and community engagement into its programming. The organization's focus on local products, including events that highlight traditional craftsmanship, agriculture, and ecological practices, positions tourism as a tool for sustainable development rather than mere amusement. This approach aligns with principles of responsible tourism, as it encourages visitors to engage thoughtfully with local resources and traditions. It addresses multiple developmental dimensions—economic, social, cultural, and ecological—while promoting sustainable tourism practices (Magnaghi, 2005) characterized by an experiential rather than observational approach (Wylie, 2007). The strategy of MbL promotes collaborative tourism, which aids in alleviating social conflicts within rural communities and facilitates the meaningful integration of natural and cultural heritage sites affected by mass tourism into the everyday lives of nearby communities, as recommended by UNESCO (2019) (Panzer-Krause 2020). By organizing diverse events that appeal to both locals and visitors, MbL fosters an inclusive form of tourism that benefits the community as a whole. This model supports the sustainable use of local resources, promotes cultural continuity, and fosters environmental stewardship, contributing to long-term tourism development that does not exploit but rather celebrates and conserves local assets. In doing so, MbL redefines tourism as a participatory and educative process, rather than a transient experience, aligning visitor activities with broader goals of cultural sustainability and community well-being.

5.4.4. Innovative Territorial Tourism Model as a Model for Community-Driven Initiatives

MbL exemplifies an innovative model of *territorial tourism* that prioritizes responsible travel, fosters local economic development based on a territory, and strengthens connections between visitors and the community, nature and culture. It has a localized and immediate focus on unique regional characteristics and on the territorial potential of a region (Goussios, 2013), aiming to connect local production with quality-conscious consumers and increase competitiveness through building trust and social capital (Quaranta et al., 2016) and addresses tourist activities as activators of tourist territory (Nunez-Torres, et al, 2023). By positioning itself as a community-centered initiative, MbL showcases how tourism can be developed in a way that empowers local stakeholders, encouraging a participatory approach to tourism governance. This model supports a circular local economy by reinvesting tourism revenue into community projects, creating jobs, and boosting local businesses involved in cultural and ecological activities. It ensures that consumption remains in the place of production and development is contextually grounded in the unique characteristics of each area (Dal Santo et al., 2017; Navajas Corral & Fernández, 2022). MbL's emphasis on creating immersive, place-based experiences establishes it as a blueprint for community-driven initiatives that value sustainability and cultural integrity. The organization's success highlights the potential of territorial tourism as a viable model for other regions aiming to develop responsible tourism practices that respect local culture, environment, and community priorities.

6. CONCLUSION

In conclusion, the effective management of local resources—ranging from indigenous seeds and agricultural products to raw materials, waste, ecosystems, energy, landscape, history, and culture—requires a holistic approach that prioritizes the existing assets within a community. The concept of an ecomuseum emphasizes that territorial development does not necessitate the construction of new infrastructure but instead focuses on leveraging current resources through networking, cooperation, and a shared vision among stakeholders. This underscores the importance of bottom-up community approaches that actively engage civil society in the decision-making process, fostering a sense of ownership and collective responsibility. Investment in locality, quality, and innovation is essential for enhancing the sustainability of these initiatives, promoting consumption in the place of production, and strengthening local economies. As the ecomuseum model emerges as a catalyst for territorial development, it can facilitate the integration of cultural heritage and environmental/ landscape stewardship, ultimately contributing to a more resilient and vibrant community, and informing new territorial policies. By recognizing and valuing the interconnectedness of local resources, ecomuseums can play a transformative role in fostering sustainable development and enhancing the cultural identity of their regions.

MbL's approach underscores the power of community engagement in safeguarding and promoting cultural heritage. Through collaborative, participatory, and sustainable practices, the organization has created a dynamic space where community members can actively participate in and celebrate their heritage, ultimately enhancing both the local identity and Messolonghi's appeal as a cultural destination. This model of engagement demonstrates the potential for community-driven initiatives to create lasting social and cultural benefits, reinforcing the interdependence between heritage preservation and community well-being.

MbL's strategy of collaboration and networking has proven to be an effective model for community-based heritage preservation. Through structured, inclusive programming and a strong network of collaborators, MbL successfully engages diverse audiences, builds a welcoming community atmosphere, and addresses a wide array of local needs. The combination of personal passion from organizers and strategic partnerships allows MbL to foster a resilient, community-centered approach to cultural and environmental stewardship. This model illustrates how effective collaboration and network-building can drive sustainable outcomes, enriching both local identity and social, cultural and economic resilience while supporting the broader goals of sustainable tourism and heritage management.

The integration of culture, education, ecology, and local resources exemplifies a comprehensive, sustainable approach to heritage management that supports both community identity and socio-economic development. By blending spatially-informed design, ecomuseum principles, and experiential learning, MbL fosters a multi-dimensional experience that resonates with participants and reinforces the community's connection to its heritage. The activation of territorial capital further illustrates how community-led heritage initiatives can promote sustainable development, fostering resilience, pride, and well-being. This model serves as a testament to the potential of locally driven, integrated heritage management in advancing both cultural preservation and regional development.

The approach to territorializing tourism effectively balances visitor experience with community benefit, providing an integrated model that leverages the unique cultural and ecological assets of Messolonghi. Through embodied visitor engagement, contextual development, sustainable practices, and community-centered innovation, MbL offers an exemplary framework for developing tourism that fosters cultural preservation, environmental stewardship, and socio-economic resilience. This model underscores the potential of tourism as a driver of sustainable development when rooted in local identity and community participation, paving the way for a redefined tourism paradigm that is aligned with the principles of territorial integrity and cultural continuity.

REFERENCES

Anthopoulou, T., & Goussios, D. (2007). Rural Geography. In Terkenli, T., Iosifidis, T., & Goussios, D. (Eds.), *Human Geography: Human, Society and Space* (pp. 234–275). Kritiki Publishing.

Beal, L., Séraphin, H., Modica, G., Pilato, M., & Platania, M. (2019). Analysing the Mediating Effect of Heritage Between Locals and Visitors: An Exploratory Study Using Mission Patrimoine as a Case Study. *Sustainability (Basel), 11*(11), 3015. DOI: 10.3390/su11113015

Bostock, J., Cooper, R., & Roberts, G. (2016). Rising to the Challenge of Sustainability: Community Events by the Community, for the Community. In Jepson, A., & Clarke, A. (Eds.), *Managing and Developing Communities, Festivals and Events*. Palgrave Macmillan., DOI: 10.1057/9781137508553_2

Campagne, P., & Pecqueur, B. (2014) Le développement territorial. Une réponse émergente à la mondialisation [Territorial Development: An Emerging Response to Globalization], Paris: Éditions Charles Léopold Mayer (ECLM). ISSN 0033-0884, DOI: DOI: 10.3917/pro.344.0095

Corsane, G., Davis, P., Elliott, S., Maggi, M., Murtas, D., & Rogers, S. (2007). Ecomuseum performance in Piemonte and Liguria, Italy: The significance of capital. *International Journal of Heritage Studies, 13*(3), 224–239. DOI: 10.1080/13527250701228148

Dal Santo, R. D., Baldi, N., Del Duca, A., & Rossi, A. (2017). The Strategic Manifesto of Italian Ecomuseums. *Museum International, 69*(1-2), 86–95. DOI: 10.1111/muse.12153

Davis, P. (2011). *Ecomuseums: a Sense of Place* (2nd ed.). Continuum.

. de Varine, H. (1996) Ecomuseum or community museum? 25 years of applied research in museology and development, Nordisk museology 2, 21-26

Doğan, M. (2015). Ecomuseum, community museology, local distinctiveness, Hüsamettindere village, Bogatepe village, Turkey. *Journal of Cultural Heritage Management and Sustainable Development, 5*(1), 61–81. DOI: 10.1108/JCHMSD-07-2013-0033

Doğan, M., & Timothy, D. J. (2019). Beyond Tourism and Taxes: The Ecomuseum and Social Development in the Ak-Chin Tribal Community. *Journal of Tourism and Cultural Change, 18*(2), 133–149. DOI: 10.1080/14766825.2019.1593994

Dragouni, M., & Fouseki, K. (2017). Drivers of community participation in heritage tourism planning: An empirical investigation. *Journal of Heritage Tourism, 13*(3), 237–256. DOI: 10.1080/1743873X.2017.1310214

. Eshuis, J. and Ripoll González, L. (2024) Conceptualising place branding in three approaches: towards a new definition of place brands as embodied experiences, Journal of Place Management and Development, Vol. ahead-of-print No. ahead-of-print. https://doi.org/DOI: 10.1108/JPMD-11-2023-0109

Everingham, P., Obrador, P., & Tucker, H. (2021). Trajectories of embodiment in Tourist Studies. *Tourist Studies, 21*(1), 70–83. DOI: 10.1177/1468797621990300

Fernandez-Gimenez, M. E., Ballard, H. L., & Sturtevant, V. E. (2008). Adaptive management and social learning in collaborative and community-based monitoring: A study of five community-based forestry organizations in the western USA. *Ecology and Society, 13*(2), 4. https://www.ecologyandsociety.org/vol13/iss2/art4/ . DOI: 10.5751/ES-02400-130204

Fox, S. E., & Le Dantec, C. A. (2014) Community historians: scaffolding community engagement through culture and heritage. *Proceedings of the 2014 conference on Designing interactive systems*, 785 - 794 DOI: 10.1145/2598510.2598563

Goussios, D. (2013) Η εδαφική προσέγγιση της ανάπτυξης της υπαίθρου [The territorial approach of rural development] Περί εντοπιότητας και ιδιοτυπίας των τροφίμων: Μια εδαφική προσέγγιση της ανάπτυξης των αγροτικών περιοχών, edited by T. Anthopoulou, 72-105. Athens: Papazisi Publications, ISBN: 978-960-02-2809-0

Goussios, D., Gaki, D., & Faraslis, I. (2023). *Ένας δρόμος ανάπτυξης για τα βουνά* [A development path for the mountains]. Evdoxos Publications.

Jamal, T. B., Taillon, J. M., & Dredge, D. (2011). Sustainable tourism pedagogy and academic-community collaboration: A progressive service-learning approach. *Tourism and Hospitality Research, 11*(2), 133–147. DOI: 10.1057/thr.2011.3

Johnston, R., & Marwood, K. (2017). Action heritage: Research, communities, social justice. *International Journal of Heritage Studies, 23*(9), 816–831. DOI: 10.1080/13527258.2017.1339111

Keramitsoglou, K. M., Koudoumakis, P., Akrivopoulou, S., Papaevaggelou, R., & Protopapas, A. L. (2023). Biodiversity as an Outstanding Universal Value for Integrated Management of Natural and Cultural Heritage. *Sustainability (Basel), 15*(11), 8540. DOI: 10.3390/su15118540

Kim, D., & Lee, K. (2022). The Impact of Community Art Activists" Authenticity on the Performance and Effect of Community Art Projects. *The Journal of Internet Electronic Commerce Research, 22*(6), 277–297. DOI: 10.37272/JIECR.2022.12.22.6.277

Koščak, M. (2016). Heritage trails through Dolenjska and Bela Krajina in Slovenia: A multi-stakeholder approach to tourism development, *PREGLEDNI ČLANAK*, Fakultet za sport i turizam, Novi Sad. *TIMS Acta, 10*(2), 109–118. DOI: 10.5937/timsact10-10519

Kyvelou, S. S., & Gourgiotis, A. (2019). Landscape as Connecting Link of Nature and Culture: Spatial Planning Policy Implications in Greece. *Urban Science (Basel, Switzerland), 3*(3), 81. DOI: 10.3390/urbansci3030081

Li, Y., & Hunter, C. (2015). Community involvement for sustainable heritage tourism: A conceptual model. *Journal of Cultural Heritage Management and Sustainable Development, 5*(3), 248–262. DOI: 10.1108/JCHMSD-08-2014-0027

Loulanski, T., & Loulanski, V. (2011). The sustainable integration of cultural heritage and tourism: A meta-study. *Journal of Sustainable Tourism, 19*(7), 837–862. DOI: 10.1080/09669582.2011.553286

Magnaghi, A. (2005). *The Urban Village: A Charter for Democracy and Local Self-Sustainable Development*. Zed Books.

Mitchell, B. (2019). Stakeholders and Partnerships. In Mitchell, B. (Ed.), *Resource and Environmental Management* (3rd ed., pp. 145–176). DOI: 10.1093/oso/9780190885816.003.0006

Morreta, V. (2021). Territorial capital in local economic endogenous development. *Regional Science Policy & Practice, 13*(1), 103–119. DOI: 10.1111/rsp3.12317

Navajas Corral, O., & Fernández, J. F. (2022). The Importance of Ecomuseums and Local Knowledge for a Sustainable Future: The La Ponte-Ecomuseu Project. In Borrelli, N., Davis, P., & Santo, P. D. (Eds.), *Ecomuseums and Climate Change* (pp. 283–203). Ledizioni.

Nuñez-Torres, A., Arones-Huarcaya, A., & Yarasca-Aybar, C. (2023). Strategies for territorial tourism planning in natural protected areas (NPAs): Alto Mayo Protected Forest (BPAM), Peru. City. *Territory and Architecture, 10*(1), 15. Advance online publication. DOI: 10.1186/s40410-023-00200-w

Pappalardo, G. (2020). Community-Based Processes for Revitalizing Heritage: Questioning Justice in the Experimental Practice of Ecomuseums. *Sustainability (Basel), 12*(21), 9270. DOI: 10.3390/su12219270

Pavlis, E. (2017). North Aegean islands as 'ecomuseums': The case of Lesvos Island. *Island Studies Journal, 12*(1), 135–145. DOI: 10.24043/isj.9

Pavlis, E. (2024). The ecomuseum as a platform for territorial development: The case of Greek ecomuseums. *International Journal of Heritage Studies, 30*(10), 1193–1210. Advance online publication. DOI: 10.1080/13527258.2024.2386695

Pavlis, E., & Terkenli, T. S. (2023). Towards a Geographical Approach of "Spatial Conscience" and its application on Landscape. *Landscape Research, 48*(8), 1004–1018. DOI: 10.1080/01426397.2023.2238639

Russo, A. P. (2005). Cultural gateways - building partnerships for sustainable development in destination regions. *Proceedings of the 45th Congress of the European Regional Science Association*, ERSA conference papers

Sloan, P., & Oliver, D. (2013). Building Trust in Multi-stakeholder Partnerships: Critical Emotional Incidents and Practices of Engagement. *Organization Studies, 34*(12), 1835–1868. DOI: 10.1177/0170840613495018

Stroja, J. (2018). My history, your history, our history: Developing meaningful community engagement within historic sites and museums. *Queensland Review, 25*(2), 300–321. DOI: 10.1017/qre.2018.29

Thrift, N. (2007). *Non-representational theory: Space, Politics, Affect*. Routledge.

Wang, C.-M. (2023). Escape to the countryside: Affects and rural eventful atmospheres. *Habitat International, 140*, 102929. DOI: 10.1016/j.habitatint.2023.102929

Waterton, E. (2014). A More-Than-Representational Understanding of Heritage? The 'Past' and the Politics of Affect. *Geography Compass, 8*(11), 823–833. DOI: 10.1111/gec3.12182

Wuisang, C., Rengkung, J., & Rondonuwu, M. (2018). Towards sustainable cultural landscape: A challenge in developing ecomuseum in Minahasa Region. 3rd International Symposium for Sustainable Landscape Development. *IOP Conference Series. Earth and Environmental Science, 179*, 012017. DOI: 10.1088/1755-1315/179/1/012017

Wylie, J. (2007). *Landscape*. Routledge. DOI: 10.4324/9780203480168

Chapter 14
The Impact of Spatial Planning on Traffic Safety:
Urban Development and Road Network Design

Ömür Kaygısız
https://orcid.org/0000-0001-9195-6339
Mugla Sitki Kocman University, Turkey

Ahmet Yıldız
https://orcid.org/0000-0002-9281-0718
Macromedia University of Applied Sciences

ABSTRACT

This chapter explores the critical relationship between spatial planning and traffic safety, focusing on how urban development patterns, land use decisions, and road network designs influence traffic accident frequency and severity. Research indicates that urban sprawl tends to reduce traffic safety, while smart growth strategies and travel demand management enhance it. Smart growth promotes compact, mixed-use, and diverse urban environments, reducing mobility demand and increasing accessibility, which significantly improves traffic safety. Similarly, strategies that reduce motor vehicle use and promote public transit, walking, and cycling contribute to fewer accidents by lowering traffic volumes and speeds. The chapter also examine how land use patterns, road network designs, traffic speed, and volume interact to affect accident rates, highlighting that speed primarily influences accident severity, while traffic volume drives accident frequency.

DOI: 10.4018/979-8-3373-1554-6.ch014

1 THE IMPACT OF SPATIAL PLANNING ON TRAFFIC SAFETY: URBAN DEVELOPMENT AND ROAD NETWORK DESIGN

Traffic safety remains a global public health challenge, with the World Health Organization estimating 1.35 million road traffic deaths annually. Spatial planning, which shapes urban development patterns, land use decisions, and road network designs, is a critical determinant of traffic safety outcomes. Urban form influences travel behavior, traffic volumes, and speeds, directly affecting accident frequency and severity. This chapter examines how spatial planning strategies—ranging from urban sprawl to smart growth, travel demand management (TDM), land use patterns, urban density, mixed land use, road network configurations, and intersection designs—can enhance traffic safety. By synthesizing empirical research and practical examples, the chapter aims to provide actionable insights for planners and policymakers to create safer urban environments, contributing to the discourse on economic geography and sustainable urban development.

1.1 Urban Development Patterns

Urban development patterns significantly affect traffic safety through road network design and travel mode choices. Garrick and Marshall (2008) studied 24 California cities, classifying them as "safe" (serious or fatal accident rates one-third of the state average) or "less safe" (near the state average). Safe cities had higher intersection density (106 vs. 63 per square mile), greater use of walking, cycling, and public transit (16% vs. 4%), and fewer traffic deaths (3.2 vs. 10.5 per 100,000 people). Older cities (pre-1950) with dense, grid-like networks were safer than newer, sprawling cities. For example, in Davis, California, pre-1940 areas with 211 intersections per square mile had half the serious accident rate of post-1970 areas with 111 intersections per square mile, alongside higher non-motorized travel (59% vs. 14% post-1980) [Garrick & Marshall, 2008].

Litman and Fitzroy (2010) explored travel demand management, finding that increased private vehicle trips raise accident rates, while reducing trips or shifting to other modes lowers risks. Redirecting traffic to less congested roads may reduce accident frequency but increase severity due to higher speeds. Smart growth strategies, promoting compact, mixed-use communities, reduce sprawl and enhance safety by lowering vehicle miles traveled (VMT) and encouraging non-motorized travel [Litman & Fitzroy, 2010]. Durning (1996) noted that sprawl increases vehicle use and speeds, raising accident frequency and severity. Jones and Jha (2009) argued that sprawl leads to high-speed road designs and non-local traffic on residential streets, increasing accidents.

Dumbaugh and Rae (2009) found that smart growth may increase accident frequency due to traffic concentration but reduces severity and fatalities by lowering speeds and VMT. Traditional safety approaches, like segregating land uses or designing hierarchical road networks, often increase accidents by promoting higher speeds [Dumbaugh & Rae, 2009]. Ewing et al. (2003) reported a 1% increase in the smart growth index reduces traffic fatalities by 1.5%, with similar effects expected at the neighborhood level [Ewing et al., 2003]. Lovegrove and Litman (2008) estimated that smart growth could reduce total accidents by 20% and severity by 29% in residential areas of Greater Vancouver.

Ewing and Dumbaugh (2009) propose a framework linking built environments to traffic safety through traffic volume and speed. Development patterns primarily affect safety via traffic volume, while road designs influence safety through permitted speeds. Traffic volume drives accident frequency, and speed determines severity [Ewing & Dumbaugh, 2009].

1.1.1 Key Impacts of Travel Demand Management [Litman & Fitzroy, 2010]:

1. Reducing vehicle trips lowers accident rates (10% distance reduction = 10–14% fewer accidents).
2. "Pay-as-you-drive" models reduce high-risk drivers' mileage, enhancing safety.
3. Shifting to public transit or car-sharing improves safety, depending on urban context.
4. Promoting walking or cycling may increase per-mile risk but reduces total accidents due to shorter trips.
5. Reducing congestion lowers accident frequency but may increase severity due to higher speeds.
6. Lowering traffic speeds in pedestrian-heavy areas reduces accident frequency and severity.
7. Smart growth increases per-mile accidents due to congestion but reduces per-capita casualties by lowering VMT and speeds.
8. Traffic restrictions reduce accidents if they lower total VMT, but have minimal impact if trips are merely rerouted.
9. Safety effects vary by demographic and geographic factors (e.g., improved cycling infrastructure reduces accidents).
10. Traffic accidents are a major societal cost, outweighing congestion or pollution; safety-focused policies are critical.

In conclusion one can say that urban sprawl increases traffic accidents by promoting car dependency and higher speeds, while smart growth and travel demand management enhance safety by reducing VMT and encouraging safer travel modes. Dense, interconnected road networks and compact urban forms are key to improving traffic safety.

1.2 Urban Land Use and Traffic Accidents

Urban land use significantly influences traffic safety by affecting traffic volume, distribution, mode choice, and accident rates. Effective land use planning can minimize traffic and trip distances to enhance safety, while unplanned large-scale developments may create chaotic traffic systems, increasing accidents [Elvik et al., 2009; Fridstrøm et al., 1995]. This section explores urban density and land use types, drawing on empirical studies to highlight their impact on traffic accidents.

1.2.1 Impact of Urban Density

Urban density, encompassing population and building density, shapes travel behavior and accident risks. Higher density generally reduces per-capita vehicle miles traveled, lowering accident rates [Litman, 2012; Elvik et al., 2009]. Newman and Kenworthy (1989) analyzed 32 cities, finding a strong negative correlation between building density and traffic volume. Similarly, Næss (1996) reported that increased building density in Norwegian settlements reduced per-capita traffic volume.

Durning (1996) found that low-density areas in Puget Sound, Washington, had four times higher per-capita traffic casualty rates than high-density areas. Elvik and Muskaug (1994) noted that urban areas have higher per-mile injury accident rates than rural areas, but are safer per capita due to lower speeds and fewer fatal crashes. In San Antonio, Texas, Dumbaugh and Rae (2009) found that a one-person increase per 0.404 acres reduced accident frequency by 0.05%, though increased vehicle trips raised accident rates by 0.75% per million miles traveled. Hadayeghi et al. (2003) observed that high-density areas near business districts face more accidents due to regional traffic and multi-lane arterials, complicating safety dynamics.

Overall, denser urban environments are safer due to reduced VMT, lower speeds, and fewer severe accidents, despite higher per-mile accident rates [Ewing & Dumbaugh, 2009; Litman, 2012].

1.2.2 Impact of Land Use Types

Different land uses generate varied trip patterns, affecting traffic volume and accident risks [Elvik et al., 2009]. Dumbaugh and Rae (2009) analyzed San Antonio, Texas, using GIS and negative binomial models, finding that arterial-focused commercial developments increased fatal accidents, while pedestrian-scale commercial areas had fewer accidents. Each additional arterial retail parcel raised total accidents by 1.3%, and each shopping center increased accidents by 6.6%, whereas pedestrian-scale retail reduced accidents by 2.2%. Managing arterial access and locating commercial uses away from high-speed roads can enhance safety.

Kim et al. (2006) studied Honolulu, Hawaii, using a 0.259 km^2 grid and linear regression models to correlate land use, population, employment, and economic output with accidents. Urban areas accounted for ten times more accidents than agricultural or conservation areas, with all accident types (bicycle, pedestrian, vehicle-vehicle) concentrated in urban zones. Population showed a positive correlation with accidents, explaining 22.3% of total accident variation, 18.8% of vehicle-vehicle, and 13.6% of pedestrian accidents. Employment had a stronger correlation, explaining 47.5% of total accidents, 41% of vehicle-vehicle, and 38.2% of pedestrian accidents. Economic output had the strongest correlation, with higher regression coefficients than population or employment.

Key findings from Kim et al. (2006) include:

- Urban areas have higher accident rates due to intense activity and dense development.
- Commercial and school areas have elevated accident rates, particularly for pedestrians and vehicle-vehicle collisions.
- Pedestrian safety in commercial and park areas requires special attention.
- Economic output and employment are more strongly linked to accidents than population.
- Bicycle accidents showed weaker correlations with economic activity.

Comparative studies in Norway and Germany provide further insights. Elvik et al. (2009) cited Blakstad (1990) and Kohler and Schwamb (1993), showing no significant accident rate differences between residential, industrial, and mixed-use areas in Sweden, but higher rates in German industrial and mixed-use areas compared to residential ones (Table 1.1). In England, Bennett and Marland (1978) found that accident risks on residential access roads increased with the number of households served, driven by higher traffic volumes. In Norway, Blakstad and Giæver (1989) reported that lowering speed limits and adding speed humps reduced accident rates on access roads, with rates dropping significantly at 30 km/h with humps.

Table 1. Comparison of urban road surroundings with the construction status and accident rates

Building Density	Type	Relative Risk	
		Sweden	Germany
Middle	Residential	1,00	1,00
	Mixed		0,97(0,84; 1,12)
	Industrial	0,86 (0,70; 1,05)	1,24 (1,04; 1,47)
Heavy	Residential	1,00	1,00
	Mixed		1,59 (1,39; 1,82)
	Industrial	1,03 (0,84; 1,27)	1,42 (1,23; 1,63)

Workplace location impacts travel behavior and safety. Strand (1993) found that relocating a company from Oslo's center to its periphery increased car commuting from 17% to 35%. In San Francisco, a similar relocation reduced public transit use from 56% to 3% and tripled car use. Lervag (1985) reported a 69% drop in public transit and a 190% increase in car use after a bank moved from Trondheim's center. Conversely, Fosli (1995) found that moving a company closer to Oslo's center increased public transit and non-motorized travel, reducing car use by 35%. Central workplace locations promote safer, less vehicle-dependent travel patterns [Elvik et al., 2009; Strand, 1993; Lervag, 1985; Fosli, 1995].

Urban land use shapes traffic safety through density and land use types. Higher density reduces per-capita accidents by lowering VMT and speeds, though per-mile rates may rise due to congestion [Litman, 2012]. Urban areas have more accidents than rural ones, but are safer per capita [Elvik & Muskaug, 1994]. Commercial and school areas face higher risks, necessitating targeted safety measures [Kim et al., 2006]. Workplace relocation to urban centers promotes public transit and reduces accidents, while peripheral locations increase car use and risks [Strand, 1993]. Strategic land use planning, focusing on density and access management, is critical for enhancing traffic safety.

1.3 Urban Road Network

Urban road networks, comprising intersections and road segments, significantly influence traffic safety. This section examines the impact of road segment characteristics on accident rates, drawing on empirical studies to highlight key findings.

1.3.1 Impact of Road Segment Characteristics

The traditional grid-based road network, prevalent before the widespread adoption of automobiles, prioritized vehicle movement but led to conflicts with pedestrians and local traffic [Dumbaugh & Rae, 2009]. Olmsted (1916) criticized the grid system for failing to meet public needs, advocating for road designs tailored to specific traffic functions to enhance safety and efficiency. An early example is New York's Central Park, where Olmsted designed a park with limited-access roads to separate transit traffic from recreational users, as shown in Figure 1.1 [Dumbaugh & Rae, 2009]. Robinson (1911) argued that roads designed beyond motor vehicle needs could reduce traffic and improve safety, particularly in residential areas.

Figure 1. Network plan of Central Park in New York

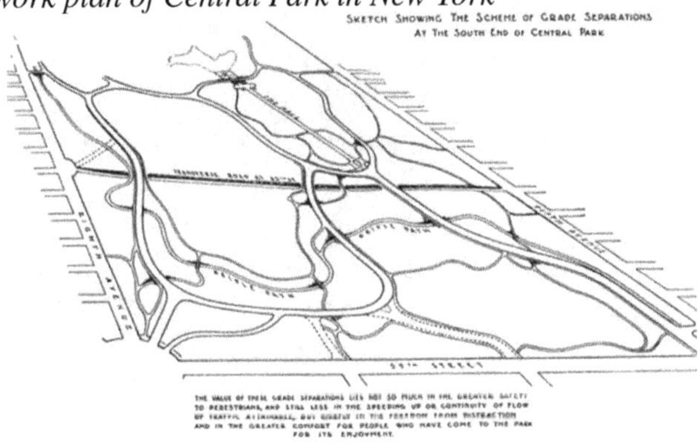

[Dumbaugh & Rae, 2009]

Early urban planning guides, such as Haldeman (1916), emphasized distinguishing primary arterials from secondary local roads to enhance safety. Wide arterials with good visibility and appropriate speeds were seen as safer for residential areas. Dumbaugh and Rae (2009) identified two safety strategies: designing arterials to allow drivers to anticipate hazards and reducing intersection numbers to lower crash risks.

Figure 2. Chicago's expanding grid system

[Currier & Ives, 1892/1991]

Grid networks, used until the 1950s, promoted walkability and development but posed safety challenges due to uniform street widths and high intersection density [Kostof, 1991]. Figure 1.2 illustrates Chicago's expanding grid system. To address these issues, modern road designs emerged in the 1950s, including warped parallel, loop, lollipop, and cul-de-sac patterns, which limit access and reduce traffic volume and speed [Garrick & Marshall, 2008]. These modern designs, shown in Figure 1.3, contrast with the traditional grid system, differing in block size, curvature, and interconnectivity [Rifaat et al., 2010; Ewing & Dumbaugh, 2009].

Figure 3. Modern approach; warped parallel, loop & lollipop and cul-de-sacs networks

a) warped parallel b) loop & lollipopp

c) warped parallel, loop & lollipop and cul-de-sacs used together

[Rifaat et al., 2010]

Rifaat et al. (2010) compared these designs in Calgary, Canada, finding that modern patterns (warped parallel, loop, lollipop, cul-de-sac) were safer than grid networks. The curves and loops encourage cautious driving, compensating for reduced visibility or control challenges. These designs also calm traffic, increasing driver awareness of pedestrians, cyclists, and children, and reduce non-local traffic, lowering accident rates. However, they can disrupt pedestrian and cyclist connectivity, potentially causing health issues [Rifaat et al., 2010].

Chatfield (1987) found that U.S. roads opened after 1967 had 20% fewer fatal accidents per 100 million vehicle-miles than pre-1967 roads, attributed to lower speeds and higher standards on busier roads. In Sweden, Björketun (1991) reported that 1970s roads had lower injury accident rates than those from the 1950s and 1960s, with 1970s roads showing 10–18% fewer accidents per million vehicle-kilometers.

Traditional safety-focused urban planning, such as aligning commercial uses along arterials, can shift safety issues rather than resolve them [Dumbaugh & Rae, 2009]. Less forgiving designs, like narrow lanes, street trees, and traffic calming measures, are more effective than wide, straight roads with high visibility, as they promote safer driving speeds [Ewing & Dumbaugh, 2009]. Litman and Fitzroy (2010) noted that such measures reduce per-mile accident rates but may increase VMT if not carefully implemented, displacing risks. Despite advances in road design and emergency services, traffic accidents remain a significant health issue, indicating limited success of these strategies.

Elvik et al. (2009) emphasized designing road networks to prioritize safety by separating access roads from transit traffic, minimizing traffic volume on access roads, and creating simple, safe systems for all users. Traffic volume and accident rates are closely linked, with studies showing varied results. Sullivan (1990) reported a 90% increase in injury accidents and 160% in property-damage accidents during peak hours in San Francisco. Persaud and Dzbik (1993) and Sandhu and Al-Kazily (1996) found accident rates doubled during peak hours in Toronto and California, respectively. However, Hall and Polanco de Hurtado (1992) found no clear link in Albuquerque.

Zhou and Sisiopiku (1997) analyzed volume-to-capacity ratios, finding that accident rates drop as ratios rise from 0.10 to 0.50 due to lower speeds, but increase beyond 0.50, particularly for property-damage accidents. Arterials have higher accident rates than local streets, with Ladrón de Guevara et al. (2004) and Lovegrove and Sayed (2006) confirming positive correlations between arterial road length and accident rates.

Ewing and Dumbaugh (2009) proposed three research areas: understanding how regional development affects trip patterns and accident exposure, reevaluating urban design's impact on crash frequency, and studying how design influences user behavior. Compact regional forms reduce VMT and per-capita accidents, but optimal implementation remains unclear. Traditional designs diverting non-local traffic to arterials increase risks for slower users like pedestrians and cyclists. Road design changes significantly affect speed and congestion, key crash factors.

In San Antonio, Texas, Dumbaugh and Rae (2009) found that each additional arterial mile increased total accidents by 15%, expressway miles in residential areas raised fatal accidents by 5%, and arterial miles increased fatal accidents by 20%. However, higher three- and four-way intersection density was linked to fewer fatal accidents, challenging traditional views that intersections are inherently dangerous. Well-designed intersections in dense networks can reduce crash frequency.

Bypass roads, separating urban and intercity traffic, enhance safety. Elvik et al. (2009) reported a 25% reduction in injury accidents and 27% in property-damage accidents after bypass construction, with effectiveness depending on:

- Higher accident rates on original roads leading to greater reductions.
- Greater traffic diversion to bypasses reducing accidents.
- Longer bypass routes reducing accident reductions.
- Safety measures on original roads (e.g., speed reduction) enhancing reductions.
- Intersection design on both old and new roads affecting accident rates.

Urban road segment characteristics, including design, function, traffic volume, and speed, significantly impact traffic safety. Modern road designs like loops and cul-de-sacs are safer than traditional grids due to lower speeds and volumes, though they may limit pedestrian connectivity [Rifaat et al., 2010]. Less forgiving designs and traffic calming measures outperform traditional high-speed arterials [Ewing & Dumbaugh, 2009]. Bypass roads reduce urban accidents, with effectiveness tied to traffic diversion and intersection design [Elvik et al., 2009]. Strategic road network planning is critical for enhancing safety and reducing accident frequency and severity.

1.3.2 Urban Intersection Characteristics and Traffic Accidents

Intersections are critical nodes in urban road networks where safety measures are essential. This section examines how intersection type, number of connections, and density affect traffic accident frequency and severity, drawing on key studies.

1.3.2.1 Intersection Design and Accident Frequency

Marks (1957) conducted an early study comparing accident rates in grid-based (traditional) and limited-access (modern) road networks at the parcel level. Grid networks, characterized by four-way (+) intersections and straight roads, allow higher vehicle speeds and access, while limited-access networks use T-intersections, cul-de-sacs, and loops to slow traffic. Key findings include:

- In grid networks, accidents are uniformly distributed but concentrated at four-way intersections, with fewer at three-way (T) intersections.
- Limited-access networks have fewer four-way intersections, and most T-intersections report no accidents.
- Over five years, grid parcels averaged 77.7 annual accidents, compared to 10.2 in limited-access parcels, largely due to the higher proportion of three-way intersections.
- Traffic calming measures, like roundabouts, may shift risks to four-way intersections.
- There is a significantly higher accident frequency at four-way intersections compared to three-way ones [Marks, 1957].

Recent studies by Lovegrove and Sayed (2006), Giaver (1990), and Sakshaug and Johannessen (2005) support Marks' findings, confirming that four-way intersections have roughly twice the accident rate of three-way intersections. Lovegrove and Sayed (2006) used a macro-level accident prediction model in Greater Vancouver, Canada, finding that areas with four-way intersections had more accidents than those with three-way intersections. Ewing and Dumbaugh (2009) noted that short-distance T-intersections effectively reduce speed, accident frequency, and severity. However, Marks (1957) has been criticized for ignoring accident severity and overall network accident rates.

1.3.2.2 Roundabouts and Safety

Converting three- or four-way intersections to modern roundabouts significantly enhances safety by reducing accident frequency and severity [Elvik et al., 2009; Robinson et al., 2000; Vanderbilt, 2008]. Roundabouts slow traffic, eliminate direct left turns, and reduce conflict points. Traditional four-way intersections have 56 potential collision points (32 vehicle-vehicle, 24 vehicle-pedestrian), while roundabouts reduce this to 16 (8 vehicle-vehicle, 8 vehicle-pedestrian). For three-way intersections, collision points drop from 21 (9 vehicle-vehicle, 12 vehicle-pedestrian) to 12 (6 vehicle-vehicle, 6 vehicle-pedestrian). Roundabouts lack high-risk crossing conflicts, with all points being merge or diverge conflicts, making crashes less severe (Figure 1.4) [Vanderbilt, 2008].

Figure 4. Potential points at intersections where vehicles may collide

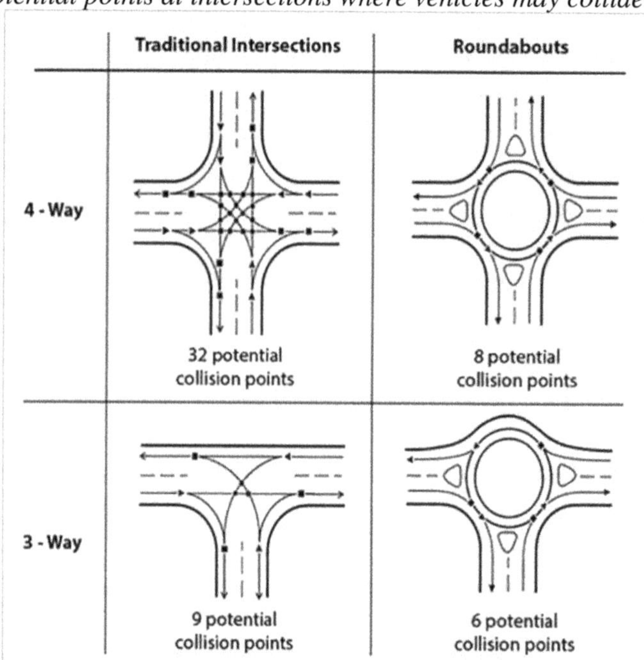

Robinson et al. (2000) analyzed 11 U.S. intersections converted to roundabouts, reporting a 37% reduction in total accidents, 51% in injury accidents, and 90% in fatal accidents (Table 3.5). Single-lane roundabouts with 30–35m central islands saw a

73% drop in injury accidents. Vanderbilt (2008) noted that roundabouts, lacking traffic lights, halve average speeds compared to signalized four-way intersections, improving pedestrian safety. Rodegerdts et al. (2010) studied roundabout conversions in six countries (Australia, France, Germany, Netherlands, UK, USA), finding injury accident reductions of 51–87% and total accident reductions of 37–61% (Table 3.6). Roundabouts prevent severe crash types like head-on or left-turn collisions by controlling traffic flow and speed.

1.3.2.3 Grade-Separated Intersections

Grade-separated (elevated) intersections, which separate traffic flows vertically, reduce accidents by eliminating crossing conflicts and minimizing lane changes [Elvik et al., 2009]. Elvik et al. (2009) compiled studies from Sweden, Germany, Finland, and Norway, finding that grade-separated intersections have lower accident rates than at-grade intersections, with the largest reductions at four-way intersections. Converting four-way intersections to fully grade-separated designs reduced injury accidents by 57% and total accidents by 42%. However, converting signalized four-way intersections with speed cameras to partial grade-separated designs increased accidents by 115%, highlighting the need for context-specific solutions (Table 3.7).

1.3.2.4 Intersection Environment and Urban Context

Jones and Jha (2009) studied four high-accident intersections in Baltimore, Maryland, from 2003 to 2007, comparing urban and suburban settings. Suburban intersections, surrounded by wider streets, lower-density buildings, and structures set farther back, had more injury and property-damage accidents than urban intersections, which featured higher buildings, shorter road segments, and buildings closer to the road. The urban environment's design facilitates safer traffic flow [Jones & Jha, 2009]. Ewing and Dumbaugh (2009) found that higher intersection density reduces speeds, leading to fewer severe accidents. In San Antonio, Texas, Dumbaugh and Rae (2009) reported that increased three- and four-way intersection frequency and population density (a 0.05% accident reduction per additional person per 0.404 acres) were linked to fewer fatal accidents.

1.3.2.5 Traffic Signals and Safety

Traffic signals reduce accidents by regulating flow, with Elvik et al. (1997) reporting a 15% reduction at three-way intersections and 30% at four-way intersections. However, Greibe (2003) analyzed 1,036 Danish intersections, finding no significant difference in accident rates between signalized and unsignalized intersections with similar traffic volumes. Signalized intersections showed only a 9% reduction in injury accidents at three-way intersections and 12% at four-way intersections, with traffic volume being the strongest predictor of accident frequency. Sakshaug and Johannessen (2005) found that four-way intersections with right-of-way priority had higher accident costs than signalized or three-way intersections, particularly at high-speed or high-volume secondary roads.

As a conclusion, urban intersection characteristics, including type, connection number, traffic signals, and surrounding land use, significantly influence accident rates. Three-way intersections are safer than four-way ones, with roundabouts reducing accidents by 37–90% by lowering speeds and conflict points [Robinson et al., 2000]. Grade-separated intersections reduce injury accidents by up to 57%, though partial conversions may increase risks [Elvik et al., 2009]. Higher intersection density and urban settings with compact designs lower accident severity [Ewing & Dumbaugh, 2009; Jones & Jha, 2009]. Traffic volume and speed remain critical factors, with signals offering mixed results [Greibe, 2003; Sakshaug & Johannessen, 2005].

2 CONCLUSION

The relationship between urban form, road networks, and traffic safety is intricate and multifaceted, as evidenced by the comprehensive studies reviewed in sections 3.2 and 3.3. Urban land use patterns significantly influence traffic safety through density, land use types, and workplace locations. Higher urban densities reduces per-capita vehicle miles traveled (VMT) and accident rates by promoting shorter trips and non-motorized travel, though it may increase per-mile accident rates due to congestion [Litman, 2012; Elvik et al., 2009]. Urban areas, despite higher per-mile accident rates, are safer per capita than rural or suburban areas due to lower speeds and fewer severe crashes [Elvik & Muskaug, 1994]. Commercial and school zones require targeted safety measures due to elevated accident risks, particularly for pedestrians [Kim et al., 2006]. Central workplace locations foster public transit use, reducing car dependency and accidents, while peripheral relocations increase VMT and risks [Strand, 1993].

Road network design further shapes safety outcomes. Modern road patterns, such as loops, lollipops, and cul-de-sacs, outperform traditional grid systems by reducing traffic volume and speed, though they may limit pedestrian connectivity [Rifaat et al., 2010]. Less forgiving designs, including narrow lanes and traffic calming measures, enhance safety by promoting cautious driving compared to wide, high-speed arterials [Ewing & Dumbaugh, 2009]. Intersection characteristics are critical, with three-way intersections and roundabouts significantly safer than four-way intersections due to fewer conflict points and lower speeds [Marks, 1957; Robinson et al., 2000]. Grade-separated intersections reduce accidents by eliminating crossing conflicts, though partial conversions may increase risks [Elvik et al., 2009]. Urban settings with higher intersection density and compact designs further reduce accident severity [Jones & Jha, 2009].

Collectively, these findings underscore that strategic urban planning and road network design can substantially enhance traffic safety. Compact urban forms, modern road patterns, and well-designed intersections mitigate accident frequency and severity by reducing VMT, controlling speeds, and minimizing high-risk conflict points. However, challenges remain, such as balancing pedestrian connectivity with vehicle safety and ensuring context-specific solutions to avoid displacing risks.

2.1 Recommendations for Planners and Policymakers

To translate these insights into actionable strategies, planners and policymakers should prioritize the following recommendations:

1. **Promote Compact Urban Development**:
    - Encourage high-density, mixed-use developments to reduce VMT and promote walking, cycling, and public transit. Zoning policies should integrate residential, commercial, and recreational spaces to minimize trip distances.
    - Incentivize central workplace locations to reduce car dependency and enhance transit use, as peripheral relocations increase accident risks.
2. **Adopt Modern Road Network Designs**:
    - Prioritize limited-access road patterns (e.g., loops, lollipops, cul-de-sacs) over grid systems in new developments to reduce traffic volume and speed. Ensure pedestrian and cyclist pathways are integrated to maintain connectivity. As connectivity increases, travel distances decrease and route options increase.
    - Avoid longer street blocks, as they encourage higher vehicle speeds due to fewer junctions that interrupt travel. Long blocks also are unsafe for pedestrians.
    - Minimize vehicle lane width to prioritize pedestrians and provide sidewalk on both sides of street wherever possible.
3. **Enhance Intersection Safety**:
    - Convert high-risk four-way intersections to three-way (T) intersections or roundabouts to reduce conflict points and accident severity. Single-lane roundabouts with 30–35m central islands are particularly effective.
    - Use grade-separated intersections in high-traffic areas to eliminate crossing conflicts, but avoid partial conversions that may increase accidents.

- o Evaluate traffic signal effectiveness case-by-case, as signals may not always reduce accidents compared to unsignalized intersections with similar volumes.

4. **Implement Targeted Safety Measures in High-Risk Zones**:
   - o Deploy enhanced pedestrian safety measures (e.g., crosswalks, pedestrian islands) in commercial and school zones, where accident risks are elevated.
   - o Use traffic calming techniques, such as speed limits and physical barriers, on residential access roads to reduce accident rates, particularly in areas with high household density.
5. **Integrate Traffic Calming and Bypass Roads**:
   - o Construct bypass roads to divert intercity traffic from urban centers, reducing congestion and enabling traffic calming on main urban roads. Effectiveness depends on diverting sufficient traffic and optimizing intersection designs.
   - o Combine bypass roads with speed reduction measures on existing urban roads to maximize accident reductions.
6. **Leverage Data-Driven Planning**:
   - o Use GIS and regression models to identify high-risk intersections and road segments, as demonstrated in San Antonio and Honolulu studies [Kim et al., 2006; Dumbaugh and Rae, 2009]. Prioritize interventions based on traffic volume, speed, and land use data.
   - o Develop macro-level accident prediction models to assess the impact of urban form and road design on safety, as shown in Vancouver [Lovegrove and Sayed, 2006].
7. **Balance Safety with Accessibility**:
   - o Address the trade-offs of limited-access designs, which enhance vehicle safety but may reduce pedestrian and cyclist mobility. Incorporate dedicated bike lanes and pedestrian pathways to ensure inclusive urban mobility.
   - o Avoid safety measures that displace risks, such as diverting traffic to high-speed arterials, which can endanger slower users.
8. **Engage Stakeholders and Update Policies**:
   - o Collaborate with urban planners, traffic engineers, and community stakeholders to align safety goals with livability and economic objectives. Public input can identify local safety concerns and improve intervention acceptance.
   - o Update zoning and transportation policies to reflect evidence-based safety practices, such as prioritizing roundabout installations and compact urban designs.

By implementing these recommendations, planners and policymakers can create safer, more livable urban environments. These strategies address the complex interplay of urban form, road design, and traffic behavior, reducing accident frequency and severity while enhancing overall urban mobility and quality of life.

REFERENCES

Bennett, G. T. & ve Marland, J., "Road accidents in traditionally designed local authority estates" TRRL Supplementary Report 394. Transport and Road Research Laboratory, Crowthorne, Berkshire, İngiltere (1978).

Björketun, U. (1991). "Linjeföring samt prediktion av olyckor utifran linjeförings data för vagar projekterade/byggda under 1950-, 1960- respektive 1970-talet", VTI-meddelande 641. Statens vag- och trafikinstitut (VTI). İsveç.

Blakstad, F. ve Giæver, T., "Ulykkesfrekvenser på vegstrekninger i tett og middels tett bebyggelse", Rapport STF63 A89005, SINTEF Samferdselsteknikk, Trondheim, Norveç (1989).

Blakstad, F. "Ulykkesfrekvenser på hovedveger i byområder", Rapport STF63 A90005, SINTEF Samferdselsteknikk, Trondheim, Norveç (1990).

Chatfield, B. V. (1987). System-Wide Safety Improvements: An Approach to Safety Consistency. In National Cooperative Highway Research Program Report 132. Transportation Research Board.

Currier, N. ve Ives, J. M.,"Detail of a bird's eye view of Chicago" The city shaped: Urban patterns and meanings through history (1991), Kostof, S. (Ed.), Bulfinch Press, New York, 116 (1892).

Dumbaugh, E., & Rae, R. (2009). ve Rae, R., "Safe urban form: Revisiting the relationship between community design and traffic safety". Journal of the American Planning Association, 75(3), 309–329. DOI: 10.1080/01944360902950349

Durning, A. T. "The Car and the City: 24 Steps to Safe Streets and Healthy Communities", New Report No.3, Northwest Environment Watch, Seattle, Washington, (1996).

Elvik, R. ve Muskaug R., "Konsekvensanalyser og trafikksikkerhet. Metode for beregning av konsekvenser for trafikksikkerheten av tiltak pa vegnettet", TØI-rapport 0281, Transportøkonomisk institutt, Oslo, Norveç (1994).

Elvik, R., & Hoye, A. ve Vaa, T., "The Handbook Of Road Safety Measures; Second Edition", Emerald Group Publishing Limited, UK, 169-206;1031-1043, (2009).

Elvik, R., Mysen, A.B. ve Vaa, T., "Trafikksikkerhetshånbok-3", Udgave, TØI Rapport, Norveç (1997).

Ewing, R., & Dumbaugh, E. (2009). ve Dumbaugh, E., "The built environment and traffic safety: A review of empirical evidence". Journal of Planning Literature, 23(4), 347–367. DOI: 10.1177/0885412209335553

Ewing, R., Schieber, R., & Zegeer, C. V. (2003). ve Zegeer, C., "Urban sprawl as a risk factor ın motor vehicle occupant and pedestrian fatalities". American Journal of Public Health, 93(9), 1541–1545. DOI: 10.2105/AJPH.93.9.1541 PMID: 12948977

Fosli, O. (1995). "Transport, arealbruk og miljø. Ein studie av dei transportmessige verknadene av IDG Norges omlokalisering fra?" Hasle til Galleri Oslo, Hovedoppgave i samfunnsgeografi, Universitetet i Oslo, Samfunnsvitenskapelig Fakultet. Norveç.

Fridstrøm, L., Ifver, J., Ingebrigtsen, S., Kulmala, R., & Thomsen, L. K. (1995). ve Thomsen, L.K., "Measuring the contribution of randomness, exposure, weather, and daylight to the variation in road accident counts.". Accident; Analysis and Prevention, 27(1), 1–20. DOI: 10.1016/0001-4575(94)E0023-E PMID: 7718070

Garrick, N. ve Marshall W., "Network, Placemaking and Sustainability" Center for Transportation and Urban Planning, University of Connecticut, Connecticut, A.B.D. (2008).

Giæver, T. (1990). "Ulykkesfrekvenser i rundkjøringer og signalregulerte kryss", SINTEF Samferdselsteknikk, STF63 A90002. Norveç.

Greibe, P. (2003). Accident prediction models for urban roads. Accident; Analysis and Prevention, 35(2), 273–285. DOI: 10.1016/S0001-4575(02)00005-2 PMID: 12504148

Hadayeghi, A., Shalaby, A. S., & Persaud, B. (2003). ve Persaud, B. N., "Macrolevel accident prediction models for evaluating safety of urban transportation systems". Transportation Research Record: Journal of the Transportation Research Board, 1840(1), 87–95. DOI: 10.3141/1840-10

Haldeman, B. A. "Transportation and main thoroughfares and street railways", City planning, J. Nolen (Ed.), Appleton & Co., New York, A.B.D, 279-313 (1916).

Hall, J. W. (1992). ve Polanco de Hurtado, M., "Effect of intersection congestion on accident rates". Transportation Research Record: Journal of the Transportation Research Board, (1376), 71–77.

Jones, D. ve Jha, M., "The effect of urban form on traffic accident incidence", CEA'10 Proceedings Of the 4th WSEAS International Conference on Computer Engineering and Applications,USA, 212-222 (2009).

Kim, K., Brunner, M., & Yamashita, E. Y. (2006). ve Yamashita, E., "Influence of land use, population, employment, and economic activity on accidents". Transportation Research Record: Journal of the Transportation Research Board, 1953(1), 56–64. DOI: 10.1177/0361198106195300107

Kostof, S. (1991). The city shaped: Urban patterns and meanings through history. Bulfinch Press.

Ladrón de Guevara, F., Washington, S. P., & Oh, J. (2004). ve Oh, J., "Forecasting travel crashes at the planning levels: Simultaneous negative binomial crash model applied in Tucson, Arizona". Transportation Research Record: Journal of the Transportation Research Board, 1897(1), 191–199. DOI: 10.3141/1897-25

Lervag, H. (1985). "Arealbruk og transport", NIBR-notat 1985:14, Norsk institutt for byog regionplanlegging. Norveç.

Litman, T. (2012). "Evaluating accessibility for transportation planning", Victoria Transport Policy Institute. Kanada.

Litman, T. ve Fitzroy, S., "Safe travels; Evaluating mobility management traffic safety impacts", Victoria Transport Policy Institute,Victoria, Kanada (2010).

Lovegrove, G., & Litman, T. "Macrolevel collision prediction models to evaluate road safety effects of mobility management strategies: New empirical tools to promote sustainable development", Transportation Research Board 87th Annual Meeting, USA, 1-24 (2008).

Lovegrove, G. R., & Sayed, T. (2006). ve Sayed, T., "Macro-level collision prediction models for evaluating neighbourhood traffic safety". Canadian Journal of Civil Engineering, 33(5), 609–621. DOI: 10.1139/l06-013

Marks, H. (1957). Subdividing for traffic safety. Traffic Quarterly, 11(3), 308–325.

Næss, P. "Miljømessig effektiv lokalisering av arbeidsplasser og boliger", In:Miljøhåndboken, (Kolbenstvedt, M., H. Silborn & T. Solheim, T. eds) Transportøkonomisk institutt, Oslo, Norveç, 1: 95-106 (1996).

Newman, P. W. G. ve Kenworthy, J. R., "Cities and Automobile Dependence: A Sourcebook", Avebury Technical, Aldershot, UK (1989).

Persaud, B. N. (1993). ve Dzbik, L., "Accident prediction models for freeways". Transportation Research Record: Journal of the Transportation Research Board, (1401), 55–60.

Rifaat, S. M., & Tay, R. (2010). ve Barros, A., "Effect of street pattern on road safety are policy recommendations sensitive to different aggregations of crashes by severity?". Transportation Research Record: Journal of the Transportation Research Board, (2147), 58–65. DOI: 10.3141/2147-08

Robinson, B. W., Rodegerdts, L., & Scarborough, W. ve Kittelson, W., "Roundabouts: An informational guide" U.S. Department of Transportation, Federal Highway Administration, Report FWHARD-00-067, Washington, ABD, 101-181 (2000).

Robinson, C. M. (1911). The width and arrangement of streets. Engineering News Publishing.

Sakshaug, K. ve Johannessen, S.,"Analyse av ulykkessteder: Verdier for normal ulykkesfrekvens og skadekostnad ved normal oggod standard", Trondheim, SINTEF Teknologi Ogsamfunn, Transportsikkerhet og –İnformatikk, Norveç (2005)

Sandhu, B. ve Al-Kazily, J., "Safety impacts of freeway traffic congestion", Transportation Research Board, 75th Annual Meeting, Washington DC., (1996).

Strand, A. (1993). "Trafikksikkerhetsmessige hensyn i arealplanleggingen: Forelesning ved EEU-kurs Trafikksikkerhet 29.10.93", Norges Tekniske Høgskole, Institutt for samferdselsteknikk. Norveç.

Sullivan, E. C. (1990). Estimating Accident Benefits of Reduced Freeway Congestion. Journal of Transportation Engineering, 116(2), 167–180. DOI: 10.1061/(ASCE)0733-947X(1990)116:2(167)

Vanderbilt, T. (2008). Traffic: Why we drive the way we do (And what it says about us). Knopf Doubleday Publishing Group.

Zhou, M., & Sisiopiku, V. P. (1997). ve Sisiopiku, V., "Relationship between volume to capacity ratios and accident rates". Transportation Research Record: Journal of the Transportation Research Board, 1581(1), 47–52. DOI: 10.3141/1581-06

Compilation of References

Abban, O. J., Rajaguru, G., & Acheampong, A. (2025). The Spillover Effect of Economic Institutions on The Environment: A Global Evidence from Spatial Econometric Analysis. Journal of Environmental Management, 373, 123645. DOI: 10.1016/j.jenvman.2024.123645 PMID: 39689530

Adams, V., Romero, J. A., & Capon, S. (2017). Making time for space: The critical role of spatial planning in adapting natural resource management to climate change. Environmental Science & Policy, 74, 57–67. DOI: 10.1016/j.envsci.2017.05.003

Ager, A., Finney, M., Kalabokidis, K., & Moore, P. (2017). Wildland fires. In Dalezios, N. R. (Ed.), Environmental Hazards Methodologies for Risk Assessment and Management (pp. 305–336). International Water Association Publishing. DOI: 10.2166/9781780407135_0307

Agriantoni Ch., (1986), The beginnings of industrialization in Greece in the 19th century, Athens [Αγριαντώνη Χ., (1986), Οι απαρχές της εκβιομηχάνισης στην Ελλάδα τον 19ο αιώνα, Αθήνα] (in Greek)

Ahlfeldt, G. M., & Feddersen, A. (2018). From periphery to core: Measuring agglomeration effects using high-speed rail. Journal of Economic Geography, 18(2), 355–390. DOI: 10.1093/jeg/lbx005

Akerlof, G. A. (1997). Social Distance and Social Decisions. Econometrica, 65(5), 1005–1027. DOI: 10.2307/2171877

Alfen. (2017). Alfen Annual Report. Almere: Alfen Beheer B.V. Retrieved 10 15, 2024, from https://alfen.com/search?search_api_fulltext=ameland

Allmendinger, P., & Haughton, G. (2010). Spatial planning, devolution, and new planning spaces. Environment and Planning. C, Government & Policy, 28(5), 803–818. DOI: 10.1068/c09163

Alrabba, M. I. (2017). The Determinants of Unemployment Rate in Jordan: A Multivariate Approach. International Journal of Economics and Finance, 9(11), 109–117. DOI: 10.5539/ijef.v9n11p109

Ameland. (2018). Sustainability. Retrieved 10 15, 2024, from https://www.vvvameland.com/practical/general/trivia/duurzaamheid

Anagnostou S., (2024), Concise History of Lesvos, 1800-1950, Mytilene, unpublished [Αναγνώστου Στρατής, (2024), Συνοπτική Ιστορία της Λέσβου, 1800-1950, Μυτιλήνη, αδημοσίευτο] (in Greek)

Anagnostou, S. (2004), The residential development of Lesvos (1462-1912), PhD Thesis, Department of Geography, University of the Aegean, Mytilene [Αναγνώστου Στρατής, (2004), Η οικιστική εξέλιξη της Λέσβου (1462-1912), Διδακτορική Διατριβή, Τμήμα Γεωγραφίας, Πανεπιστήμιο Αιγαίου, Μυτιλήνη] (in Greek)

Anagnostou, S., Tataris, G., Sidiropoulos, G., Pelekou, E., & Gougoulios, N. 2024. Spatial development planning. In: Kalabokidis K. (Ed.). Impact study and expertise for addressing problems in the affected area of Vrissa-Vatera-Stavros in Lesvos Island, Greece, after the fire of July 23, 2022. University of the Aegean. ISBN: 978-618-86408-1-8. Pp. 206-223. (In Greek).

Anagnostou, S., Kalabokidis, K., & Sidiropoulos, G. (2019, Fall). Vatera of Lesvos in the next decade: The need for a plan for regeneration, reconstruction and development. [In Greek]. Echo of Vrissa, (66), 19–22.

Andam, K. S., Ferraro, P. J., Pfaff, A., Sanchez-Azofeifa, G. A., & Robalino, J. A. (2008). Measuring the effectiveness of protected area networks in reducing deforestation. Proceedings of the National Academy of Sciences of the United States of America, 105(42), 16089–16094. DOI: 10.1073/pnas.0800437105

Angrist, J. D., & Pischke, J.-S. (2009). Mostly harmless econometrics: An empiricist's companion. Princeton university press. DOI: 10.1515/9781400829828

Anselin, L., Gallo, J. L., & Jayet, H. (2008). Spatial Panel Econometrics. In The Econometrics of Panel Data: Fundamentals and Recent Developments in Theory and Practice, 625-660.

Anselin, L. (1988). Spatial Econometrics: Methods and Models. Springer Science & Business Media. DOI: 10.1007/978-94-015-7799-1

Anselin, L. (1992). Spatial Data Analysis With GIS: An Introductıon to Application. In The Social Sciences. National Center for Geographic Information and Analysis.

Anselin, L. (2005). Exploring Spatial Data with GeoDa TM: A Workbook. Center for Spatially Integrated Social Science.

Anselin, L., Florax, R., & Rey, S. J. (Eds.). (2013). Advances in Spatial Econometrics: Methodology, Tools and Applications. Springer Science & Business Media.

Anselin, L., & Hudak, S. (1992). Spatial Econometrics in Practice: A Review of Software Options. Regional Science and Urban Economics, 22(3), 509–536. DOI: 10.1016/0166-0462(92)90042-Y

Anthopoulou, T., & Goussios, D. (2007). Rural Geography. In Terkenli, T., Iosifidis, T., & Goussios, D. (Eds.), Human Geography: Human, Society and Space (pp. 234–275). Kritiki Publishing.

Antonova, N., Abramova, S., & Polyakova, V. (2020). "cities are good for us": Assessment of urban attractiveness by students. [IOP Publishing.]. IOP Conference Series. Earth and Environmental Science, 459(3), 032090. DOI: 10.1088/1755-1315/459/3/032090

Apaydın, Ş. (2019). Türkiye'de İşsizliğin Kaynakları: Bir Yapısal Vektör Hata Düzeltme Modeli. Manas Sosyal Araştırmalar Dergisi, 8(1), 1002–1016. DOI: 10.33206/mjss.464857

Aral, N., & Aytaç, M. (2018). Türkiye'de İşsizliğin Mekansal Analizi. Marmara Üniversitesi Öneri Dergisi, 13(49), 1–20.

Arianoutsou, M., & Kazanis, D. 2024. Assessment of the potential for natural regeneration of Pinus brutia forests in the burned area of Vrissa-Vatera-Stavros in southern Lesvos. In: Kalabokidis K. (Ed.). Impact study and expertise for addressing problems in the affected area of Vrissa-Vatera-Stavros in Lesvos Island, Greece, after the fire of July 23, 2022. University of the Aegean. ISBN: 978-618-86408-1-8. Pp. 93-104. (In Greek).

Aschauer, D. A. (1989). Is public expenditure productive? Journal of Monetary Economics, 23(2), 177–200. DOI: 10.1016/0304-3932(89)90047-0

Asian Development Bank. (2011): Framework of Inclusive Growth Indicators Key Indicators for Asia and the Pacific 2011 Special Supplement https://sustainabled evelopment.un.org/content/documents/739indicators.pdf

Asian Development Bank. (2021): Inclusive growth. Retrieved, https://www.adb .org/themes/social, development/ poverty, reduction/inclusive, growth

Atmaca, Y. (2020). An Evaluation on the Contribution of Development Agencies to Local Development Policies''. Journal of Public Administration and Policy, 1(2), 66–81.

Au, C.-C., & Henderson, J. V. (2006). Are chinese cities too small? The Review of Economic Studies, 73(3), 549–576. DOI: 10.1111/j.1467-937X.2006.00387.x

Azolibe, C. B., Dimnwobi, S. K., & Uzochukwu-Obi, C. P. (2022). The Determinants of Unemployment Rate in Developing Economies: Does Banking System Credit Matter? Journal of Economic and Administrative Sciences.

Babaoğlu, C., & Kocaoğlu, M. (2017). The İssue of Urban İmmigrants and Local Governments. Turkish Administration Magazine, 485, 497–516.

Baccaro, L., & Rei, D. (2007). Institutional Determinants of Unemployment in OECD Countries: Does the Deregulatory View Hold Water? International Organization, 61(3), 527–569. DOI: 10.1017/S0020818307070221

Bacchetta, M., Cerra, V., Piermartini, R., & Smeets, M. (2021). Trade and Inclusive Growth. International Monetary Fund.

Bachantourian, M., Kalabokidis, K., Palaiologou, P., & Chaleplis, K. (2023). Optimizing fuel treatments allocation to protect the wildland–urban Interface from large-scale wildfires in Greece. Fire (Basel, Switzerland), 6(2), 75. DOI: 10.3390/fire6020075

Badinger, H., & Url, T. (2002). Determinants of Regional Unemployment: Some Evidence from Austria. Regional Studies, 36(9), 977–988. DOI: 10.1080/0034340022000022558

Bae, C.-H. C., Sandlin, G., Bassok, A., & Kim, S. (2007). The exposure of disadvantaged populations in freeway air-pollution sheds: A case study of the Seattle and Portland. Environment and Planning. B, Planning & Design, 34(1), 154–170. DOI: 10.1068/b32124

Bakalis, C. (2016). Tobacco Networks in the Aegean Islands. Advances in Historical Studies, 5(2), 63–79. DOI: 10.4236/ahs.2016.52007

Balkaya, D., Kutluay, T. F., & Tutar, N. F. (2022). Think Global Move Local: Micro Based Sustainable Local Development''. International Academic Social Resources Journal, 7(35), 299–309. DOI: 10.29228/ASRJOURNAL.57789

Baltagi, B. H. (2005). Econometric Analysis of Panel Data. John Wiley & Sons.

Banzhaf, H. S., & Walsh, R. P. (2008). Do people vote with their feet? an empirical test of tiebout's mechanism. The American Economic Review, 98(3), 843–863. DOI: 10.1257/aer.98.3.843

Batista e Silva, F., Barranco, R., Proietti, P., Pigaiani, C., & Lavalle, C. (2021). A new European regional tourism typology based on hotel location patterns and geographical criteria. Annals of Tourism Research, 89, 103077. DOI: 10.1016/j.annals.2020.103077

Bayrak, R., & Tatlı, H. (2018). The Determinants of Youth Unemployment: A Panel Data Analysis of OECD Countries. The European Journal of Comparative Economics, 15(2), 231–248.

Beal, L., Séraphin, H., Modica, G., Pilato, M., & Platania, M. (2019). Analysing the Mediating Effect of Heritage Between Locals and Visitors: An Exploratory Study Using Mission Patrimoine as a Case Study. Sustainability (Basel), 11(11), 3015. DOI: 10.3390/su11113015

Beck, U. (2013). World at Risk. Polity Press.

Bennett, G. T. & ve Marland, J., "Road accidents in traditionally designed local authority estates" TRRL Supplementary Report 394. Transport and Road Research Laboratory, Crowthorne, Berkshire, İngiltere (1978).

Beriatos, E. (2005). Planning and management of protected areas in Greece. Sustainablc Dcvelopment and Planning II, Vol. 1, WIT Transactions on Ecology and the Environment, 84, 389. WIT Press. ISSN 1743-3541

Bernard, A. B., Moxnes, A., & Saito, Y. U. (2019). Production networks, geography, and firm performance. Journal of Political Economy, 127(2), 639–688. DOI: 10.1086/700764

Bianchi, R. (2018). The political economy of tourism development: A critical review. Annals of Tourism Research, 70, 88–102. DOI: 10.1016/j.annals.2017.08.005

Biesbroek, G. R., Swart, R. J., & van der Knaap, W. G. M. (2009). The mitigation-adapration dichotomy and the role of spatial panning. Habitat International, 33(3), 230–237. DOI: 10.1016/j.habitatint.2008.10.001

Björketun, U. (1991). "Linjeföring samt prediktion av olyckor utifran linjeförings data för vagar projekterade/byggda under 1950-, 1960- respektive 1970-talet", VTI-meddelande 641. Statens vag- och trafikinstitut (VTI). İsveç.

Blakstad, F. "Ulykkesfrekvenser på hovedveger i byområder", Rapport STF63 A90005, SINTEF Samferdselsteknikk, Trondheim, Norveç (1990).

Blakstad, F. ve Giæver, T., "Ulykkesfrekvenser på vegstrekninger i tett og middels tett bebyggelse", Rapport STF63 A89005, SINTEF Samferdselsteknikk, Trondheim, Norveç (1989).

Bolund, P., & Hunhammar, S. (1999). Ecosystem services in urban areas. Ecological Economics, 29(2), 293–301. DOI: 10.1016/S0921-8009(99)00013-0

Bonilla-Bedoya, S., Mora, A., Vaca, A., Estrella, A., & Herrera, M. Á. (2020). Modelling the relationship between urban expansion processes and urban forest characteristics: An application to the Metropolitan District of Quito. Computers, Environment and Urban Systems, 79, 101420. DOI: 10.1016/j.compenvurbsys.2019.101420

Borg, N. (2023, August 18) Fact-check: Does Malta have the highest population density in Europe? Times of Malta. Available at: https://timesofmalta.com/article/factcheck-does-malta-highest-population-density-europe.1049809

Boshoff, D. G. (2016). Principles of Highest and Best Use Valuation;A South African Legal Perspective with Practical Implications. Researchgate. DOI: 10.18638/quaesti.2016.4.1.283

Bosker, M., Brakman, S., Garretsen, H., & Schramm, M. (2012). Relaxing hukou: Increased labor mobility and china's economic geography. Journal of Urban Economics, 72(2-3), 252–266. DOI: 10.1016/j.jue.2012.06.002

Bostock, J., Cooper, R., & Roberts, G. (2016). Rising to the Challenge of Sustainability: Community Events by the Community, for the Community. In Jepson, A., & Clarke, A. (Eds.), Managing and Developing Communities, Festivals and Events. Palgrave Macmillan., DOI: 10.1057/9781137508553_2

Boulanger, P. (1991), "L'Ile de Mytilène et le négoce français au XVIIIe siècle", in: Les Villes dans l'Empire ottoman: activités et sociétés, CNRS, Paris, 1991, t. 1, p. 273-298 (Greek translation in: Λεσβιακά, Δελτίον της Εταιρείας Λεσβιακών Μελετών, τ. XVI, Μυτιλήνη, 1996, σ. 176-200) (in French/Greek)

Boz, Y., & Cay, T. (2024). How Smart and Sustainable are the Cities in Turkiye? -National Policies and the Enthusiasm Level of the Local Governments. Heliyon, 10(4), e26002. DOI: 10.1016/j.heliyon.2024.e26002 PMID: 38390120

BRITANNICA. (2019, January 3). Helgoland island, Germany. Retrieved 10 15, 2024, from https://www.britannica.com/place/Helgoland

Bruin, K., Goosen, H., Ierland, E. v., & Groeneveld, R. (2013). Costs and benefits of adapting spatial planning to climate change: lessons learned from a large-scale urban development project in the Netherlands. Springer.

Campagne, P., & Pecqueur, B. (2014) Le développement territorial. Une réponse émergente à la mondialisation [Territorial Development: An Emerging Response to Globalization], Paris: Éditions Charles Léopold Mayer (ECLM). ISSN 0033-0884, DOI: DOI: 10.3917/pro.344.0095

Campos, J., & De Rus, G. (2009). Some stylized facts about high-speed rail: A review of hsr experiences around the world. Transport Policy, 16(1), 19–28. DOI: 10.1016/j.tranpol.2009.02.008

Canetti, A., Cordeiro, M., Póvoa, P., Mattos, D., & Muñoz, E. (2017). Understanding multi-temporal urban forest cover using high-resolution images. Urban Forestry & Urban Greening, 29, 106–112. DOI: 10.1016/j.ufug.2017.10.020

Caporale, G. M., & Gil-Alana, L. (2014). Youth Unemployment in Europe: Persistence and Macroeconomic Determinants. Comparative Economic Studies, 59(4), 581–591. DOI: 10.1057/ces.2014.29

Carten'ı, A., Pariota, L., & Henke, I. (2017). Hedonic value of high-speed rail services: Quantitative analysis of the students' domestic tourist attractiveness of the main italian cities. Transportation Research Part A, Policy and Practice, 100, 348–365. DOI: 10.1016/j.tra.2017.04.018

Çelebi, H. İ., & Tek, M. (2022). The İmportance of Municipal Cooperative Relations for Local Development: The Case of Hatay Metropolitian Municipality''. Journal of Social Sciences, 58(58), 190–209. DOI: 10.29228/SOBIDER.62659

Çelebi, İ., & Alas, B. (2024). Urban Transformation Strategy Proposition in Local Governments - Şişli Example. JENAS Journal of Environmental and Natural Studies, 6(1), 37–51.

Cellmer, R., Senetra, A., & Szczepanska, A. (2012). The Effect of Environmental Factors on Real Estate Value. Knowing to manage the territory, protect the environment, evaluate the cultural heritage, 1-13.

Chang, Z., Diao, M., Jing, K., & Li, W. (2021). High-speed rail and industrial movement: Evidence from china's greater bay area. Transport Policy, 112, 22–31. DOI: 10.1016/j.tranpol.2021.08.013

Chan, K. W. (2015). Five decades of the chinese hukou system. In Handbook of Chinese migration (pp. 23–47). Edward Elgar Publishing. DOI: 10.4337/9781783476640.00009

Chan, K. W., & Buckingham, W. (2008). Is china abolishing the hukou system? The China Quarterly, 195, 582–606. DOI: 10.1017/S0305741008000787

Chape, S., Harrison, J., Spalding, M., & Lysenk, I. (2005). Measuring extent and effectiveness of protected areas as an indicator for meeting global biodiversity targets. Philosophical Transactions of the Royal Society of London. Series B, Biological Sciences, 360(1454), 433–455. DOI: 10.1098/rstb.2004.1592

Charitatos, P. (2004). Unexplored Spetses. Road.

Chatfield, B. V. (1987). System-Wide Safety Improvements: An Approach to Safety Consistency. In National Cooperative Highway Research Program Report 132. Transportation Research Board.

Chatzi, E., Derdemezi, E.-T., & Tsilimigkas, G. (2025). The impact of built-up area dispersion on the cultural heritage of the region of the South Aegean, Greece. ISPRS International Journal of Geo-Information, 14(3), 97. DOI: 10.3390/ijgi14030097

Chatzis, N., Kkallas, C., Papazachos, C., Anthymidis, M., Rovithis, E., Karakostas, C., & Papaioannou, C. 2019. Stochastic simulation of seismic motion and site-effects studies of ambient noise and seismic data: The case of the Vrisa settlement and the 2017 M = 6.3 Lesvos earthquake. 15th International Congress of the Geological Society of Greece, Athens, 22-24 May 2019 | Harokopio University of Athens, Greece. Bulletin of the Geological Society of Greece, Sp. Pub. 7 Ext. Abs. GSG2019-250. Pp. 24–25.

Chen, C., & Fan, C. C. (2016). China's hukou puzzle: Why don't rural migrants want urban hukou? China Review, 16(3), 9–39.

Chen, G., & Silva, J. A. e. (2013). Regional impacts of high-speed rail: A review of methods and models. Transportation Letters, 5(3), 131–143. DOI: 10.1179/1942786713Z.00000000018

Chen, M., Liu, W., & Tao, X. (2013). Evolution and assessment on china's urbanization 1960–2010: Under-urbanization or over-urbanization? Habitat International, 38, 25–33. DOI: 10.1016/j.habitatint.2012.09.007

Cho, M. Y. (2013). The specter of" the people": urban poverty in northeast China. Cornell University Press. DOI: 10.7591/cornell/9780801451652.001.0001

Chorianopoulos, I., Tsilimigkas, G., Koukoulas, S., & Balatsos, T. (2014). The shift to competitiveness and a new phase of sprawl in the Mediterranean city: Enterprises guiding growth in Messoghia–Athens. Cities (London, England), 39, 133–143. DOI: 10.1016/j.cities.2014.03.005

Chowdhury, M. S., & Hossain, M. (2014). Determinants of Unemployment in Bangladesh: A Case Study. Developing Country Studies, 4(3).

Christiansen, L. E., Lin, J. M., Pereira, P., & Topalova, R. T. (2017): Unlocking the Potential of Greater Female Employment in Europe, in: Intereconomics, Review of European Economic Policy, 52(1), 2017, pp. 5, 16

Cilliers, E. J., & Timmermans, W. (2013). Approaching value added planning in the green environment. Journal of place Management and development, 144-157.

Ciriminna, R., Pagliaro, M., Meneguzzo, F., & Pecoraino, M. (2016, April 9). Solar energy for Sicily's remote islands: On the route from fossil to renewable energy. Retrieved 10 15, 2024, from https://www.sciencedirect.com/science/article/pii/S2212609015301254

Clark, X., Dollar, D., & Micco, A. (2004). Port efficiency, maritime transport costs, and bilateral trade. Journal of Development Economics, 75(2), 417–450. DOI: 10.1016/j.jdeveco.2004.06.005

Clean Energy for EU Islands. (2019). EU ISLANDS IN THE ENERGY TRANSITION A CATALOGUE OF GOOD PRACTICES. Clean Energy for EU Islands Secretariat.

CNBS. (2023). China Statistical Yearbook 2023, by China National Bureau of Statistics. CNBS.

CNN. (2018, August 21). Τήλος, το πρώτο ενεργειακά αυτόνομο νησί στην Ελλάδα. Retrieved 10 15, 2024, from https://www.cnn.gr/perivallon/story/143481/tilos-to-proto-energeiaka-aytonomo-nisi-stin-ellada

CNRB. (2023). Statistical Bulletin of National Railways of China 2022, by China National Railway Bureau. CNRB.

CNRG (2020). In Outline of Planning for Railway Pioneering in the New Era of a Strong Transportation State, by China National Railway Group (CNRG).

Conze, A. (1865). Reise auf der Insel Lesbos. Rümpler. (in German)

Cook, S. J., Hays, J., & Franzese, R. (2015). Model Specification and Spatial Interdependence. Research Methods in Political Science and International Relations, 1-30.

Cook, B. (2010). Knowledges, controversies and floods: national-scale flood management in Bangladesh. Durham University.

Corsane, G., Davis, P., Elliott, S., Maggi, M., Murtas, D., & Rogers, S. (2007). Ecomuseum performance in Piemonte and Liguria, Italy: The significance of capital. International Journal of Heritage Studies, 13(3), 224–239. DOI: 10.1080/13527250701228148

Cotteleer, G., & Peerlings, J. H. (2011). Spatial planning procedures and property prices: The role of expectations. Landscape and Urban Planning, 100(1-2), 77–86. DOI: 10.1016/j.landurbplan.2010.11.010

Council of the European Union. (1992). Directive 92/43/EEC of 21 May 1992 on the conservation of natural habitats and of wild fauna and flora. Official Journal of the European Union, L, 206(7).

Cuinet V., (1892), La Turquie d'Asie: géographie administrative, statistique, descriptive et raisonnée de chaque province de l'Asie-Mineure, (4 vol.), Paris (in French)

Currier, N. ve Ives, J. M.,"Detail of a bird's eye view of Chicago" The city shaped: Urban patterns and meanings through history (1991), Kostof, S. (Ed.), Bulfinch Press, New York, 116 (1892).

DAFNI –Network of Sustainable Greek Islands. (2021, July 20). Sustainable Islands Prosperous people. Retrieved 10 15, 2024, from https://dafninetwork.gr/ylopoiisi-ergou-ksi/

Dal Santo, R. D., Baldi, N., Del Duca, A., & Rossi, A. (2017). The Strategic Manifesto of Italian Ecomuseums. Museum International, 69(1-2), 86–95. DOI: 10.1111/muse.12153

Dalezios, N., Kalabokidis, K., Koutsias, N., & Vasilakos, C. 2018. Chapter 10. Wildfires and Remote Sensing: An Overview. In: Remote Sensing of Hydrometeorological Hazards, G.P. Petropoulos and T. Islam (eds.). CRC Press, Taylor & Francis Group, Boca Raton, Florida, US. Pp. 211-236.

Danova, M., & Vozarov, I. K. (2020). Regional Disparities in the Relationship Between Economic Growth and Unemployment. Scientific Papers of the University of Pardubice, 28(2).

Darques, R., Sidiropoulos, G., & Kalabokidis, K. (Eds.). (2024). The Geography of Greece - Managing Crises and Building Resilience. World Regional Geography Book Series. Springer. DOI: 10.1007/978-3-031-29819-6

Davies, R., Barbosa, O., Fuller, R., Tratalos, J., Burke, N., Lewis, D., & Gaston, K. (2008). City-wide relationships between green spaces, urban land. Springer Science and Business Media publications.

Davis, P. (2011). Ecomuseums: a Sense of Place (2nd ed.). Continuum.

Davoodi, R. H., Montiel, J. P., & Martirosyan, T. A. (2021): Macroeconomic Stability and Inclusive Growth. IMF https://www.imf.org/en/Publications/WP/Issues/2021/03/19/Macroeconomic, Stability, and, Inclusive, Growth, 50259

Davoudi, S., & Strange, I. (2009). conceptions of space and place in strategic spatial planning. Abingdon: Routledge.

De Mello, D. L., Mark A. (2012): Promoting Inclusive Growth: Challenges and Policies. Paris, OECD and the World Bank. https://openknowledge.worldbank.org/handle/10986/16948 License: CC BY, NC, ND 3.0 IGO.

Delenda, E. (2013). The impacts of tourism on small communities: The case of Spetses (Student project). Retrieved from: http://digilib.teiemt.gr/jspui

Department of economic and social affairs. (2014). World urbanization prospects: The 2014 revision. Population Devision, United Nation.

Derdemezi, E.-T., Tsilimigkas, G., & Kizos, T. (2021). Mining activity and island landscape issues: Evidence from Cyclades islands, Greece. European Planning Studies, 30(2), 384–404. DOI: 10.1080/09654313.2021.1958172

Desmet, K., & Rossi-Hansberg, E. (2013). Urban accounting and welfare. The American Economic Review, 103(6), 2296–2327. DOI: 10.1257/aer.103.6.2296

Dimopoulos, P., Drakou, E., Kokkoris, I. P., Katsanevakis, S., Kallimanis, A., Tsiafouli, M., Bormpoudakis, D., Kormas, K., & Arends, J. (2017). The need for the implementation of an ecosystem services assessment in Greece: Drafting the national agenda. One Ecosystem, 2, e13714. DOI: 10.3897/oneeco.2.e13714

Doğan, M. (2015). Ecomuseum, community museology, local distinctiveness, Hüsamettindere village, Bogatepe village, Turkey. Journal of Cultural Heritage Management and Sustainable Development, 5(1), 61–81. DOI: 10.1108/JCHMSD-07-2013-0033

Doğan, M., & Timothy, D. J. (2019). Beyond Tourism and Taxes: The Ecomuseum and Social Development in the Ak-Chin Tribal Community. Journal of Tourism and Cultural Change, 18(2), 133–149. DOI: 10.1080/14766825.2019.1593994

Donaldson, D. (2018). Railroads of the raj: Estimating the impact of transportation infrastructure. The American Economic Review, 108(4-5), 899–934. DOI: 10.1257/aer.20101199

Donaldson, D., & Hornbeck, R. (2016). Railroads and american economic growth: A "market access" approach. The Quarterly Journal of Economics, 131(2), 799–858. DOI: 10.1093/qje/qjw002

Dragouni, M., & Fouseki, K. (2017). Drivers of community participation in heritage tourism planning: An empirical investigation. Journal of Heritage Tourism, 13(3), 237–256. DOI: 10.1080/1743873X.2017.1310214

Du, J., Zhu, X., Li, X., & Ünal, E. (2024). What Drives the Green development Behavior of Local Governments? A Perspective of Grounded Theory''. Heliyon, 10(6), 1–18. DOI: 10.1016/j.heliyon.2024.e27744 PMID: 38509992

Dumbaugh, E., & Rae, R. (2009). ve Rae, R., "Safe urban form: Revisiting the relationship between community design and traffic safety". Journal of the American Planning Association, 75(3), 309–329. DOI: 10.1080/01944360902950349

Dunn, C. (2010). The Appraisal of real estate 3rd canadian edition. SAUDER; school of Business .

Duranton, G., & Puga, D. (2019). Urban growth and its aggregate implications. Technical report. National Bureau of Economic Research. DOI: 10.3386/w26591

Durning, A. T. "The Car and the City: 24 Steps to Safe Streets and Healthy Communities", New Report No.3, Northwest Environment Watch, Seattle, Washington, (1996).

Dustmann, C., & Glitz, A. (2011). Migration and education. In Handbook of the Economics of Education (Vol. 4, pp. 327–439). Elsevier. DOI: 10.1016/B978-0-444-53444-6.00004-3

E.U. (2018): Towards Sustainable Development: Mainstreaming Environment and Climate Change into Development, E.U. SOCIAL PROTECTION pp 1, 16

EADO. (2023). Resilience of the Greek Forest Ecosystems to Climate Change. Committee for the Resilience of the Greek Forest Ecosystems to Climate Change (EADO) [M. Arianoutsou, C. Zerefos, K. Kalabokidis, A. Poupkou, F. Aravanopoulos (Eds.)]. Academy of Athens. (In Greek)

Economic Commission for Europe. (2008). Spatial planning; key instrument for development and effective governance with spacial reference to countries in Transition. United Nations.

EEA (European Environment Agency). (2018). Impervious Built-up 2018. Copernicus Land Monitoring Service.

Efthymiopoulos, I. (2016). Smart Islands Projects and Strategies. Friedrich-Ebert-Stiftung.

Eggenberger, M., & Partidario, M. (2014). Development of a framework to assist the integration of environmental, social and economic issues in spatial planning. Impact Assessment and Project Appraisal. Advance online publication. DOI: 10.3152/147154600781767448

Eita, J. H., & Ashipala, J. (2010). Determinants of Unemployment in Namibia. International Journal of Business and Management, 5(10). Advance online publication. DOI: 10.5539/ijbm.v5n10p92

el Piketty T. (2014): Capital in the 21st century, Cambridge, MA: Harvard University

Elhorst, J. P. (2010). Applied Spatial Econometrics: Raising The Bar. Spatial Economic Analysis, 5(1), 9–28. DOI: 10.1080/17421770903541772

Elhorst, J. P. (2014). Spatial Econometrics: From Cross-Sectional Data to Spatial Panels, 479. Springer. DOI: 10.1007/978-3-642-40340-8

Elmqvist, T., Zipperer, W. C., & Güneralp, B. (2015). Urbanization, habitat loss, and biodiversity decline: Solution pathways to break the cycle. In Mounsey, H., & Tomlinson, R. (Eds.), Building Databases for Global Science (1st ed.). Routledge., DOI: 10.4324/9781315849256

EL-STAT (Hellenic Statistical Authority). (2020). Retrieved from https://www.statistics.gr/en/home

Elvik, R. ve Muskaug R., "Konsekvensanalyser og trafikksikkerhet. Metode for beregning av konsekvenser for trafikksikkerheten av tiltak pa vegnettet", TØI-rapport 0281, Transportøkonomisk institutt, Oslo, Norveç (1994).

Elvik, R., & Hoye, A. ve Vaa, T., "The Handbook Of Road Safety Measures; Second Edition", Emerald Group Publishing Limited, UK, 169-206;1031-1043, (2009).

Elvik, R., Mysen, A.B. ve Vaa, T., "Trafikksikkerhetshånbok-3", Udgave, TØI Rapport, Norveç (1997).

ENERGIAKADEMIET. (2019). Samsoe Version 2.0. Retrieved 10 15, 2024, from https://energiakademiet.dk/en/uafhaengig-af-fossile-braendsler/

Enshassi, A., Kochendoerfer, B., & Rizq, E. (2014). An evaluation of environmental impacts of construction projects. Revista ingeieria de construccion, 234-254.

e-poleodomia. GIS platform for urban and settlement boundaries. Retrieved from http://gis.epoleodomia.gov.gr/v11#/

Erdem, A. D., & Bilgili, A. (2023). Climate Change Combat Tools in Turkey: National Smart Cities Strategy and Action Plan. Journal of Contemporary Local Governments, 32(1), 51–78.

Eroğlu, H. T. (2010). Sustainable Local Development and City Councils. Dumlupınar University Journal of Social Sciences, 26(2), 44–55.

Eser, B. Y., & Terzi, H. (2008). Türkiye'de İşsizlik Sorunu ve Avrupa İstihdam Stratejisi. Erciyes Üniversitesi İktisadi ve İdari Bilimler Dergisi, (30), 229–250.

Estevens, A., Cocola-Gant, A., López-Gay, A., & Pavel, F. (2023). The role of the state in the touristification of Lisbon. Cities (London, England), 137, 104275. DOI: 10.1016/j.cities.2023.104275

EU. (2002). European Parliament and Council Recommendation on ICZM (2002/413/EC). European Commission.

EUNICE. (2020). Tilos Project. Retrieved 10 15, 2024, from EUNICE: https://eunice-group.com/gr/projects/tilos-project/

European Commission. (2011). Our life insurance, our natural capital: An EU biodiversity strategy to 2020. European Commission.

European Commission. (2020). EU biodiversity strategy for 2030: Bringing nature back into our lives. COM, 2020. European Commission.

European Commission. (2020). The Natura 2000 network. Retrieved from https://ec.europa.eu/environment/nature/natura2000/index_en.htm

European Parliament & Council of the European Union. (2009). Directive 2009/147/EC of 30 November 2009 on the conservation of wild birds (replacing Council Directive 79/409/EEC). Official Journal of the European Union, L, 20(7).

European Parliament & Council of the European Union. (2009). Directive 2009/147/EC of the European Parliament and of the Council of 30 November 2009 on the conservation of wild birds (codified version). Official Journal of the European Union, L, 20, 7–25. https://data.europa.eu/eli/dir/2009/147/oj

European Union. (1992). Council Directive 92/43/EEC of 21 May 1992 on the conservation of natural habitats and of wild fauna and flora. Official Journal of the European Communities, L 206/7.

Eurostat (2023) Tourism Satellite Accounts in Europe – 2023 edition. Statistical Reports. Available at: https://ec.europa.eu/eurostat/web/products-statistical-reports/w/ks-ft-22-011

Evans, D. (2006). The habitats of the European Union Habitats Directive. Biology and Environment, 106B(3), 167–173. DOI: 10.1353/bae.2006.0032

Everingham, P., Obrador, P., & Tucker, H. (2021). Trajectories of embodiment in Tourist Studies. Tourist Studies, 21(1), 70–83. DOI: 10.1177/1468797621990300

Ewing, R., & Dumbaugh, E. (2009). ve Dumbaugh, E., "The built environment and traffic safety: A review of empirical evidence". Journal of Planning Literature, 23(4), 347–367. DOI: 10.1177/0885412209335553

Ewing, R., Schieber, R., & Zegeer, C. V. (2003). ve Zegeer, C., "Urban sprawl as a risk factor ın motor vehicle occupant and pedestrian fatalities". American Journal of Public Health, 93(9), 1541–1545. DOI: 10.2105/AJPH.93.9.1541 PMID: 12948977

Faber, B. (2014). Trade integration, market size, and industrialization: Evidence from china's national trunk highway system. The Review of Economic Studies, 81(3), 1046–1070. DOI: 10.1093/restud/rdu010

Fan, C. C., Sun, M., & Zheng, S. (2011). Migration and split households: A comparison of sole, couple, and family migrants in beijing, china. Environment & Planning A, 43(9), 2164–2185. DOI: 10.1068/a44128

Fang, Y. and Shi, Z. (2018). Children of migrant parents: Migrating together or left behind.

Fang, Y., & Shi, Z. (2018, June). Children of migrant parents: Migrating together or left behind. Habitat International, 76, 62–68. DOI: 10.1016/j.habitatint.2018.05.001

Fernandez-Gimenez, M. E., Ballard, H. L., & Sturtevant, V. E. (2008). Adaptive management and social learning in collaborative and community-based monitoring: A study of five community-based forestry organizations in the western USA. Ecology and Society, 13(2), 4. https://www.ecologyandsociety.org/vol13/iss2/art4/ . DOI: 10.5751/ES-02400-130204

Fernando, A., Rosete Verges, F., & Larruga, J. (2024). Coastal Laws evolution in Spain: An analysis from the public policy cycle. Marine Policy, 162, 106052. Advance online publication. DOI: 10.1016/j.marpol.2024.106052

Ferrer-i-Carbonell, A., & Frijters, P. (2004). How Important is Methodology for the Estimates of the Determinants of Happiness? Economic Journal (London), 114(497), 641–659. DOI: 10.1111/j.1468-0297.2004.00235.x

Filindra, M. "Spetses," in D. Philippidis (Ed.), Greek Traditional Architecture, Volume 4: Peloponnese A (pp. 245-248). Athens: Melissa, 1987.

Filiztekin, A. (2009). Regional Unemployment in Turkey. Papers in Regional Science, 88(4), 863–878. DOI: 10.1111/j.1435-5957.2009.00237.x

Fischer, A. P., Spies, T. A., Steelman, T. A., Moseley, C., Johnson, B. R., Bailey, J. D., Ager, A. A., Bourgeron, P., Charnley, S., Collins, B. M., Kline, J. D., Leahy, J. E., Littell, J. S., Millington, J. D. A., Nielsen-Pincus, M., Olsen, C. S., Paveglio, T. B., Roos, C. I., Steen-Adams, M. M., & Bowman, D. M. J. S. (2016). Wildfire risk as a socioecological pathology. Frontiers in Ecology and the Environment, 14(5), 276–284. DOI: 10.1002/fee.1283

Fischer, M. M., & Wang, J. (2011). Spatial Data Analysis: Models, Methods and Techniques. Springer Science & Business Media. DOI: 10.1007/978-3-642-21720-3

Fleischhauer, M. (2008). The Role of Spatial Planning in Strengthening Urban Resilience. Springer. DOI: 10.1007/978-1-4020-8489-8_14

Folawewo, A. O., & Adeboje, O. M. (2017). Macroeconomic Determinants of Unemployment: Empirical Evidence from Economic Community of West African States. African Development Review, 29(2), 197–210. DOI: 10.1111/1467-8268.12250

Forest maps and wildlife park data. Retrieved from http://geodata.gov.gr

Fosli, O. (1995). "Transport, arealbruk og miljø. Ein studie av dei transportmessige verknadene av IDG Norges omlokalisering fra?" Hasle til Galleri Oslo, Hovedoppgave i samfunnsgeografi, Universitetet i Oslo, Samfunnsvitenskapelig Fakultet. Norveç.

Fox, S. E., & Le Dantec, C. A. (2014) Community historians: scaffolding community engagement through culture and heritage. Proceedings of the 2014 conference on Designing interactive systems, 785 - 794 DOI: 10.1145/2598510.2598563

Franzitta, V., & Curto, D. (2017). Sustainability of the Renewable Energy Extraction Close to the Mediterranean Islands. Retrieved 10 15, 2024, from https://www.mdpi.com/1996-1073/10/3/283/htm

Fretz, S., Parchet, R., and Robert-Nicoud, F. (2017). Highways, market access, and spatial sorting.

Fridstrøm, L., Ifver, J., Ingebrigtsen, S., Kulmala, R., & Thomsen, L. K. (1995). ve Thomsen, L.K., "Measuring the contribution of randomness, exposure, weather, and daylight to the variation in road accident counts.". Accident; Analysis and Prevention, 27(1), 1–20. DOI: 10.1016/0001-4575(94)E0023-E PMID: 7718070

Gaganis, P., Kontos, T., & Anastasiou, M. 2024. Soil erosion risk assessment of the Vatera area of Lesvos Island as a consequence of the disastrous fire of July 2022. In: Kalabokidis K. (Ed.). Impact study and expertise for addressing problems in the affected area of Vrissa-Vatera-Stavros in Lesvos Island, Greece, after the fire of July 23, 2022. University of the Aegean. ISBN: 978-618-86408-1-8. Pp. 106-120. (In Greek).

Gakenheimer, R. (1999). Urban mobility in the developing world. Transportation Research Part A, Policy and Practice, 33(7-8), 671–689. DOI: 10.1016/S0965-8564(99)00005-1

Galettas, I., & Mpouzoumpardi, M. (2004). Spetses: History - Ethnography (Vol. I). Spetsiot Union.

Garmendia, M., de Uren˜a, J. M., Ribalaygua, C., Leal, J., & Coronado, J. M. (2008). Urban residential development in isolated small cities that are partially integrated in metropolitan areas by high speed train. European Urban and Regional Studies, 15(3), 249–264. DOI: 10.1177/0969776408090415

Garrick, N. ve Marshall W., "Network, Placemaking and Sustainability" Center for Transportation and Urban Planning, University of Connecticut, Connecticut, A.B.D. (2008).

Gayawan, E., Mpandiko, A. D., & Ojo, O. (2025). Modelling the Spatio-Temporal Dynamics of Youth Unemployment in Africa: A Bayesian Approach. Regional Science Policy & Practice, 17(1), 100149. DOI: 10.1016/j.rspp.2024.100149

Geldmann, J., Barnes, M., Coad, L., Craigie, I. D., Hockings, M., & Burgess, N. D. (2013). Effectiveness of terrestrial protected areas in reducing habitat loss and population declines. Biological Conservation, 161, 230–238. DOI: 10.1016/j.biocon.2013.02.018

Giæver, T. (1990). "Ulykkesfrekvenser i rundkjøringer og signalregulerte kryss", SINTEF Samferdselsteknikk, STF63 A90002. Norveç.

Giang, D. T., & Pheng, L. S. (2011). Role of construction in economic development; review of key concepts in the past 40 years. Habitat International, 35(1), 118–125. DOI: 10.1016/j.habitatint.2010.06.003

Gibson, C. (2021). Critical tourism studies: New directions for volatile times. Tourism Geographies, 23(4), 659–677. DOI: 10.1080/14616688.2019.1647453

Glaeser, E. L., & Mare, D. C. (2001). Cities and skills. Journal of Labor Economics, 19(2), 316–342. DOI: 10.1086/319563

González-García, A., Palomo, I., Arboledas, M., González, J. A., Múgica, M., Mata, R., & Montes, C. (2022). Protected areas as a double-edged sword: An analysis of factors driving urbanization in their surroundings. Global Environmental Change, 74, 102522. DOI: 10.1016/j.gloenvcha.2022.102522

Gospodini, A. (2005). Urban development, redevelopment and regeneration encouraged by transport infrastructure projects: The case study of 12 european cities. European Planning Studies, 13(7), 1083–1111. DOI: 10.1080/09654310500242121

Gourzis, K., Alexandridis, G., Gialis, S., & Caridakis, G. (2019). Studying the spatialities of short-term rentals' sprawl in the urban fabric: The case of Airbnb in Athens, Greece. In Artificial Intelligence Applications and Innovations: AIAI 2019 IFIP WG 12.5 International Workshops: MHDW and 5G-PINE 2019, Hersonissos, Crete, Greece, May 24–26, 2019 [Springer International Publishing.]. Proceedings, 15, 196–207.

Gourzis, K., Herod, A., Chorianopoulos, I., & Gialis, S. (2022). On the recursive relationship between gentrification and labour market precarisation: Evidence from two neighbourhoods in Athens, Greece. Urban Studies (Edinburgh, Scotland), 59(12), 2545–2564. DOI: 10.1177/00420980211031775

Goussios, D. (2013) Η εδαφική προσέγγιση της ανάπτυξης της υπαίθρου [The territorial approach of rural development] Περί εντοπιότητας και ιδιοτυπίας των τροφίμων: Μια εδαφική προσέγγιση της ανάπτυξης των αγροτικών περιοχών, edited by T. Anthopoulou, 72-105. Athens: Papazisi Publications, ISBN: 978-960-02-2809-0

Goussios, D., Gaki, D., & Faraslis, I. (2023). Ένας δρόμος ανάπτυξης για τα βουνά [A development path for the mountains]. Evdoxos Publications.

Greek Ministry of Environment and Energy. Natura 2000 areas. Retrieved from http://geoportal.ypen.gr/

Green, A. E., Hogarth, T., & Shackleton, R. E. (1999). Longer distance commuting as a substitute for migration in britain: A review of trends, issues and implications. International Journal of Population Geography, 5(1), 49–67. DOI: 10.1002/(SICI)1099-1220(199901/02)5:1<49::AID-IJPG124>3.0.CO;2-O

Greening the islands foundation. (2019, May 1). Germany's Helgoland island calls for legislative changes to support its pioneering role in sustainability and zero-emission ambition. Retrieved 10 15, 2024, from https://greeningtheislands.org/germanys-helgoland-island-calls-for-legislative-changes-to-support-its-pioneering-role-in-sustainability-and-zero-emission-ambition/

Greibe, P. (2003). Accident prediction models for urban roads. Accident; Analysis and Prevention, 35(2), 273–285. DOI: 10.1016/S0001-4575(02)00005-2 PMID: 12504148

Grydehøj, A., & Kelman, I. (2016). Island Smart Eco-Cities: Innovation,Secessionary Enclaves, and the Selling of Sustainability. Urban Island Studies.

Guirao, B. (2013). Spain: Highs and lows of 20 years of hsr operation. Journal of Transport Geography, 31, 201–206. DOI: 10.1016/j.jtrangeo.2013.05.010

Guirao, B., Casado-Sanz, N., & Campa, J. L. (2020). Labour opportunities provided by spanish high-speed rail (hsr) commuting services in a period of financial crisis: An approach based on regional wage disparities and housing rental prices. Regional Studies, 54(4), 539–549. DOI: 10.1080/00343404.2018.1530751

Günaydın, D., & Çetin, M. (2015). Genç İşsizliğin Temel Makroekonomik Belirleyicileri: Ampirik Bir Analiz. Pamukkale Üniversitesi Sosyal Bilimler Enstitüsü Dergisi, (22), 17–34.

Gündoğdu, S., & Boyalı, H. (2018). ''The Role of Local Governments in Local Development'', Talas, M. and Hilal, E., (eds), Civil Society, Political System and Political Governance, İksad Publishing House, Turkey, 60-80.

Guo, Y., Cao, L., Song, Y., Wang, Y., & Li, Y. (2022). Understanding the formation of city-hsr network: A case study of yangtze river delta, china. Transport Policy, 116, 315–326. DOI: 10.1016/j.tranpol.2021.12.014

Gupta, J., & Vegelin, C. (2016): "Sustainable development goals and inclusive development", Int Environ Agreements 16, pp. 433–448 https://link.springer.com/content/pdf/10.1007/s10784, 016, 9323, z.pdf

Gupta, J., Pouw, N., & Tonen, R. M. (2015). Towards an Elaborated Theory of Inclusive Development'. European Journal of Development Research, 27(4), 541–559.

Haas, A. and Osland, L. (2014). Commuting, migration, housing and labour markets: Complex interactions.

Hadayeghi, A., Shalaby, A. S., & Persaud, B. (2003). ve Persaud, B. N., "Macrolevel accident prediction models for evaluating safety of urban transportation systems". Transportation Research Record: Journal of the Transportation Research Board, 1840(1), 87–95. DOI: 10.3141/1840-10

Hainmueller, J. (2012). Entropy balancing for causal effects: A multivariate reweighting method to produce balanced samples in observational studies. Political Analysis, 20(1), 25–46. DOI: 10.1093/pan/mpr025

Haldeman, B. A. "Transportation and main thoroughfares and street railways", City planning, J. Nolen (Ed.), Appleton & Co., New York, A.B.D, 279-313 (1916).

Hall, J. W. (1992). ve Polanco de Hurtado, M., "Effect of intersection congestion on accident rates". Transportation Research Record: Journal of the Transportation Research Board, (1376), 71–77.

Hanley, D., Li, J., and Wu, M. (2022). High-speed railways and collaborative innovation.

Hanley, D., Li, J., & Wu, M. (2022, March). High-speed railways and collaborative innovation. Regional Science and Urban Economics, 93, 103717. DOI: 10.1016/j.regsciurbeco.2021.103717

Hassamancıoğlu, V. M. (2021). An Evaluation of Urban Green Space Management Strategies of Local Governments in Turkey. OPUS İnternational Journal of Society Researches, 18(39), 627–656.

He, D., Chen, Z., Pei, T., & Zhou, J. (2023). Analysis of structural evolution and its influencing factors of the high-speed railway network in china's three urban agglomerations. Cities (London, England), 132, 104063. DOI: 10.1016/j.cities.2022.104063

Hellenic Ornithological Society. (2022). Important Bird Areas data. Retrieved from https://ornithologiki.gr/el/

Hellenic Statistical Authority (ESYE-ELSTAT), Population Censuses: 1920, 1928, 1940, 1951, 1961, 1971, 1981, 1991, 2001, 2011, 2021 [ΕΣΥΕ-ΕΛΣΤΑΤ, Απογραφές Πληθυσμού: 1920, 1928, 1940, 1951, 1961, 1971, 1981, 1991, 2001, 2011, 2021] (in Greek)

Herod, A., Gialis, S., Psifis, S., Gourzis, K., & Mavroudeas, S. (2022). The impact of the COVID-19 pandemic upon employment and inequality in the Mediterranean EU: An early look from a Labour Geography perspective. European Urban and Regional Studies, 29(1), 3–20. DOI: 10.1177/09697764211037126

Hersperger, A., Oliveira, E., Pagliarin, S., Palka, G., Verburg, P., Bolliger, J., & Gradinaru, S. (2018). Urban land-use change: The role of strategic spatial planning. Global Environmental Change, 51, 32–42. DOI: 10.1016/j.gloenvcha.2018.05.001

Hodgson, C. (2018). The effect of transport infrastructure on the location of economic activity: Railroads and post offices in the american west. Journal of Urban Economics, 104, 59–76. DOI: 10.1016/j.jue.2018.01.005

Holling, C. (1973). Resilience and Stability of ecological systems. Annual Review of Ecology and Systematics, 4(1), 1–23. DOI: 10.1146/annurev.es.04.110173.000245

Hong, Z. and Sun, Y. (2020). Power, capital, and the poverty of farmers' land rights in china.

Hong, Z., & Sun, Y. (2020, March). Power, capital, and the poverty of farmers' land rights in China. Land Use Policy, 92, 104471. DOI: 10.1016/j.landusepol.2020.104471

Huang, X., & Chen, M. (2022). Understanding the role of housing in rural migrants' intention to settle in cities: Evidence from china. Habitat International, 128, 102650. DOI: 10.1016/j.habitatint.2022.102650

Huang, Y.. (2019). The correlation between hsr construction and economic development– empirical study of chinese cities. Transportation Research Part A, Policy and Practice, 126, 24–36. DOI: 10.1016/j.tra.2019.05.017

İlhan, A., & Akdeniz, C. (2024). The Effect of Real Wages on Employment after the Global Financial Crisis: The Case of the Turkish Manufacturing Industry. Sosyoekonomi, 32(61), 147–166. DOI: 10.17233/sosyoekonomi.2024.03.08

IMF (2024). World Economic Outlook Databases.

IMF. (2013): Inclusive Growth: Measurement and Determinants, IMF Working paper publ, WP/13/135

IMF. (2017): Fiscal Monitor: Tackling Inequality, IMF Working paper publ, https://www.imf.org/en/Publications/FM/Issues/2017/10/05/fiscal, monitor, october, 2017

IMF. (2017): FOSTERING INCLUSIVE GROWTH G, 20 Leaders, Summit, July 7, 8, Hamburg, Germany https://www.imf.org/external/np/g20/pdf/2017/062617.pdf

IMF. (2019): Fiscal Policy and Development: Human, Social, and Physical Investments for the SDGs, Working paper SDN/19/03 https://www.imf.org/en/Publications/Staff, Discussion, Notes/Issues/2019/01/18/Fiscal, Policy, and, Development, Human, Social, and, Physical, Investments, for, the, SDGs, 46444

IMF. (2019): Inequality of Opportunity, Inequality of Income and Economic Growth, IMF Working paper WP/19/34 publ, https://www.imf.org/en/Publications/WP/Issues/2019/02/15/Inequality, of, Opportunity, Inequality, of, Income, and, Economic, Growth, 46566

IMF. (2021): Regional Disparities, Growth, and Inclusiveness, IMF Working paper WP/21/39 https://www.imf.org/en/Publications/WP/Issues/2021/02/13/Regional, Disparities, Growth, and, Inclusiveness, 50076

Irewole, O. E. (2019). Inflatıon and Unemployment Relationshıp: A Dynamic Report Of Nıgeria and Mexico in The Perspective of Phillips Curve From 1991-2016. International Journal of New Economics and Social Sciences, 2(10), 11–30. DOI: 10.5604/01.3001.0013.8085

Ivanovskiy, I., & Azmitov, R. (2015). Economic Depreciation in the Property Objects Valuation. Mediterranean Journal of Social Sciences, •••, 512–518. DOI: 10.5901/mjss.2015.v6n1s3p512

Jamal, T. B., Taillon, J. M., & Dredge, D. (2011). Sustainable tourism pedagogy and academic-community collaboration: A progressive service-learning approach. Tourism and Hospitality Research, 11(2), 133–147. DOI: 10.1057/thr.2011.3

James, L., Halkier, H., Sanz-Ibáñez, C., & Wilson, J. (2023). Advancing evolutionary economic geographies of tourism: Trigger events, transformative moments and destination path shaping. Tourism Geographies, 25(8), 1819–1832. DOI: 10.1080/14616688.2023.2285324

Janssen, M., & Scheffer, M. (2004). Overexploitation of Renewable Resources by Ancient Societies and the Role of Sunk-Cost Effects. Ecology and Society, 9(1), art6. DOI: 10.5751/ES-00620-090106

Jayalath, G., & Jayalath, S. (2019). Landform process centered approach for planning the build environment with particular reference to managing the impacts of floods and landslides.

Jim, C. Y., & Chen, W. Y. (2009). Value of scenic views: Hedonic assessment of private housing in Hong Kong. Landscape and Urban Planning, 91(4), 226–234. DOI: 10.1016/j.landurbplan.2009.01.009

Johnston, R., & Marwood, K. (2017). Action heritage: Research, communities, social justice. International Journal of Heritage Studies, 23(9), 816–831. DOI: 10.1080/13527258.2017.1339111

Jones, D. ve Jha, M., “The effect of urban form on traffic accident incidence”, CEA'10 Proceedings Of the 4th WSEAS International Conference on Computer Engineering and Applications,USA, 212-222 (2009).

Kabaklarlı, E., & Gür, M. (2011). Türkiye’de Genç İşsizlik Sorunu ve Ekonomik Belirleyicilerinin Uzun Dönem Eş-Bütünleşme Analizi. 14. Uluslararası İktisat Öğrencileri Kongresi.

Kadiroğlu, A., & Türko, E. S. (2024). “Actors of Local Economic Development in Türkiye”. Turkish Studies – Economics, Finance. Politics, 19(1), 207–224.

Kaitatzidis M., (2009), Aegean Steam Navigation P.M. Courtgis & Co, 1883-1911, Patras (bilingual edition, Greek/English)

Kalabokidis, K. (Ed.). 2024. Impact study and expertise for addressing problems in the affected area of Vrissa-Vatera-Stavros in Lesvos Island, Greece, after the fire of July 23, 2022. University of the Aegean. ISBN: 978-618-86408-1-8. 376 p. (In Greek).

Kalabokidis, K., Palaiologou, P., & Finney, M. 2014. Fire behavior simulation in Mediterranean forests using the Minimum Travel Time algorithm. In: Wade DD & Fox RL (Eds.), Robinson ML (Comp.). Proceedings of 4th Fire Behavior and Fuels Conference, 18-22 February 2013, Raleigh, NC, USA and 1-4 July 2013, St. Petersburg, Russia. Published by the International Association of Wildland Fire: Missoula, MT, USA. Pp. 468-492.

Kalabokidis, K., Roussou, O., Vasilakos, C., Cristaki, M., Zikoulis, A., Palaiologou, P., Kypriotellis, P., & Gougoulios, N. 2024. Fire ecology and management. In: Kalabokidis K. (Ed.). Impact study and expertise for addressing problems in the affected area of Vrissa-Vatera-Stavros in Lesvos Island, Greece, after the fire of July 23, 2022. University of the Aegean. ISBN: 978-618-86408-1-8. Pp. 18-70. (In Greek).

Kalabokidis, K. 1996. Wildland fire safety in residential areas. In Proceedings International Fire Safety Conference, 24-25 May 1996, Santorini, Greece. Department of Civil Engineering, Aristotle University of Thessaloniki, Greece. Pp. 267-278.

Kalabokidis, K., Iliopoulos, N., & Gliglinos, D. (2013). Fire Meteorology and Forest Fire Behavior in a Changing Climate. ION Publications. (In Greek)

Kalabokidis, K., Iosifides, T., Henderson, M., & Morehouse, B. (2008). Wildfire policy and use of science in the context of human–environment system on the Aegean Archipelago. Environmental Science & Policy, 11(5), 408–421. DOI: 10.1016/j.envsci.2008.01.006

Kalabokidis, K., Korres, G., Soulakellis, N., & Feidas, H. (Eds.). (2021). Social Sciences and Geography: Theory, Methods and Techniques of Spatial Analysis. In memoriam to Professor Pavlos Kanaroglou. University of the Aegean, Department of Geography. (In Greek)

Kalabokidis, K., & Palaiologou, P. (2019). Mediterranean Forest Fuels. In Manzello, S. (Ed.), Encyclopedia of Wildfires and Wildland-Urban Interface (WUI) Fires. Springer. DOI: 10.1007/978-3-319-51727-8_29-1

Kallimanis, A. S., Touloumis, K., Tzanopoulos, J., Mazaris, A. D., Apostolopoulou, E., Stefanidou, S., Scott, A. V., Potts, S. G., & Pantis, J. D. (2015). Vegetation coverage change in the EU: Patterns inside and outside Natura 2000 protected areas. Biodiversity and Conservation, 24(3), 579–591. DOI: 10.1007/s10531-014-0837-9

Kampouris Z.P., (1978), Natural disasters in Lesvos in the 19th century, Mytilene [Καμπούρης Z., (1978), Θεομηνίες στη Λέσβο τον 19ο αιώνα, Μυτιλήνη] (in Greek)

Kampouris, Z. P. (1962), "The last years of the Ottoman period in Lesvos", Lesviaka, vol. 4, Mytilene, (pp. 102-188) [Καμπούρης Z., (1962), «Τα τελευταία χρόνια της Τουρκοκρατίας στη Λέσβο», Λεσβιακά τ.4, Μυτιλήνη, σ. 102-188] (in Greek)

Karaalp-Orhan, H. S., & Gülel, F. E. (2016). The Impact of The Global Financial Crises on Regional Unemployment in Turkey. International Journal of Economic Perspectives, 10(1), 151–160.

Karpat K., (1985), Ottoman population 1850-1914, Demographic and Social Characteristics, London

Kati, V., Hovardas, T., Dieterich, M., Ibisch, P. L., Mihok, B., & Selva, N. (2015). The challenge of implementing the European network of protected areas Natura 2000. Conservation Biology, 29(1), 260–270. DOI: 10.1111/cobi.12366

Kavroudakis, D. 2024. Spatial analysis. In: Kalabokidis K. (Ed.). Impact study and expertise for addressing problems in the affected area of Vrissa-Vatera-Stavros in Lesvos Island, Greece, after the fire of July 23, 2022. University of the Aegean. ISBN: 978-618-86408-1-8. Pp. 194-203. (In Greek).

Kazanis, D., Roussou, O., Bazos, I., Kalabokidis, K., & Arianoutsou, M. 2023. Estimating the post-fire regeneration potential of Pinus brutia forests at south Lesvos island, Greece. Poster paper in Proceedings of the 11th HELECOS Conference. Patras, Greece, 4-7 October 2023. Pp. 191-192.

Kelly, E., Tuda, D., Whelan, A., Gourzis, K., Avagianou, A., Krommyda, V., . . . Judge, C. (2022) Baseline Study on Impact of Youth Employment Policies. Cowork4YOUTH Report. Accessed at: https://www.cowork4youth.org/wp-content/uploads/2022/11/Cowork4YOUTH_Output-3_Baseline-Study_Final-1_compressed.pdf

Keramitsoglou, K. M., Koudoumakis, P., Akrivopoulou, S., Papaevaggelou, R., & Protopapas, A. L. (2023). Biodiversity as an Outstanding Universal Value for Integrated Management of Natural and Cultural Heritage. Sustainability (Basel), 15(11), 8540. DOI: 10.3390/su15118540

Kesebir, M. (2018). 2008 Küresel Finansal Kriz ve Bu Krizin Türkiye'deki Ekonomik Göstergelere Etkisi. Anemon Muş Alparslan Üniversitesi Sosyal Bilimler Dergisi, 6(4), 531–538. DOI: 10.18506/anemon.378086

Kim, D., & Lee, K. (2022). The Impact of Community Art Activists" Authenticity on the Performance and Effect of Community Art Projects. The Journal of Internet Electronic Commerce Research, 22(6), 277–297. DOI: 10.37272/JIECR.2022.12.22.6.277

Kim, K. S. (2000). High-speed rail developments and spatial restructuring: A case study of the capital region in south korea. Cities (London, England), 17(4), 251–262. DOI: 10.1016/S0264-2751(00)00021-4

Kim, K., Brunner, M., & Yamashita, E. Y. (2006). ve Yamashita, E., "Influence of land use, population, employment, and economic activity on accidents". Transportation Research Record: Journal of the Transportation Research Board, 1953(1), 56–64. DOI: 10.1177/0361198106195300107

Kim, K., Park, S. J., & Kweon, Y.-J. (2007). Highway traffic noise effects on land price in an urban area. Transportation Research Part D, Transport and Environment, 12(4), 275–280. DOI: 10.1016/j.trd.2007.03.002

Kizos, T., Koulouri, M., Vakoufaris, H., & Psarrou, M. (2010). Preserving characteristics of the agricultural landscape through agri-environmental policies: The case of cultivation terraces in Greece. Landscape Research, 35(6), 577–593. DOI: 10.1080/01426397.2010.519434

Kizos, T., Tsilimigkas, G., & Karampela, S. (2017). What drives built-up area expansion on islands? Using soil sealing indicators to estimate built-up area patterns on Aegean islands, Greece. Tijdschrift voor Economische en Sociale Geografie, 108(1), 35–52. DOI: 10.1111/tesg.12244

Klasen, S. (2010): Measuring and Monitoring inclusive growth: Multiple Definitions, Open Questions and some Constructive Proposals, Sustainable Development Working Paper Series, Asian Development Bank, Manila

Kobani, D., & Douglas, P. (2021). Indigeneous people, Multiculturalism and inclusiveness in portharcourt city developmnt planning. 2-5.

Kolodny, E., & Darques, R. (2004), "Turcs, Grecs et réfugiés dans l'île de Lesbos au XXe siècle", Méditerranée 103 (pp. 65-74) (in French)

Koresawa, A., & Konvitz, J. (2001). Towards a New Role for Spatial Planning. Organisation for Economic Co-operation and Development.

Koščak, M. (2016). Heritage trails through Dolenjska and Bela Krajina in Slovenia: A multi-stakeholder approach to tourism development, PREGLEDNI ČLANAK, Fakultet za sport i turizam, Novi Sad. TIMS Acta, 10(2), 109–118. DOI: 10.5937/timsact10-10519

Kostof, S. (1991). The city shaped: Urban patterns and meanings through history. Bulfinch Press.

Koutsopoulou, A., & Coccossis, C. (2012). DYNAMICS OF ISLAND SYSTEMS: TOWARDS A STRATEDY FOR INTEGRATED DEVELOPMENT OF THE SMALL GREEK ISLANDS. University of Thessaly, Department of Planning and Regional Development. (In Greek)

Kovács, E., Kelemen, K., Kalóczkai, A., Margóczi, K., Pataki, G., Gébert, J., Málovics, G., Balázs, B., Roboz, A., Krasznai Kovács, E., & Mihók, B. (2015). Understanding the links between ecosystem service trade-offs and conflicts in protected areas. Ecosystem Services, 12, 117–127. DOI: 10.1016/j.ecoser.2014.09.012

Krajewska, M., Zrobek, S., & Kovac, M. S. (2014). The role of spatial planning in the investment process in poland and slovenia. De Gruyter. DOI: 10.2478/remav-2014-0017

Kutluay Tutar, F., Canbulut, S., & Tutar, N. F. (2022). Smart cities for a sustainable life in local economic''. International Journal of Disciplines Economics & Administrative Scienves Studies, 8(43), 472–495.

Kyvelou, S. S., & Gourgiotis, A. (2019). Landscape as Connecting Link of Nature and Culture: Spatial Planning Policy Implications in Greece. Urban Science (Basel, Switzerland), 3(3), 81. DOI: 10.3390/urbansci3030081

Ladrón de Guevara, F., Washington, S. P., & Oh, J. (2004). ve Oh, J., "Forecasting travel crashes at the planning levels: Simultaneous negative binomial crash model applied in Tucson, Arizona". Transportation Research Record: Journal of the Transportation Research Board, 1897(1), 191–199. DOI: 10.3141/1897-25

Lahouri, D., Kalloniatis, C., Kalabokidis, K., Tzoraki, O., Roussou, O., Dimou, I., Kopsacheilis, V., Andreadis, O., Zikoulis, A., & Krallis, F. 2024. Communication and promotion issues. In: Kalabokidis K. (Ed.). Impact study and expertise for addressing problems in the affected area of Vrissa-Vatera-Stavros in Lesvos Island, Greece, after the fire of July 23, 2022. University of the Aegean. ISBN: 978-618-86408-1-8. Pp. 311-340. (In Greek).

Lai, S., & Zoppi, C. (2017). The influence of Natura 2000 sites on land-taking processes at the regional level: An empirical analysis concerning Sardinia (Italy). Sustainability (Basel), 9(2), 259. DOI: 10.3390/su9020259

Landow, G. P. (2012, 03 25). Retrieved from The Victorian Web: http://www .victorianweb.org/science/darwin/geochange.html

Lane, R., Kronsell, A., Reynolds, D., Raven, R., & Lindsay, J. (2024). Role of Local Governments and Households in Low-Waste City Transitions''. Environmental Innovation and Societal Transitions, 52, 100879. DOI: 10.1016/j.eist.2024.100879

Lantitsou, K. (2017). Eco development and environmental spatial planning. Fresenius Environmental Bulletin, 1291–1300.

Launay L. de, (1897), Chez les Grecs de Turquie, Paris (in French) Mantran R., (1989), Histoire de l'Empire Ottoman, Paris (in French)

Lervag, H. (1985). "Arealbruk og transport", NIBR-notat 1985:14, Norsk institutt for byog regionplanlegging. Norveç.

LeSage, J. P. (1999). The Theory and Practice of Spatial Econometrics. University of Toledo, 28(11), 1–39.

Levantis, E. (2001). Greek National Report. Revue juridique de l'Environnement Année 2001 H-S pp., pp. 239-254.

Ley 197/1963. (1963). Ley 197/1963 sobre Centros y Zonas de Inter´es Turístico Nacional. Boletín Oficial del Estado, 313,.

Lin, C., Liu, J., & Li, W. (2023). Influence of the high-speed railway (hsr) construction on industrial structure transformation. Enterprise Information Systems, 17(2), 1942998. DOI: 10.1080/17517575.2021.1942998

Li, R., Xu, M., & Zhou, H. (2023). Impact of high-speed rail operation on urban economic resilience: Evidence from local and spillover perspectives in china. Cities (London, England), 141, 104498. DOI: 10.1016/j.cities.2023.104498

Litman, T. ve Fitzroy, S., "Safe travels; Evaluating mobility management traffic safety impacts", Victoria Transport Policy Institute,Victoria, Kanada (2010).

Litman, T. (2012). "Evaluating accessibility for transportation planning", Victoria Transport Policy Institute. Kanada.

Liu, Z., Zeng, S., Jin, Z., & Shi, J. J. (2022). Transport infrastructure and industrial agglomeration: Evidence from manufacturing industries in china. Transport Policy, 121, 100–112. DOI: 10.1016/j.tranpol.2022.04.001

Li, Y., & Hunter, C. (2015). Community involvement for sustainable heritage tourism: A conceptual model. Journal of Cultural Heritage Management and Sustainable Development, 5(3), 248–262. DOI: 10.1108/JCHMSD-08-2014-0027

Loulanski, T., & Loulanski, V. (2011). The sustainable integration of cultural heritage and tourism: A meta-study. Journal of Sustainable Tourism, 19(7), 837–862. DOI: 10.1080/09669582.2011.553286

Loungani, P. (2017). The Power of Two: Inclusive Growth and the IMF. Review of European Economic Policy, 52(2), 92–99.

Lovegrove, G., & Litman, T. “Macrolevel collision prediction models to evaluate road safety effects of mobility management strategies: New empirical tools to promote sustainable development”, Transportation Research Board 87th Annual Meeting, USA, 1-24 (2008).

Lovegrove, G. R., & Sayed, T. (2006). ve Sayed, T., “Macro-level collision prediction models for evaluating neighbourhood traffic safety”. Canadian Journal of Civil Engineering, 33(5), 609–621. DOI: 10.1139/l06-013

Lowicki, D., & Piotrowska, S. (2015). Monetary valuation of road noise. Residential property prices as an indicator of the acoustic climate quality. Ecological Indicators, 52, 472–479. DOI: 10.1016/j.ecolind.2015.01.002

Luck, G. W., Ricketts, T. H., Daily, G. C., & Imhoff, M. (2004). Alleviating spatial conflict between people and biodiversity. Proceedings of the National Academy of Sciences of the United States of America, 101(1), 182–186. DOI: 10.1073/pnas.2237148100

Madison Kolodny E., (1974), La population des îles de la Grèce. Essai de géographie insulaire en Méditerranée Orientale, 3 vol., Aix-en-Provence (in French)

Magnaghi, A. (2005). The Urban Village: A Charter for Democracy and Local Self-Sustainable Development. Zed Books.

Ma, L., Niu, D., & Sun, W. (2021). Transportation infrastructure and entrepreneurship: Evidence from high-speed railway in china. China Economic Review, 65, 101577. DOI: 10.1016/j.chieco.2020.101577

Malta Ministry of Tourism. (n.d.). National Tourism Policy 2015-2020. Retrieved from https://tourism.gov.mt/wp-content/uploads/2023/04/National-Tourism-POlicy-2015-2020.pdf

Ma, M., & Hassink, R. (2013). An evolutionary perspective on tourism area development. Annals of Tourism Research, 41, 89–109. DOI: 10.1016/j.annals.2012.12.004

Mammides, C., Zotos, S., & Martini, F. (2024). Quantifying the amount of land lost to artificial surfaces in European habitats: A comparison inside and outside Natura 2000 sites using a quasi-experimental design. Biological Conservation, 293, 110556. DOI: 10.1016/j.biocon.2024.110556

Mantzaris, A. (1928). Our tobaccos. Pyrsos. [Μάντζαρης Α., (1928), Τα καπνά μας, Πυρσός, Αθήνα] (in Greek)

Maqbool, M. S., Mahmood, T., Sattar, A., & Bhalli, M. (2013). Determinants of Unemployment: Empirical Evidences From Pakistan. Pakistan Economic and Social Review, 51(2), 191–208.

Marks, H. (1957). Subdividing for traffic safety. Traffic Quarterly, 11(3), 308–325.

Marmaras, E. (2019). Cycladic aspects of settlement development and new human geography in the islands. Gutenberg. (In Greek)

Martínez Cordero, J. (2013). Modificaciones conceptuales de la ribera del mar,en la reforma de la Ley de costas. In E. Perez Galvez, & A. Aleman Monterreal, Costas y urbanismo. El litoral tras la Ley 2/2013, de protecci'on y uso sostenible del litoral y de modificaci'on de la Ley de Costas (pp. 173-224). Wolters Kluwer Espana.

Matos, A. T., Santos, C. J. F., & Chevalier, D. R. (2002). Fire and restoration of the largest urban forest of the world in Rio de Janeiro City, Brazil. Urban Ecosystems, 6(3), 151–161. DOI: 10.1023/A:1026164427792

McDonald, R. I., Forman, R. T. T., Kareiva, P., Neugarten, R., Salzer, D., & Fisher, J. (2009). Urban effects, distance, and protected areas in an urbanizing world. Landscape and Urban Planning, 93(1), 63–75. DOI: 10.1016/j.landurbplan.2009.06.002

McDonald, R. I., Kareiva, P., & Forman, R. T. T. (2008). The implications of current and future urbanization for global protected areas and biodiversity conservation. Biological Conservation, 141(6), 1695–1703. DOI: 10.1016/j.biocon.2008.04.025

Mcentire, D. A. (2001). Triggering agents, vulnerabilities and disaster reduction: Towards a holistic paradigm. Disaster Prevention and Management, 10(3), 189–196. DOI: 10.1108/09653560110395359

McKenna, D., Naumann, S., McFarland, K., Graf, A., & Evans, D. (2014). Literature review, the ecological effectiveness of the Natura 2000 Network. ETC/BD report to the EEA, 30 pp. DOI: 10.13140/RG.2.1.4358.9288

McKinley, T. (2010): Inclusive Growth Criteria and Indicators: An Inclusive Growth Index for Diagnosis of Country Progress, ADB Sustainable Development Working Paper Series, No. 14. Asian Development Bank, Mandaluyong City, Philippines

Mendola, D., & Volo, S. (2017). Building composite indicators in tourism studies: Measurements and applications in tourism destination competitiveness. Tourism Management, 59, 541–553. DOI: 10.1016/j.tourman.2016.08.011

Menemenlis, D., Palaiologou, P., & Kalabokidis, K. (2023). Wildfire-residential risk analysis using building characteristics and simulations to enhance structural fire resistance in Greece. Fire (Basel, Switzerland), 6(10), 403. DOI: 10.3390/fire6100403

Mergos, G., Papadaskalopoulos, A., & Christofakis, E. (2004). Economic Characteristics and Development Strategy of Greek Islands. Academy of Athens, Economic Studies Office. (In Greek)

Meschede, H. (2015, November). SMART ENERGY SYSTEMS FOR USING RES UP TO 100%-THE COMBINATION OF 100% RES AND TOURISM EXAMPLES FROM "LA GOMERA" AND "JUIST" ISLANDS. Colimbari, Crete: University of Kassel. DOI: 10.13140/RG.2.2.21071.18088

Mills, B., Yessis, J., Boyd, D., & Andrey, J. (2001). The urban environment a hazard source and sink. Environments, 17–38.

Ministry of Environment and Energy, Directorate of Natural Environment Management and Biodiversity. (2024). Prioritized Action Framework (PAF) for Natura 2000 in Greece for the Multiannual Financial Framework period 2021–2027. Greece.

Ministry of Environment and Urbanization Climate Change. (2011). 2024-2030 National Smart Cities Strategy and Action Plan (2023). Retrieved from, https://www.akillisehirler.gov.tr/wp-content/uploads/2023/12/Taslak-Eylem_Plani.pdf

Ministry of Environmental Protection. (2020, 08 17). Ministry of Environmental Protection. Retrieved from Protecting Israel's Coast: https://www.gov.il/en/pages/coastal_environment_preservation

Ministry of the Environment. D. (2001). Spatial planning as an instrument for promoting sustainable development in the Nordic countries; Action programme for 2001-2004. Denmark: Nordisk Miljomaerkning.

Mitchell, B. (2019). Stakeholders and Partnerships. In Mitchell, B. (Ed.), Resource and Environmental Management (3rd ed., pp. 145–176). DOI: 10.1093/oso/9780190885816.003.0006

Monte, F., Redding, S. J., & Rossi-Hansberg, E. (2018). Commuting, migration, and local employment elasticities. The American Economic Review, 108(12), 3855–3890. DOI: 10.1257/aer.20151507

Morehouse, B. J., Henderson, M., Kalabokidis, K., & Iosifides, T. (2011). Wildland fire governance: Perspectives from Greece. Journal of Environmental Policy and Planning, 13(4), 349–371. DOI: 10.1080/1523908X.2011.611678

Morreta, V. (2021). Territorial capital in local economic endogenous development. Regional Science Policy & Practice, 13(1), 103–119. DOI: 10.1111/rsp3.12317

Mpia, N. (2018). Socio-economic transformations on the island of Spetses and their impacts on the natural environment: A diachronic investigation (Master's thesis). Retrieved from: https://apothesis.eap.gr/archive/item/155521

Mücher, S., Gerard, F., Olschofsky, K., Hazeu, G., Luque, S., Pino, J., Gregor, M., Wachowicz, M., Halada, L., Tompo, E., Kohler, R., Petit, S., Smith, G., & Kolar, J. (2006). Spatial impact of conservation sites (Natura 2000) on land cover changes. Proceedings of the 2nd Workshop of the EARSeL SIG on Land Use and Land Cover.

Nadin, V., & Stead, D. (2008). European Spatial Planning Systems, Social Models and Learning. disP - The Planning Review, 35-47.

Nadin, V. (2007). The emergence of the spatial planning approach in England. Planning Practice and Research, 22(1), 43–62. DOI: 10.1080/02697450701455934

Næss, P. "Miljømessig effektiv lokalisering av arbeidsplasser og boliger", In:Miljøhåndboken, (Kolbenstvedt, M., H. Silborn & T. Solheim, T. eds) Transportøkonomisk institutt, Oslo, Norveç, 1: 95-106 (1996).

Nagopoulos, N., Paraskevopoulos, D., Maragoudakis, M., Kouroutzas, C., ... 2024. Qualitative social research on the protection of the natural environment as a key factor of local development and social cohesion. In: Kalabokidis K. (Ed.). Impact study and expertise for addressing problems in the affected area of Vrissa-Vatera-Stavros in Lesvos Island, Greece, after the fire of July 23, 2022. University of the Aegean. ISBN: 978-618-86408-1-8. Pp. 242-285. (In Greek).

Natarajan, L. (2015). Socio-spatial learing: A case study of community knowledge in participatory spatial planning. progress planning.

Navajas Corral, O., & Fernández, J. F. (2022). The Importance of Ecomuseums and Local Knowledge for a Sustainable Future: The La Ponte-Ecomuseu Project. In Borrelli, N., Davis, P., & Santo, P. D. (Eds.), Ecomuseums and Climate Change (pp. 283–203). Ledizioni.

NCMA (National Cadastre and Mapping Agency). (2023). Viewing orthophotos. Greek Cadastre. Retrieved from https://www.ktimatologio.gr

Nel, H. (2001). A Project Management Approach to the Implementation of Development Programmes within the Local Government Sphere: An Emprical Analysis. Development Southern Africa, 18(5), 605–624. DOI: 10.1080/03768350120097450

Netto, A. (2009). What is tourism? Definitions, theoretical phases and principles. In J. Tribe, Philosophical Issues in Tourism (pp. 43-61). Channel View Publications. DOI: 10.21832/9781845410988-004

Newman, P. W. G. ve Kenworthy, J. R., "Cities and Automobile Dependence: A Sourcebook", Avebury Technical, Aldershot, UK (1989).

Newton, C. T. (1865), Travels and discoveries in the Levant, London

Nikellis, E. 2017. Personal communication on local civil protection issues in the aftermath of the Vrissa 2017 earthquake. Vrissa-Vatera resident and administrator of a local webpage (i.e. https://vatera.gr/platanos/, in Greek), Lesvos Island, Greece.

Nikoloas, K., Dimitra, V., & Agapi, X. (2014). Real estate values and environment: A case study on the effect of the environment on residential real estate values. International Journal of Academic Research, 861–866.

Nuñez-Torres, A., Arones-Huarcaya, A., & Yarasca-Aybar, C. (2023). Strategies for territorial tourism planning in natural protected areas (NPAs): Alto Mayo Protected Forest (BPAM), Peru. City. Territory and Architecture, 10(1), 15. Advance online publication. DOI: 10.1186/s40410-023-00200-w

O'Gorman, K., & Maclntosh, R. (2015). Research Methods for Business & Management (2nd ed.). Goodfellow Publishers Limited. DOI: 10.23912/978-1-910158-51-7-2736

OECD. (2013): Together We Stand: Inclusive Growth, Speech by OECD Secretary General Angel Gurria, at OECD Workshop on inclusive growth, 3 April 2013, Paris

OECD. (2018): Beyond GDP: Measuring What Counts for Economic and Social Performance, OECD Publishing, Paris https://policydialogue.org/files/publications/papers/FINAL, Beyond, GDP_, Measuring, What, Counts, for, Economic, and, Social, Performance.pdf

OECD. (2018): Joining Forces to Leave No One Behind Multilingual Summaries Development Co, operation Report, OECD Publishing, Paris, https://www.oecd, ilibrary.org/sites/53d35549, el/index.html?itemId=/content/component/53d35549

OECD. (2018): Opportunities for all: a framework for policy action on inclusive growth, OECD Publishing, Paris https://doi.org/, en.DOI: 10.1787/9789264301665

OECD. (2018): The Framework For Policy Action On Inclusive Growth, OECD Publishing, Paris https://www.oecd.org/mcm/2018/documents/C, MIN, 2018, 5, EN.pdf

OECD. (2018): The Productivity, Inclusiveness Nexus, OECD Publishing, Paris, https://doi.org/, en. https://www.oecd.org/publications/the, productivity, inclusiveness, nexus, 9789264292932, en.htmDOI: 10.1787/9789264292932

OECD. (2020): How's Life? 2020 Measuring Well, being, OECD Publishing, Paris, http://www.oecd.org/statistics/how, s, life, 23089679.htm

OECD. (2020): Regional Policy for Greece Post-2020, OECD Publishing, Paris, https://www.oecd, ilibrary.org/urban, rural, and, regional, development/regional, policy, for, greece, post, 2020_cedf09a5, en

OECD/European Union/EC-JRC. (2008). Handbook on Constructing Composite Indicators: Methodology and User Guide. OECD Publishing., DOI: 10.1787/9789264043466-

Ogbeide, F. I., Kanwanye, H., & Kadiri, S. (2016). Revisiting the Determinants of Unemployment in Nigeria: Do Resource Dependence and Financial Development Matter? African Development Review, 28(4), 430–443. DOI: 10.1111/1467-8268.12222

OGG (Official Government Gazette) A' 245/2020, Law No. 4759/2020 on Spatial Planning, Urban Planning, and Other Provisions, Greek Official Gazette, 245, Part I, 2020. Available at: https://www.et.gr/

Ojeda, A. B., & Kieffer, M. (2020). Touristification. Empty concept or element of analysis in tourism geography? Geoforum, 115, 143–145. DOI: 10.1016/j.geoforum.2020.06.021

Olatunji, S. O., Olawumi, T. O., & Awodele, O. A. (2017). Achieving value for money in construction projects. Civil and environmental Research, 54-61.

Oniore, J. O., Bernard, A., & Gyang, E. (2015). Macroeconomic determinants of unemployment in Nigeria. International Journal of Economics. Commerce & Management, 3(10), 215–230.

Özden, P. P. (2006). ''Proposals on Applicability of Urban Regeneration in Turkey''. Istanbul University Faculty of Political Sciences Journal, 35, 215-233.

Özden, M., Kantar, G., & Çınar, S. (2021). The Importance of Local Governments in the Urban Branding Process. Journal of Yasar University, 16(62), 1005–1021.

Özsabuncuoğlu, İ. H., & Direkçi, T. (2012). Şeçilmiş Bazı Makroekonomik Değişkenlerin Türkiye'de Bölgesel İstihdama Etkisi. Turkish Economic Association, Discussion Paper, 2012/86.

Palaiologou, P., & Kalabokidis, K. 2024. Fuel breaks and estimated burning conditions with the aim of breaking the continuity of forest vegetation in the wider area of Vatera, Vrissa, Stavros and Polichnitos. In: Kalabokidis K. (Ed.). Impact study and expertise for addressing problems in the affected area of Vrissa-Vatera-Stavros in Lesvos Island, Greece, after the fire of July 23, 2022. University of the Aegean. ISBN: 978-618-86408-1-8. Pp. 71-82. (In Greek).

Palaiopanis, R., Pappas, I., Kizos, A., & Tscheulin, T. 2024. Agricultural issues in the affected area of Vrissa- Vatera-Stavros of Lesvos by the fire of 23 July 2022. In: Kalabokidis K. (Ed.). Impact study and expertise for addressing problems in the affected area of Vrissa-Vatera-Stavros in Lesvos Island, Greece, after the fire of July 23, 2022. University of the Aegean. ISBN: 978-618-86408-1-8. Pp. 170-192. (In Greek).

Pantazis, D., Moussas, V., Murgante, B., Daverona, A., Stratakis, P., Vlissidis, N., . . . Gadolou, E. (2017, October). SMART SUSTAINABLE ISLANDS VS SMART SUTAINABLE CITIES. Athens: ISPRS.

Papageorgiou, K., & Vogiatzakis, I. N. (2006). Nature protection in Greece: An appraisal of the factors shaping integrative conservation and policy effectiveness. Environmental Science & Policy, 9(5), 476–486. DOI: 10.1016/j.envsci.2006.03.003

Papatheodorou, A. (2004). Exploring the evolution of tourism resorts. Annals of Tourism Research, 31(1), 219–237. DOI: 10.1016/j.annals.2003.10.004

Pappalardo, G. (2020). Community-Based Processes for Revitalizing Heritage: Questioning Justice in the Experimental Practice of Ecomuseums. Sustainability (Basel), 12(21), 9270. DOI: 10.3390/su12219270

Paraskevaidis P., (2013), The refugees of the second persecution in Lesvos, (2nd edition)

Pavlis, E. (2017). North Aegean islands as 'ecomuseums': The case of Lesvos Island. Island Studies Journal, 12(1), 135–145. DOI: 10.24043/isj.9

Pavlis, E. (2024). The ecomuseum as a platform for territorial development: The case of Greek ecomuseums. International Journal of Heritage Studies, 30(10), 1193–1210. Advance online publication. DOI: 10.1080/13527258.2024.2386695

Pavlis, E., & Terkenli, T. S. (2023). Towards a Geographical Approach of "Spatial Conscience" and its application on Landscape. Landscape Research, 48(8), 1004–1018. DOI: 10.1080/01426397.2023.2238639

Peeters, P., Gössling, S., Klijs, J., Milano, C., Novelli, M., Dijkmans, C., . . . Postma, A. (2018) Overtourism: impact and possible policy responses. Research for TRAN Committee. Available at: https://www.europarl.europa.eu/RegData/etudes/STUD/2018/629184/IPOL_STU(2018)629184_EN.pdf

Peker, O., & Bölükbaş, M. (2012). Türkiye'de İşsizliği Belirleyen Faktörlerin Ampirik Bir Analizi. Atatürk Üniversitesi İktisadi ve İdari Bilimler Dergisi, 26(3-4), 33–43.

Pellach, C., & Carmon, D. (2016). Mare Nostrum Project-Final Report: Legal-Institutional Instruments for Integrated Coastal Zone Management (ICZM) in the Mediterranean. Mare Nostrum Project.

Persaud, B. N. (1993). ve Dzbik, L., "Accident prediction models for freeways". Transportation Research Record: Journal of the Transportation Research Board, (1401), 55–60.

Perugini, F. (2024). Local Government Efficiency and Economic Growth: The İtalian Case. Socio-Economic Planning Sciences, 91, 101775. DOI: 10.1016/j.seps.2023.101775

Post, J., & Lundin, C. (1996). Guidelines for integrated coastal zone management. The International Bank for Reconstruction and Development/The World Bank. DOI: 10.1596/0-8213-3735-1

Pouw, N., & McGregor, A. (2014): An Economics of Wellbeing: What Would Economics Look Like if it were Focused on Human Wellbeing? Institute of Development Studies Working Paper No. 436, London: IDS. Press

Princé, K., Rouveyrol, P., Pellissier, V., Touroult, J., & Jiguet, F. (2021). Long-term effectiveness of Natura 2000 network to protect biodiversity: A hint of optimism for common birds. Biological Conservation, 253, 108871. DOI: 10.1016/j.biocon.2020.108871

Qi, Y., Yang, Y., & Jin, F. (2013). China's economic development stage and its spatiotemporal evolution: A prefectural-level analysis. Journal of Geographical Sciences, 23(2), 297–314. DOI: 10.1007/s11442-013-1011-0

Rabianski, J. (2003). Primary and secondary data: Concepts, concerns, errors, and issues. The Appraisal Journal.

Raifu, I. A. (2017). On The Determinants of Unemployment in Nigeria: What Are The Roles of Trade Openness and Current Account Balance? Review of Innovation and Competitiveness: A Journal of Economic and Social Research, 3(4), 5-30.

Ranieri, R., & Ramos, R. A. (2013): After All, What is Inclusive Growth? International Policy Centre for Inclusive Growth, Brasilia http://www.ipc, undp.org/pub/IPCOnePager188.pdf

Ranieri, R., & Ramos, R. A. (2013): Inclusive growth: building up a concept, IPC, IG Working Paper Number 104, International Policy Centre for Inclusive Growth, Brasilia

Ranieri, R., & Ramos, R. A. (2013): Inclusive Growth: The Building up of a concept, International Policy Centre for Inclusive Growth, Brasilia http://www.ipc, undp.org/pub/IPCWorkingPaper104.pdf

Redman, C. (1999). human impact on ancient environments. university of Arizona press.

Reid, A., Teamey, K., & Dillon, J. (2007). Valuing and utilizing traditional ecological knowledge: Tensions in the context of education and the environment. Environmental Education Research, 1–19.

Ren, W., Xue, B., Yang, J., & Lu, C. (2020). Effects of the northeast china revitalization strategy on regional economic growth and social development. Chinese Geographical Science, 30(5), 791–809. DOI: 10.1007/s11769-020-1149-5

Resurreccion, B. P., Sajor, E. E., & Fajber, E. (2008). Climate Adaptation in Asia: Knowledge Gaps and Research Issues in South East Asia. ISET-International and ISET.

Rifaat, S. M., & Tay, R. (2010). ve Barros, A., "Effect of street pattern on road safety are policy recommendations sensitive to different aggregations of crashes by severity?". Transportation Research Record: Journal of the Transportation Research Board, (2147), 58–65. DOI: 10.3141/2147-08

Robinson, B. W., Rodegerdts, L., & Scarborough, W. ve Kittelson, W., "Roundabouts: An informational guide" U.S. Department of Transportation, Federal Highway Administration, Report FWHARD-00-067, Washington, ABD, 101-181 (2000).

Robinson, C. M. (1911). The width and arrangement of streets. Engineering News Publishing.

Romero, A. C., Issii, T. M., Pereira-Silva, E. F. L., & Hardt, E. (2018). Effects of urban sprawl on forest conservation in a metropolitan water source area. Revista Árvore, 42(1), e420114. DOI: 10.1590/1806-90882018000100014

Rougon F., (1892), Smyrne, Paris (in French)

Russo, A. P. (2005). Cultural gateways - building partnerships for sustainable development in destination regions. Proceedings of the 45th Congress of the European Regional Science Association, ERSA conference papers

Sağlam, C., & Kaya, H. F. (2020). Bilgi ve İletişim İktisadi Faaliyetinin İşsizlik Oranı Üzerindeki Kısa ve Uzun Dönem Etkileri. Uluslararası Ekonomi ve Yenilik Dergisi, 6(2), 261–275.

Şahin, D., & Güven, A. (2011). ''Local Development and The Fight Against Poverty of Local Governments''. 26. Turkish Finance Symposium, 281-304.

Sakshaug, K. ve Johannessen, S.,"Analyse av ulykkessteder: Verdier for normal ulykkesfrekvens og skadekostnad ved normal oggod standard", Trondheim, SINTEF Teknologi Ogsamfunn, Transportsikkerhet og –İnformatikk, Norveç (2005)

Sandhu, B. ve Al-Kazily, J., "Safety impacts of freeway traffic congestion", Transportation Research Board, 75th Annual Meeting, Washington DC., (1996).

Scharp, K. M., Paxman, C. G., & Thomas, L. J. (2016). "i want to go home" homesickness experiences and social-support-seeking practices. Environment and Behavior, 48(9), 1175–1197. DOI: 10.1177/0013916515590475

Sekulo, E. (2018, March). Sustainable Island Tourism: the Future of the Island. Dubrovnik: RIT Croatia Digital repository - Rochester Institute of Technology. Retrieved from https://repository.acmt.hr/islandora/object/acmt%3A33/-datastream/PDF/view

Senaratne, A. (2019). Ensuring Sustainable Development in post-conflict Sri Lanka. In Managing Domestic and International Challenges and opportunities in post conflict development. DOI: 10.1007/978-981-13-1864-1_5

Seto, K. C., Golden, J. S., Alberti, M., & Ii, B. L. T. (2017). Sustainability in an urbanizing planet. Proceedings of the National Academy of Sciences of the United States of America, 114(34), 8935–8938. DOI: 10.1073/pnas.1606037114

Sever, E., & İğdeli, A. (2018). The Determining Factors of Youth Unemployment in Developing Countries: The Case of Turkey. International Journal of Social and Economic Sciences, 8(1), 75–83.

Sezer, Ö., & Vural, T. (2009). ''Within the Framework of Globalization-localization İnteraction, Development-oriented Local Social Policies in Local Governments''. National Development and Local Governments: Proceedings of the 4th National Local Governments Symposium, 2.

Shi, H., & Goulias, K. (2024). Understanding The Zero-Emission Vehicle Market Spatial Diffusion and Its Determinants from 2019 to 2022 Using Spatial Econometric Models. Energy, 313, 133607. DOI: 10.1016/j.energy.2024.133607

Sifneou, E. (1996a), Lesvos. Economic and Social History, 2nd Edition, 2016, Aeolida, Mytilene [Σιφναίου Ε., (1996), Λέσβος. Οικονομική και Κοινωνική Ιστορία 2η έκδοση, 2016, Αιολίδα, Μυτιλήνη] (in Greek)

Sifneou, E. (1996b), “The industrialization of Lesvos during the late Ottoman period”, Lesviaka, Bulletin of the Society of Lesvian Studies, vol. XVI, Mytilene, p. 454-478 [Σιφναίου Ε., (1996), «Η εκβιομηχάνιση της Λέσβου κατά την ύστερη Τουρκοκρατία», Λεσβιακά, Δελτίον της Εταιρείας Λεσβιακών Μελετών, τ. ΙΣΤ' Μυτιλήνη, σ. 454-478] (in Greek)

Sifneou, E. (Ed.). (2002), Soapmaking in Lesvos. A memento, Livanis, Athens (bilingual edition, Greek/English)

Sifneou, E. (2007). Lesvos-Ayvalik-Adramytti: “Expansionism” or equal exchange in the trade and manufacture of olive-oil? In Kitromilides, P., & Michaelaris, P. (Eds.), Mytilene and Ayvalik: A bilateral historical relationship in the north-eastern Aegean (pp. 249–260). [Σιφναίου Ε., (2007), “Λέσβος-Κυδωνίες-Αδραμύτι: Οικονομικός επεκτατισμός ή ισότιμη συναλλαγή στην εμπορία και τη μεταποίηση του ελαιολάδου;” στο: Κιτρομηλίδης Π., Μιχαηλάρης Π., (Επιμ.), Μυτιλήνη και Αϊβαλί (Κυδωνίες): Μια αμφίδρομη σχέση στο βορειοανατολικό Αιγαίο, (σελ. 249-260), Αθήνα] (in Greek)

Silva, P. d. (2008, 02 07). On Relativity of knowledge. Retrieved from Sri lanka Guardian: http://www.srilankaguardian.org/2008/02/on-relativity-of-knowledge.html

Singh, P. K. (2019). Spatial planning for socio-economic developoment . International Journal of Geography and Regional Planning, 102-110.

Sloan, P., & Oliver, D. (2013). Building Trust in Multi-stakeholder Partnerships: Critical Emotional Incidents and Practices of Engagement. Organization Studies, 34(12), 1835–1868. DOI: 10.1177/0170840613495018

SMART ISLAND. (2015, January). Retrieved 10 15, 2024, from http://www.smartisland.eu/en/replicability/pantelleria.html

Sobieraj, J. (2017). Impact of spatial planning on the pre-investment phase of the development process in the residential construction field. Archives of civil engineering, 113-131.

Song, Y. (2014). What should economists know about the current chinese hukou system?

Song, Y. (2014, June). What should economists know about the current Chinese hukou system? China Economic Review, 29, 200–212. DOI: 10.1016/j.chieco.2014.04.012

Soulakellis, N., Vasilakos, C., Tataris, G., Papadopoulou, E., Proestakis, S., Hatzivassiliou, T., & Chroni, A. 2024. Web cartography. In: Kalabokidis K. (Ed.). Impact study and expertise for addressing problems in the affected area of Vrissa-Vatera-Stavros in Lesvos Island, Greece, after the fire of July 23, 2022. University of the Aegean. ISBN: 978-618-86408-1-8. Pp. 287-309. (In Greek).

Stiglitz, J. (2012). The price of inequality: how today's divided society endangers our future. Penguin.

Strand, A. (1993). "Trafikksikkerhetsmessige hensyn i arealplanleggingen: Forelesning ved EEU-kurs Trafikksikkerhet 29.10.93", Norges Tekniske Høgskole, Institutt for samferdselsteknikk. Norveç.

Strategy and Budget Presidency. (2018). 2024-2028 Twelfth Development Plan (2023). Retrieved from, https://www.sbb.gov.tr/kalkinma-planlari/

Stratigea, A., & Katsoni, V. (2015). A strategic policy scenario analysis framework for the sustainable tourist development of peripheral small island areas: The case of Lefkada-Greece Island. European Journal of Futures Research, 3(1), 5. DOI: 10.1007/s40309-015-0063-z

Strauss, W., & Howe, N. (1991). Generations: The history of America's future, 1584 to 2069. William Morrow.

Strohmeier, M. (2010), "Economy and society in the Aegean province of the Ottoman empire, 1840-1912", Turkish Historical Review, 1.2, (pp. 164-195) DOI: 10.1163/187754610X538825

Stroja, J. (2018). My history, your history, our history: Developing meaningful community engagement within historic sites and museums. Queensland Review, 25(2), 300–321. DOI: 10.1017/qre.2018.29

Subramaniam, T., & Baharumshah, A. Z. (2011a). Determinants of Unemployment in the Philippines. The Empirical Economics Letters, 10(12), 1248–1257.

Subramaniam, T., & Baharumshah, A. Z. (2011b). Unemployment and Speed of Adjustment in ASEAN-3 Economies: A Cointegration Analysis. The Singapore Economic Review, 56(03), 327–347. DOI: 10.1142/S0217590811004353

Sullivan, E. C. (1990). Estimating Accident Benefits of Reduced Freeway Congestion. Journal of Transportation Engineering, 116(2), 167–180. DOI: 10.1061/(ASCE)0733-947X(1990)116:2(167)

Sun, X., Yan, S., Liu, T., & Wang, J. (2023). The impact of high-speed rail on urban economy: Synergy with urban agglomeration policy. Transport Policy, 130, 141–154. DOI: 10.1016/j.tranpol.2022.11.004

Sutanta, H., Rajabifard, A., & Bishop, I. D. (2010). Integrating Spatial Planning and Disaster Risk Reduction at the Local Level in the Context of Spatially Enabled Government. Spatially Enabling Society Research, Emerging Trends and Critical Assessment, 55-68.

Sutanta, H., Rajabifard, A., & Bishop, I. (2013). Disaster risk reduction using acceptable risk measures for spatial planning. Journal of Environmental Planning and Management, 56(6), 761–785. DOI: 10.1080/09640568.2012.702314

Tartaruga, Mytilene [Παρασκευαΐδης Π., (2013), Η προσφυγιά του β' διωγμού στη Λέσβο, Ταρταρούγα, Μυτιλήνη] (in Greek)

Terkenli, T. (2005). Human activity in landscape seasonality: The case of tourism in Crete. Landscape Research, 30(2), 221–239. DOI: 10.1080/01426390500044408

The world Bank. (2022, October 6). The world Bank in Sri Lanka. Retrieved from worldbank.org: https://www.worldbank.org/en/country/srilanka/overview

Theodora, Y. (2019). Aegean Sea: Challenges and dilemmas in management and planning for local development in fragmented insular regions. Heritage, 2(3), 108. Advance online publication. DOI: 10.3390/heritage2030108

Thrift, N. (2007). Non-representational theory: Space, Politics, Affect. Routledge.

Tiwari, I., & Gupta, B. (2023). Environment and Sustainable Development. 213-215.

Trimurti, C. P., & Komalasari, Y. (2014). Determinants of Unemployment: Empirical Evidences From 7 Province in Indonesia. [SCIRJ]. Scientific Research Journal, 2(8), 5–9.

Tsilimigkas, G., & Derdemezi, E.-T. (2020). Unregulated built-up area expansion on Santorini Island, Greece. European Planning Studies, 28(9), 1790–1811. DOI: 10.1080/09654313.2019.1687656

Tsilimigkas, G., & Gourgiotis, A. (2023). Unregulated urban sprawl and spatial distribution of fire events: Evidence from Greece. Environmental Hazards, 17(5), 436–455. DOI: 10.1080/17477891.2018.1430554

Tsilimigkas, G., Gourgiotis, A., & Derdemezi, E.-T. (2022). Spatial planning incompetence to discourage urban sprawl on Greek islands: Evidence from Paros, Greece. Journal of Coastal Conservation, 26(2), 11. Advance online publication. DOI: 10.1007/s11852-022-00859-2

Tsilimigkas, G., Kizos, T., & Derdemezi, E.-T. (2021). Urbanization and land cover change in Aegean islands: Assessing the socio-economic impact of tourism development. Land (Basel), 10(5), 545. DOI: 10.3390/land10050545

Tuğaç, Ç. (2021). "Duties and Responsibilities of Local Governments from Urban Resilience Perspective". İdealkent. Journal of Urban Studies, 28(3), 984–1019.

Türk, S. M. (2024). Local Governments and Democracy. Conservative Thought Magazine, 20(66), 430–444.

TURKSTAT. (2024). Bölgesel İstatistikler Veri Tabanı. Access Date: 12.05.2024, Access Link: https://biruni.tuik.gov.tr/bolgeselistatistik/anaSayfa.do?dil=tr

Tzoraki, O., Siarkos, E., Hatzivassiliou, T., & Tsoulibrk, A. 2024. Assessment of flood risk and expected cost of damage. In: Kalabokidis K. (Ed.). Impact study and expertise for addressing problems in the affected area of Vrissa-Vatera-Stavros in Lesvos Island, Greece, after the fire of July 23, 2022. University of the Aegean. ISBN: 978-618-86408-1-8. Pp. 121-167. (In Greek).

UNDP. (2014): Human Development Report Sustaining Human Progress: Reducing Vulnerabilities and Building Resilience https://www.undp.org/content/undp/en/home/librarypage/hdr/2014, human, development, report.html

UNDP. (2017): Strategy for Inclusive and Sustainable Growth, UNDP's Strategy for Inclusive and Sustainable Growth | UNDP

UNDP. (2020): The 2020 Human Development Report The next frontier Human development and the Anthropocene http://hdr.undp.org/sites/default/files/hdr2020.pdf

UNWTO. (2022) Measuring the Sustainability of Tourism (MST) Estimating Tourism Direct GDP with limited data. UNWTO Reports. Available at: https://webunwto.s3.eu-west-1.amazonaws.com/s3fs-public/2022-12/Measuring_TDGDP.pdf

UNWTO World Tourism Organization. (2024). World Tourism Barometer 2024- Volume 22 • Issue 1. https://www.unwto.org/publications. Retrieved from https://webunwto.s3.eu-west-1.amazonaws.com/s3fs-public/2024-01/UNWTO_Barom24_01_January_Excerpt.pdf?VersionId=IWu1BaPwtlJt66kRIw9WxM9L.y7h5.d1

Uyar Bozdağlıoğlu, E. Y. (2008). Türkiye'de İşsizliğin Özellikleri ve İşsizlikle Mücadele Politikalari. Manas Üniversitesi Sosyal Bilimler Dergisi, 10(20), 45–65.

Uysal, Y. (2022). The role of local governments in combating climate change and global warming: Findings and recommendations''. The Journal of Kesit Academy, 8(30), 324–354. DOI: 10.29228/kesit.57382

Vafeas, N., & Mandamadiotou, M. (2023), The islands of North-Eastern Aegean. From the Ottoman Empire to the Greek State, Ed. Alexandria, Athens [Βαφέας Ν., Μανδαμαδιώτου Μ., (2023), Τα νησιά του βορειοανατολικού Αιγαίου. Από την Οθωμανική Αυτοκρατορία στο Ελληνικό Κράτος, Εκδ. Αλεξάνδρεια, Αθήνα] (in Greek)

Vale, L. (2005). The resilient city: How modern cities recover from disaster. Oxford university press. DOI: 10.1093/oso/9780195175844.001.0001

Vanderbilt, T. (2008). Traffic: Why we drive the way we do (And what it says about us). Knopf Doubleday Publishing Group.

Vasilakos, C., & Verykios, V. (2024). Burned olive trees identification with a deep learning approach in Unmanned Aerial Vehicle images. Remote Sensing (Basel), 16(23), 4531. DOI: 10.3390/rs16234531

Vassi, A., & Christodoulopoulou, G. (2014). Planning for sustainability at the destination level through limiting accessibility on the island of Spetses, Greece. WIT Transactions on Ecology and The Environment. Sustainable Tourism, VI, 187–198. DOI: 10.2495/ST140151

Vega, S. H., & Elhorst, J. (2015). The SLX model. Journal of Regional Science, 55(3), 1–25.

Vileniske, i. g., & Arturas, N. (2012). environmental ethics and sustainability in housing design. 1-13.

Vlachakis, G., Vlachaki, E., & Lourenço, P. B. (2020). Learning from failure: Damage and failure of masonry structures, after the 2017 Lesvos earthquake (Greece). Engineering Failure Analysis, 117, 104803. DOI: 10.1016/j.engfailanal.2020.104803

Vogiatzis, A. (Ed.). (2012). Lesvos in 1912. Detailed depiction of the state of the island by the first Greek Administration. Aeolida. [Βογιατζής Α. (Επιμ.), (2012), Η Λέσβος το 1912. Λεπτομερής αποτύπωση της εικόνας του νησιού από την πρώτη Ελληνική Διοίκηση, Αιολίδα, Μυτιλήνη] (in Greek)

von Scherzer, K. (1873), Smyrna, Vienna (Greek Edition, 1995, Istoritis, Athens) [Scherzer K., Σμύρνη, Ιστορητής, Αθήνα] (in Greek)

Vortherms, S. A., & Liu, G. G. (2022). Hukou as benefits: Demand for hukou and wages in china. Urban Studies (Edinburgh, Scotland), 59(15), 3167–3183. DOI: 10.1177/00420980221074911

Wang, C.-M. (2023). Escape to the countryside: Affects and rural eventful atmospheres. Habitat International, 140, 102929. DOI: 10.1016/j.habitatint.2023.102929

Wang, C., & Shen, J. (2022). Settlement intention of migrants in urban china: The effects of labor-market performance, employment status, and social integration. Applied Geography (Sevenoaks, England), 147, 102773. DOI: 10.1016/j.apgeog.2022.102773

Wang, F., Liu, Z., Xue, P., & Dang, A. (2022). High-speed railway development and its impact on urban economy and population: A case study of nine provinces along the yellow river, china. Sustainable Cities and Society, 87, 104172. DOI: 10.1016/j.scs.2022.104172

Wang, Z., & Chen, L. (2019). Destination choices of chinese rural–urban migrant workers: Jobs, amenities, and local spillovers. Journal of Regional Science, 59(3), 586–609. DOI: 10.1111/jors.12444

Waterton, E. (2014). A More-Than-Representational Understanding of Heritage? The 'Past' and the Politics of Affect. Geography Compass, 8(11), 823–833. DOI: 10.1111/gec3.12182

Watson, J. E. M., Dudley, N., Segan, D. B., & Hockings, M. (2014). The performance and potential of protected areas. Nature, 515(7525), 67–73. DOI: 10.1038/nature13947

Wave for Energy. (2022, November 9). Pantelleria. Retrieved 10 15, 2024, from https://www.waveforenergy.com/news/portfolio-items/pantelleria/

Wegelin, E. A., & Borgman, K. (1995). Options for Municipal Interventions in Urban Poverty Alleviation''. Environment and Urbanization, 7(2), 131–151. DOI: 10.1177/095624789500700219

Wijesundara, J. (2015). Re-invigorating the spirit of place of kandy; urban Design and city planning strategies. Built- environment: Sri lanka, 1-11.

WISEGRID. (2024). Kythnos. Retrieved 10 15, 2024, from https://www.wisegrid.eu/pilot-sites/kythnosgr

Word Bank. (2017): Environmental and Social Framework, International Bank for Reconstruction and Development, World Bank Washington DC https://pubdocs.worldbank.org/en/837721522762050108/Environmental, and, Social, Framework.pdf

World Bank. (2009): What is Inclusive Growth? World Bank, Washington DC

World Bank. (2012): Inclusive Green Growth: The Pathway to Sustainable Development, World Bank Washington DC https://openknowledge.worldbank.org/handle/10986/6058

World Bank. (2014): Strategic Framework For Mainstreaming Citizen Engagement In World Bank Group Operations, World Bank Washington DC https://consultations .worldbank.org/sites/default/files/materials/consultation, template/engaging, citizens, improved, resultsopenconsultationtemplate/materials/finalstrategicframeworkforce_4.pdf

World Bank. (2017): Portraits Of Labor Market Exclusion 2.0 Country Policy Paper (CPP) for Greece, International Bank for Reconstruction and Development, World Bank Washington DC http://documents1.worldbank.org/curated/en/118191522094488076/pdf/Country, Policy, Paper, for, Greece.pdf

World Bank. (2017): Promoting inclusive growth by creating opportunities for the urban poor (Philippines Urbanization Review Policy Note). Washington, DC: World Bank Group. http://documents.worldbank.org/curated/en/90447149580 8486974/ Promoting, inclusive, growth, by, creating opportunitiesfor, the, urban, poor

World Bank. (2019): Breaking the Cycle of Roma Exclusion in the Western Balkans, World Bank Washington DC https://www.worldbank.org/en/region/eca/publication/ breaking, cycle, of, roma, exclusion, in, western, balkans

Wu, J., & Loucks, O. (1995). from balace of nature to hierarchical patch dynamics; A paradigm shift in ecology. 1-2.

Wuisang, C., Rengkung, J., & Rondonuwu, M. (2018). Towards sustainable cultural landscape: A challenge in developing ecomuseum in Minahasa Region. 3rd International Symposium for Sustainable Landscape Development. IOP Conference Series. Earth and Environmental Science, 179, 012017. DOI: 10.1088/1755-1315/179/1/012017

Wylie, J. (2007). Landscape. Routledge. DOI: 10.4324/9780203480168

Xanthopoulos, G. 2024. Wildland-urban interface area. In: Kalabokidis K. (Ed.). Impact study and expertise for addressing problems in the affected area of Vrissa-Vatera-Stavros in Lesvos Island, Greece, after the fire of July 23, 2022. University of the Aegean. ISBN: 978-618-86408-1-8. Pp. 83-92. (In Greek).

Xiong, Y. (2015). The broken ladder: Why education provides no upward mobility for migrant children in china. The China Quarterly, 221, 161–184. DOI: 10.1017/S0305741015000016

Yeşildal, A. (2020a). Local Government in the Struggle against Poverty: New Trends from Global to Local. Journal of Social Policy Conferences, 78, 455–481.

Yeşildal, A. (2020b). Political Economy of Environment and Sustainable Development; Local Governments and Multilevel Governance. Kocaeli University Journal of Social Sciences, 1(39), 189–208.

Yıldız, M. and Gümüş, E. (2023). ''Digital Transformation in Local Governments: E-Municipality Applications''. Journal of Economics, Business, Politics and International Relations, 9(2-1), 201-221.

Yılgör, M. (2019). Geoda, ArcGis, Matlab Uygulamalı Mekansal Ekonometri. Der Yayınevi.

You, H., Yang, J., Xue, B., Xiao, X., Xia, J., Jin, C., & Li, X. (2021). Spatial evolution of population change in northeast china during 1992–2018. The Science of the Total Environment, 776, 146023. DOI: 10.1016/j.scitotenv.2021.146023

Yuan, H., Guan, Z., Jiang, C., & Yang, H. (2023). Peer effect in the construction of china's high-speed rail stations: Empirical evidence from spatial econometric analysis. Transportation Research Part A, Policy and Practice, 173, 103720. DOI: 10.1016/j.tra.2023.103720

Zammit Pace, M., Bray, M., Baily, B., & Potts, J. (2018). Beach management review of the Maltese Islands. Proceedings of the Thirtneeth International MEDCOAST Congress on coastal and marine sciences, engineering, management and conservation, (pp. 203-214).

Zengin, E., Başkurt, M., & Es, M. (2014). Local goverments and local development''. Manas Social Journal of Research, 3(2), 95–124.

Zeren, F. (2010). Mekansal Etkileşim Analizi. İstanbul Üniversitesi İktisat Fakültesi Ekonometri ve İstatistik Dergisi, (12), 18-39.

Zhang, B., & Qi, R. (2021). Transportation infrastructure, innovation capability, and urban economic development. Transformations in Business & Economics, •••, 20.

Zhang, J., Huang, J., Wang, J., & Guo, L. (2020a). Return migration and hukou registration constraints in chinese cities. China Economic Review, 63, 101498. DOI: 10.1016/j.chieco.2020.101498

Zhang, L., Sharpe, R. V., Li, S., & Darity, W. A.Jr. (2016). Wage differentials between urban and rural-urban migrant workers in china. China Economic Review, 41, 222–233. DOI: 10.1016/j.chieco.2016.10.004

Zhang, P., Zhao, Y., Zhu, X., Cai, Z., Xu, J., & Shi, S. (2020b). Spatial structure of urban agglomeration under the impact of high-speed railway construction: Based on the social network analysis. Sustainable Cities and Society, 62, 102404. DOI: 10.1016/j.scs.2020.102404

Zhou, M., & Sisiopiku, V. P. (1997). ve Sisiopiku, V., "Relationship between volume to capacity ratios and accident rates". Transportation Research Record: Journal of the Transportation Research Board, 1581(1), 47–52. DOI: 10.3141/1581-06

Zhou, X., Lu, X., Lian, H., Chen, Y., & Wu, Y. (2017). Construction of a spatial planning system at city level: Case study of "integration of multi-planning" in Yuanqing Wu. Habitat International, 65, 32–48. DOI: 10.1016/j.habitatint.2017.04.015

Zou, J., Chen, J., & Chen, Y. (2022). Hometown landholdings and rural migrants' integration intention: The case of urban china. Land Use Policy, 121, 106307. DOI: 10.1016/j.landusepol.2022.106307

Zouros, N., Valiakos, E., Lambrakopoulos, A., & Antonakis, E. 2024. Display of geological – natural heritage of Vrisa area. In: Kalabokidis K. (Ed.). Impact study and expertise for addressing problems in the affected area of Vrissa-Vatera-Stavros in Lesvos Island, Greece, after the fire of July 23, 2022. University of the Aegean. ISBN: 978-618-86408-1-8. Pp. 226-241. (In Greek).

Zouros, N. (2007). Geomorphosite assessment and management in protected areas of Greece. Case study of the Lesvos Island – coastal geomorphosites. Geographica Helvetica, 62(3), 169–180. DOI: 10.5194/gh-62-169-2007

Zrobek, S., Trojanek, M., Sokolnik, A. Z., & Trojanek, R. (2015). The influence of environmental factors on property buyers' choice of residential location in Poland. Journal of International Students, 164–174.

Zsolnai, L. (2011). Ethical Principles and Economic Transformation – A Buddhist Approach, Issues in Business Ethics. Springer Science+Business Media .

About the Contributors

George M. Korres holds a doctorate in Economic Sciences (Economics of Innovation) from University of London (QMW) and an M.Sc from Essex University. Currently, he is a Professor at the University of the Aegean, Department of Geography, School of Social Science. He was also a visiting fellow, for the academic period 2010-20111, at the University of Newcastle, Centre from Urban Regional & Development Studies, (CURDS), School of Geography, Sociology and Politics. He was visiting scholar at the University of Leeds during 2006-2007. Over the last 10 years, during both his academic and professional career, he has gained an in depth experience in the fields of economics, business and politics in the area of European Economics and Innovation Systems. His main areas of interest and specialisation include: European Integration, Social Policy and Social Inclusion, Economic and Innovation Policies, EU enlargement and the Euro-Mediterranean partnership. He has collaborated with many European universities, Institutions, NGOs as well as academia. George Korres has also been involved in various trans-European and international projects and he has published widely in this area in many International Journals, such as Applied Econometrics, Regional and Sectoral Economic Studies, Southwestern Economic Review, Pensions, Sage, Management Sciences and Regional Development etc, also he has published a number of books with international publishers, such as MacMillan-Palgrave Press, Springer, Edward and Elgar, Ashgate Publishers and Nova (New York) Publishers.

Maria P. Michailidis is an Professor and the ex-Dean of the School of Business at the University of Nicosia, Cyprus. She is a Community-Social Psychologist, a Psychotherapist and a Vocational Counselor. She is the author of several book chapters, journal articles and editor of several books. Dr. Michailidis has a long industrial experience as she worked in the United States and Cyprus, as well as teaching experience from teaching in tertiary education both in the United States and Cyprus. She is a researcher and a regular speaker and presenter at International conferences. Her research interests include occupational health, illness and wellness, gender and intercultural studies, unemployment issues, working women and stress, occupational stress, women entrepreneurs. Most of her research work has been supported by grants; she has also been a research partner in European Union projects such as Leonardo da Vinci; Socrates/Erasmus Thematic Networks and Grundtvig Multilateral Projects.

Coşkun Akdeniz, Ph.D. Associate Professor in the Economics Department at TNKU University, Department of Economics in Tekirdağ, Turkey, specializing in economics, particularly Turkish manufacturing.

Spyros Anagnostou is an Associate Professor at the University of the Aegean-Department of Geography. He has obtained a Ph.D from the Department of Geography at the University of the Aegean, also he has a DEA, Urban and Spatial Management from the Université d'Aix-Marseille III DESS, Law and Urban Planning – Université d'Aix-Marseille III and a B.A. degree (Maîtrise) Regional and Urban Planning – Université d'Aix-Marseille III. His research interests focusing in the fields of City Planning, Regional Development with Emphasis on Island Space, Culture and Local Development, Transport Infrastructures, Geoeconomics, Geopolitics, Urban History, and Historical Geography. He has also collaborated in research projects, has presented her work in a number of international conferences and has published in established international journals.

Efstratia Chatzi is a PhD candidate in Spatial and Urban Planning at the University of the Aegean, Department of Geography. She holds a bachelor's degree in Geography and two master's degrees from the University of the Aegean, specializing in applied geoinformatics, natural hazards, and integrated coastal zone management. Her doctoral research focuses on terrestrial and marine spatial planning tools, with a case study of the South Aegean Region. Her research interests include spatial and urban planning, marine and coastal spatial planning, geoinformatics, cultural and natural heritage, and sustainable development of island and coastal areas.

Despina Dimelli is an architect (NTUA) with a master's degree from the postgraduate program "Architecture – Spatial Design" and a second Master's in environmental design of cities and buildings. She also holds a PhD from the School of Architecture, NTUA. She has served as an advisor to the Secretary General for Spatial Planning and Urban Environment at the Ministry of Environment, responsible for urban and spatial planning issues. She has been a member of working groups on urban and spatial planning reform, informal industrial zones (e.g., in Oinofyta), and metropolitan planning councils. She participates in international scientific bodies and committees, reviews journal articles, and has been involved in national and European research projects. Her work includes numerous publications and conference presentations. She has curated educational material, edited student work volumes, and written teaching notes. She has also received awards in national and international architectural competitions.

Despina Dimelli is an architect (NTUA) with a master's degree from the postgraduate program "Architecture – Spatial Design" and a second Master's in environmental design of cities and buildings. She also holds a PhD from the School of Architecture, NTUA. She has served as an advisor to the Secretary General for Spatial Planning and Urban Environment at the Ministry of Environment, responsible for urban and spatial planning issues. She has been a member of working groups on urban and spatial planning reform, informal industrial zones (e.g., in Oinofyta), and metropolitan planning councils. She participates in international scientific bodies and committees, reviews journal articles, and has been involved in national and European research projects. Her work includes numerous publications and conference presentations. She has curated educational material, edited student work volumes, and written teaching notes. She has also received awards in national and international architectural competitions.

Stelios Gialis is a Professor at the Department of Geography of the University of the Aegean in Greece. His research interests include the spatial divisions of flexible/atypical/informal employment, urban restructuring, and labor markets.

Kostas Gourzis is a postdoctoral researcher in the Department of Geography at the University of the Aegean, and an Assistant Professor (elect) in the Department of Economic and Regional Development at Panteion University. He has published in numerous international scientific journals of high impact factor including Antipode (Wiley), the International Journal of Urban and Regional Research (Wiley-Blackwell), European Urban and Regional Studies (SAGE), and Urban Studies (JSTOR). He is a Fulbright Alumni, while he has been awarded with the 2022 Jim Lewis Prize for the most innovative research paper of the year. In 2015, he received the Herta Macht Thesis Prize for the highest quality masters thesis submitted to a Dutch university in the field of Spatial Sciences.

Lishuang Huang is a Ph.D. Candidate in Jinan University, her research focus on Labor Economics and Taxation.

Charuka Deshan is a passionate researcher and Research Assistant in the Department of Building Economics, at the University of Moratuwa. He is currently pursuing an MSc in Research, focusing on sustainable urban mobility and its impact on societal well-being.

Kostas D. Kalabokidis teaches geography of natural disasters and wildland fires (hazards geography) with an emphasis on the use of geographic information systems (GIS) at the University of the Aegean in Mytilene, Greece, where he has been also director of graduate studies in "Geography and Applied Geo-Informatics."

Shixi Kang is a PhD student at University of Barcelona.

Shengming Li is a graduate student at the School of Economics of Guizhou University.

Alexandra Ntousi is an architect from Athens and a graduate Technical University of Crete.

Evangelos Pavlis is Assistant Professor at the Department of Agricultural Economics and Rural Development at the Agricultural University of Athens in Greece with specialization in social analysis of rural space. He conducted a post-doc in social/rural geography (Panteion University, Greece, 2016–2020), and holds a Ph.D. in Cultural/Landscape Geography with emphasis on rural landscape (University of the Aegean, Greece, 2006–2012), an M.Sc. in Rural Change (Coventry University, UK, 2001–2002) and a bachelor's in Countryside Change and Management (Coventry University, UK, 1997–2000). His research, publishing and teaching experience and interests regard social/rural geography and cultural/landscape geography. He has experience as an administrator and a consultant in environmental/landscape, local development and social economy acts and initiatives and as a member of organizational and scientific committees of conferences, events, summer schools and workshops. He is a Member of the Board of the Hellenic Association of Geographers. His main research topics and publishing material involve issues such as landscape change, landscape perception, landscape values, the transformation and multifunctionality of rural space, the challenges of post-productive rural areas, ecomuseums, social economy and commons and sustainable/territorial rural development.

Eleni Prifti is an architect from Athens, Greece, and a graduate of the Technical University of Crete. Her work focuses on cultural heritage and the transformation of the built environment, exploring how architecture can respond to historical, social, and spatial contexts in both urban and rural settings.

Dimitris Psarologos is a junior researcher at the Department of Geography of the University of the Aegean in Greece. His interests include regional and labour inequalities, WebGIS applications, and data analysis.

George Sykas is a PhD candidate at the Department of Geography of the University of the Aegean in Greece. His interests include regional and labour inequalities, investment flows from industrial production to the built environment, and the interplay between the sectors of construction and tourism.

Jingwen Tan is a PhD candidate at University of Barcelona, his main research is in field of Development and Labor Economics.

Aylin Tepecik received his bachelor's degree from Erciyes University, Department of Urban and Regional Planning in 2021 and his master's degree from the same university and department in 2023. His research area focused on topics such as Rural Development, Regional Planning, Sustainable Planning and Urban Sustainability and aimed to contribute to the field.

Georgios Tsilimigkas is a Professor of Urban and Spatial Planning at the University of the Aegean, Department of Geography. He holds a PhD in Spatial Planning from the University of Thessaly (2007) and postgraduate degrees from the University of Thessaly, Université Paul Valéry–Montpellier III, and the Mediterranean Agronomic Institute of Montpellier (CIHEAM). He is a graduate engineer in Spatial, Urban and Regional Planning and also holds a diploma in Architecture from the National Technical University of Athens. His research interests include urban and spatial planning, regional and local development, housing and settlement planning, and planning in rural, island, and mountainous areas. He also focuses on marine and coastal spatial planning, cultural landscapes, and the protection and sustainable use of cultural and natural heritage.

Meng Wei is a graduate student at the School of Economics at Huaqiao University.

Index

C

D

E

G

H

I

L

M

N

P

R

S

T

U

V

W

www.ingramcontent.com/pod-product-compliance
Lightning Source LLC
LaVergne TN
LVHW081544300926
849161LV00006B/1513

* 9 7 9 8 3 3 7 3 1 5 5 4 6 *